MONOGRAPHS ON THE PHYSICS AND CHEMISTRY OF MATERIALS

MONOGRAPHS ON THE PHYSICS AND CHEMISTRY OF MATERIALS

Theory of dielectrics M. Frohlich
Strong solids (Third edition) A. Kelly and N. H. Macmillan
Optical spectroscopy of inorganic solids B. Henderson and G. F. Imbusch
Quantum theory of collective phenomena G. L. Sewell
Principles of dielectrics B. K. P. Scaife
Surface analytical techniques J. C. Rivière
Basic theory of surface states Sydney G. Davison and Maria Steslicka
Acoustic microscopy Andrew Briggs
Light scattering: principles and development W. Brown
Quasicrystals: a primer (Second edition) C. Janot
Interfaces in crystalline materials A. P. Sutton and R. W. Balluffi
Atom probe field ion microscopy M. K. Miller, A. Cerezo, M. G. Hetherington, and G. D. W. Smith
Rare-earth iron permanent magnets J. M. D. Coey
Statistical physics of fracture and breakdown in disordered systems B. K. Chakrabarti and L. G. Benguigui
Electronic processes in organic crystals and polymers (Second edition) M. Pope and C. E. Swenberg
NMR imaging of materials B. Blümich
Statistical mechanics of solids L. A. Girifalco
Experimental techniques in low-temperature physics (Fourth edition) G. K. White and P. J. Meeson
High-resolution electron microscopy (Third edition) J. C. H. Spence
High-energy electron diffraction and microscopy L.-M. Peng, S. L. Dudarev, and M. J. Whelan
The physics of lyotropic liquid crystals: phase transitions and structural properties A. M. Figueiredo Neto and S. Salinas

THE PHYSICS OF LYOTROPIC LIQUID CRYSTALS

PHASE TRANSITIONS AND STRUCTURAL PROPERTIES

ANTÔNIO M. FIGUEIREDO NETO and SILVIO R. A. SALINAS

Instituto de Fisica
Universidade de São Paulo
São Paulo, Brazil

OXFORD
UNIVERSITY PRESS

OXFORD
UNIVERSITY PRESS

Great Clarendon Street, Oxford OX2 6DP

Oxford University Press is a department of the University of Oxford.
It furthers the University's objective of excellence in research, scholarship,
and education by publishing worldwide in

Oxford New York

Auckland Bangkok Buenos Aires Cape Town Chennai
Dar es Salaam Delhi Hong Kong Istanbul Karachi Kolkata
Kuala Lumpur Madrid Melbourne Mexico City Mumbai Nairobi
São Paulo Shanghai Taipei Tokyo Toronto

Oxford is a registered trade mark of Oxford University Press
in the UK and in certain other countries

Published in the United States
by Oxford University Press Inc., New York

First published 2005

A catalogue record for this title is available from the British Library

Library of Congress Cataloguing in Publication Data
(Data available)

ISBN 0 19 85 2550 8

10 9 8 7 6 5 4 3 2 1

Typeset by Newgen Imaging Systems (P) Ltd., Chennai, India
Printed in India
on acid-free paper by
Thomson Press (India) Ltd.

PREFACE

Soaps are among the most interesting molecules. Soap-making was known as early as 2800 BC. A soap-like material has been found in clay cylinders from excavations in ancient Babylon. Inscriptions on these cylinders indicate that fats were boiled with ashes, which is a method of making soap. The purpose of this product, however, has not been clearly established by archeologists. In the Ebers Papyrus (1500 BC), Egyptians describe the combination of animal and vegetable oils with alkaline salts in order to form a soap-like material, which was then used for washing and for therapeutic procedures in skin diseases.

The use of soaps for washing is directly related to some fundamental concepts at the level of molecular length scales: *self-assembling* and *ordering*. Soaps belong to the class of *amphiphilic* molecules. An amphiphile or surfactant molecule is formed by a hydrophilic, water-soluble, part, chemically bounded to a hydrophobic, oil-soluble, part. Mixtures of amphiphilic molecules and solvents, under suitable conditions of temperature, pressure and relative concentrations of the different components, are known to display a host of *lyotropic mesophases*. The basic units of these mesophases are molecular aggregates, spontaneously formed mainly due to hydrophobic–hydrophilic effects.

Lyotropic systems give spectacular examples of polymorphism and phase transformations depending on changes of temperature, pressure and other physico-chemical parameters.

The use of amphiphilic molecules in everyday life was originally due to the empirical properties of mixtures of these molecules with polar and non-polar solvents. In the last decades, however, there was an enormous improvement of experimental techniques, as the scattering and diffraction of light, neutrons, and X-rays, nuclear magnetic resonance, electron microscopy and fluorescence, atomic force microscopy, nonlinear optical techniques, which are among the most powerful tools of condensed matter physics. These techniques lead to the establishment of additional and more precise information on the structure, local ordering, and phase transitions, of the phase diagrams of lyotropic mixtures. The Landau–Ginzburg theory of phase transitions, as well as many-body and renormalization-group techniques, which were important advances of statistical physics, have provided a number of models and concepts for accounting to the experimental features of phase diagrams and critical behavior in lyotropic systems. There is today a unifying view of different sorts of "self-assembled" systems (lyotropics, microemulsions, polymers, gels, membranes, thin films), which are forming the new area of "complex fluids."

In the beginning of the twentieth century, on the basis of investigations of the behavior of physico-chemical parameters (detergency, electric conductivity, and interfacial tension) of a mixture of amphiphiles and water, McBain proposed the idea of a micelle as an aggregate of surfactant molecules. In 1949, Debye recognized the existence of a critical micellar concentration, and the groups of Ekwall, Luzzati, and Winsor, performed outstanding investigations of a number of basic phase diagrams, and established the main features of the structure of lyotropic phases. These investigations, summarized in a review by Ekwall in 1975, were stimulated by the practical application of amphiphilic compounds in the production of cosmetics, in the pharmaceutical and oil industries, and also as an interface with biological membranes in living cells. Connections and analogies were established with microemulsions (isotropic mixtures of amphiphiles, water, and oil), surfactant layers (as Langmuir–Blodgett films), biological membranes, block copolymers, colloidal suspensions, among several other systems. The interface with biology was deeply emphasized by the modelling of cell membranes as amphiphilic bilayers.

The discovery of a nematic phase in a lyotropic mixture of sodium decylsulfate and water, by Lawson and Flautt in 1967, opened up the opportunity to use similar concepts for analyzing different sorts of liquid crystalline systems, thermotropics, and lyotropics.

Although the physics of thermotropic liquid crystals is vastly discussed in the literature, for example, in the outstanding book of de Gennes, the physics of lyotropic liquid crystals has not been sufficiently discussed. We then believe that it is relevant to have a text describing the basic structures and phase transitions in lyotropic mesophases, and collecting information from different experimental techniques, which were fundamental for the characterization of molecular self-assembled structures. This book is planned to give a unifying presentation of the structures and physical properties of lyotropic liquid crystalline systems. We present a comprehensive set of experimental results, published so far in several specialized journals, and we discuss the characterization of different structures and the corresponding phase transitions.

This book contains eight chapters. In Chapter 1, we present the main experimental facts and techniques related to the characterization of the lyotropic mesophases. All of the structures of these systems are discussed on the basis of complementary experimental results, obtained by several groups and using different techniques. Besides introducing the basic nomenclature and properties of lyotropic mixtures, we also refer to technological applications and to the interface with biology. In Chapter 2, we present a pedagogical discussion of basic theoretical notions of phase transitions and critical phenomena in simple magnetic and liquid crystalline systems. We take advantage of simple models, and of standard mean-field calculations and Landau expansions, for providing an overview and some illustrations of the main concepts in this area. In Chapter 3, we discuss phase diagrams and the Gibbs phase rule, and present the main experimental phase diagrams of binary, ternary and multicomponent lyotropic mixtures. We also refer to theoretical attempts to account for the phase diagrams of a binary mixture. In Chapter 4, we discuss phase diagrams and phase transitions in lyotropic liquid crystals from the point of view of the symmetry transformations between periodically ordered mesophases. This chapter was written in collaboration with Dr Bruno Mettout, to whom we are deeply grateful. In Chapter 5, we present the isotropic micellar and bicontinuous phases, their main features, structure and location in the experimental phase diagrams. We also mention some models and theoretical calculations for the sponge phase. In Chapter 6, we discuss nematic and cholesteric phases. We present experimental phase diagrams and phase structures, as well as an overview of some calculations, with emphasis on the need of introducing an additional non-critical order parameter in order to account for the experimental phase diagrams. In Chapter 7, we present experimental results for one-, two-, and three-dimensionally ordered lyotropic structures. Finally, in Chapter 8, we refer to some recent extensions and neighboring topics of the general area of lyotropic mixtures. We include brief surveys of research on ferrofluids, microemulsions, diblock copolymers, and Langmuir–Blodgett films.

This book comes from years of collaboration among the authors and many colleagues at different laboratories and theoretical groups around the world in the areas of lyotropic liquid crystals and phase transitions in condensed matter physics. We hope that these collaborators, which are deeply acknowledged, have been suitably quoted in the extensive bibliography at the end of each chapter. We wish to express our special indebtedness to Dr Bruno Mettout, who helped us to write Chapter 4, and to Professor Pierre Tolédano, who encouraged us in the early stages of this project. We are also indebted to Dr Sonke Adlung, from the Oxford University Press, who gave us strong support during all of the stages of the project, and to Mr Carlos E. Siqueira and Mr Carlos R. Marques, for helping us draw most of the figures. Our research work has been supported by the Brazilian agencies Fapesp and CNPq.

Antônio M. Figueiredo Neto and
Silvio R. A. Salinas
São Paulo, May 2004.

CONTENTS

1

LYOTROPIC SYSTEMS: MAIN EXPERIMENTAL FACTS AND TECHNIQUES

1.1 Introduction

Liquid crystals [1] are intermediate states of matter or mesophases, halfway between an isotropic liquid and a solid crystal. In nature, some substances, or even mixtures of substances, present these mesomorphic states. This picture leads to the concept of *ordering.* In a solid crystal, the basic units display translational long-range order, with the center of mass of atoms or molecules located on a crystal lattice; in some cases, the basic units also display orientational order. In an isotropic liquid, the basic units do not present either positional or orientational long-range order. From one ordering limit (solid crystal) to the other (isotropic liquid), there may exist many different situations. In plastic crystals, the basic units (globular molecules, e.g.) are located on a lattice but without any orientational order. In liquid crystals, the basic units display orientational order and even positional order along some directions. These materials flow like an isotropic fluid and have characteristic optical properties of solid crystals. Liquid crystals were firstly classified as thermotropics and lyotropics, depending on the physico-chemical parameters responsible for the phase transitions.

In *thermotropic* liquid crystals the basic units are molecules, and phase transitions depend on temperature and pressure. A pronounced shape anisotropy (in other words, the anisometry) is the main feature of the molecules which give rise to a thermotropic mesophase. Rods, disks, and banana-shaped are examples of molecular geometries associated with thermotropic liquid crystals. Besides pure substances, mixtures of molecules can also present thermotropic mesomorphic properties. Thermotropics are widely used in displays of low energy cost and in many sensor devices.

Lyotropic liquid crystals, shortly called *lyotropics* or *lyomesophases*, are mixtures of amphiphilic molecules and solvents at given temperature and relative concentrations. The mesomorphic properties change with temperature, pressure and the relative concentrations of the different components of the mixture. An important feature of lyotropics, turning them different from thermotropics, is the self-assembly of the amphiphilic molecules as supermolecular structures, which are the basic units of these mesophases. Although there are not many devices based on lyotropics, their physico-chemical properties have an interesting interface with biology, and the understanding of these properties has been relevant for improving some technological aspects of cosmetics, soaps, food, crude oil recovery, and detergent production.

It is interesting to point out that there is a family of complex isotropic fluids, which have been called *microemulsions* [2], whose characteristics [3], in some respects, overlap with those of lyotropics. Microemulsions are mixtures of oil, water and amphiphile molecules, which behave as an optically isotropic and thermodynamically stable liquid solution [4]. These systems differ from the emulsions, which are *kinetically stable.* In microemulsions, the typical size of the basic units (self-assembled molecular aggregates) is about 10 nm, which makes the mixture transparent to visible light. On the other hand, emulsions diffuse visible light, displaying a milky or cloudy aspect, which indicates that their basic units are larger, typically, of micrometer dimensions. The conceptual boundaries between lyotropics, in particular the isotropic phases, and microemulsions are not sharp; sometimes, the isotropic phases of the same mixture, with oil as one of the components, are included in different sides of this border. In order to differentiate them, we point out that microemulsions are two-phase systems and lyotropics are one-phase systems. In this book, we always refer to lyotropics and use their nomenclature to describe the isotropic micellar and bicontinuous phases, even if oil is present in the mixture.

Another family involving characteristics of lyotropics and thermotropics has been recently investigated. These systems are made of a mixture of thermotropic liquid crystals and solvents. This mixture *does not present* molecular aggregates, as micelles or other supermolecular structures, but the polymorphism of the phase diagram depends on temperature *and* the relative concentrations of the different components. Since new phases appear as a function of the concentration of the solvent, these mixtures are different from those which give rise to the swelled thermotropic phases. They will be discussed in Chapter 8 of this book.

1.1.1 *The hydrophobic and hydrophilic effects*

Water is present in almost all of the lyotropic mixtures. The behavior of a molecule of a given substance with respect to the water molecules plays a crucial role in the formation of a lyomesophase.

In the field of complex and supermolecular fluids, the concepts *hydrophobic* (hates water) and *hydrophilic* (loves water) refer to the affinity of a particular molecule with respect to the water molecules. Sometimes these effects are treated as *interactions*, but this is not the case. The involved interactions are of electrostatic nature, since water molecules have a permanent dipole moment [5] $p = 6.2 \times 10^{-30}$ C m. From the point of view of electrostatic dipole–dipole interactions, similar molecules, or even parts of molecules, tend to be together. Therefore, polar molecules are easily dissolved in water, and non-polar substances (e.g., paraffin) are difficult to be dissolved in water.

The mechanism of ordering the water molecules, based on the hydrogen bonds, plays an essential role in these effects [6]. At room temperatures (~25°C), the water molecules arrange themselves as an isotropic liquid. A distortion of this structural arrangement, which costs energy, takes place upon the introduction of a solute. If the solute is polar, some energy compensation occurs and the dilution becomes possible. On the other hand, if the solute is nonpolar, no energy compensation occurs and the dilution is difficult.

1.1.2 *Amphiphilic molecules*

Amphiphilic molecules are always present in the composition of lyotropic liquid crystals. They may be synthesized for different purposes, ranging from interests in basic science to technological applications in various branches of industry.

The name amphiphilic comes from the Greek prefix *amphi*, which means *both* or *double*, and the word *phile*, which means *like* or *love*. This word is applied to a compound that displays a double "preference," "loving both," from the electrostatic point of view. It is used to name a molecule with a polar water-soluble group attached to a water-insoluble hydrocarbon chain. An example of this type of molecule, sodium decylsulfate (NadS or SdS), is illustrated in Fig. 1.1. These molecules are *surfactants* (from surface active agent), since they can modify the properties of surfaces and interfaces between different media, as solid–liquid or liquid–gas interfaces.

There are different types of natural and chemically synthesized amphiphilic molecules: anionic amphiphiles (soaps of fatty acids; e.g., potassium laurate), detergents (e.g. sodium decylsulfate); cationic amphiphiles (e.g. hexadecyl trimethylammonium bromide); nonionic amphiphiles (e.g. pentaethyleneglycol dodecyl ether); and zwitterionic amphiphiles (which develop an electric dipole in the presence of water; e.g., lysolecithin). In Fig. 1.2, we sketch some examples. Another type of surfactant molecules that give rise to a lyotropic mesophase are the *anelydes*. These molecules are able to selectively complex some metallic ions [7], which are then incorporated in their structure.

In addition to these so-called classical amphiphiles, there are molecules with a more complex topology, with more than one polar group, which also give rise to lyotropic mesophases. For example, we mention the gemini surfactants [8], the rigid spiro-tensiles, and phospholipids [9], with molecules of the hydrophilic group grafted in a position lateral to a rod-like rigid core [10]. The *facial* amphiphiles [11] (Fig. 1.2(g)) are block molecules in which two alkyl chains are placed in both sides of a calamitic core and the polar group is attached to the core, perpendicular to the stick-like molecule. In the *bolaamphiphiles* (Fig. 1.2(h)), there are two polar heads in both sides of the stick-like molecule

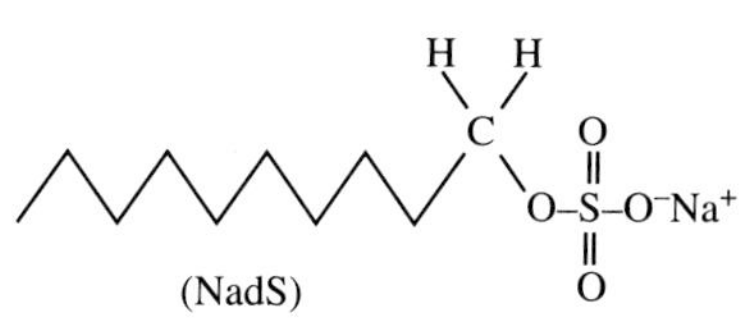

FIG. 1.1. Amphiphilic molecule of sodium decylsulfate.

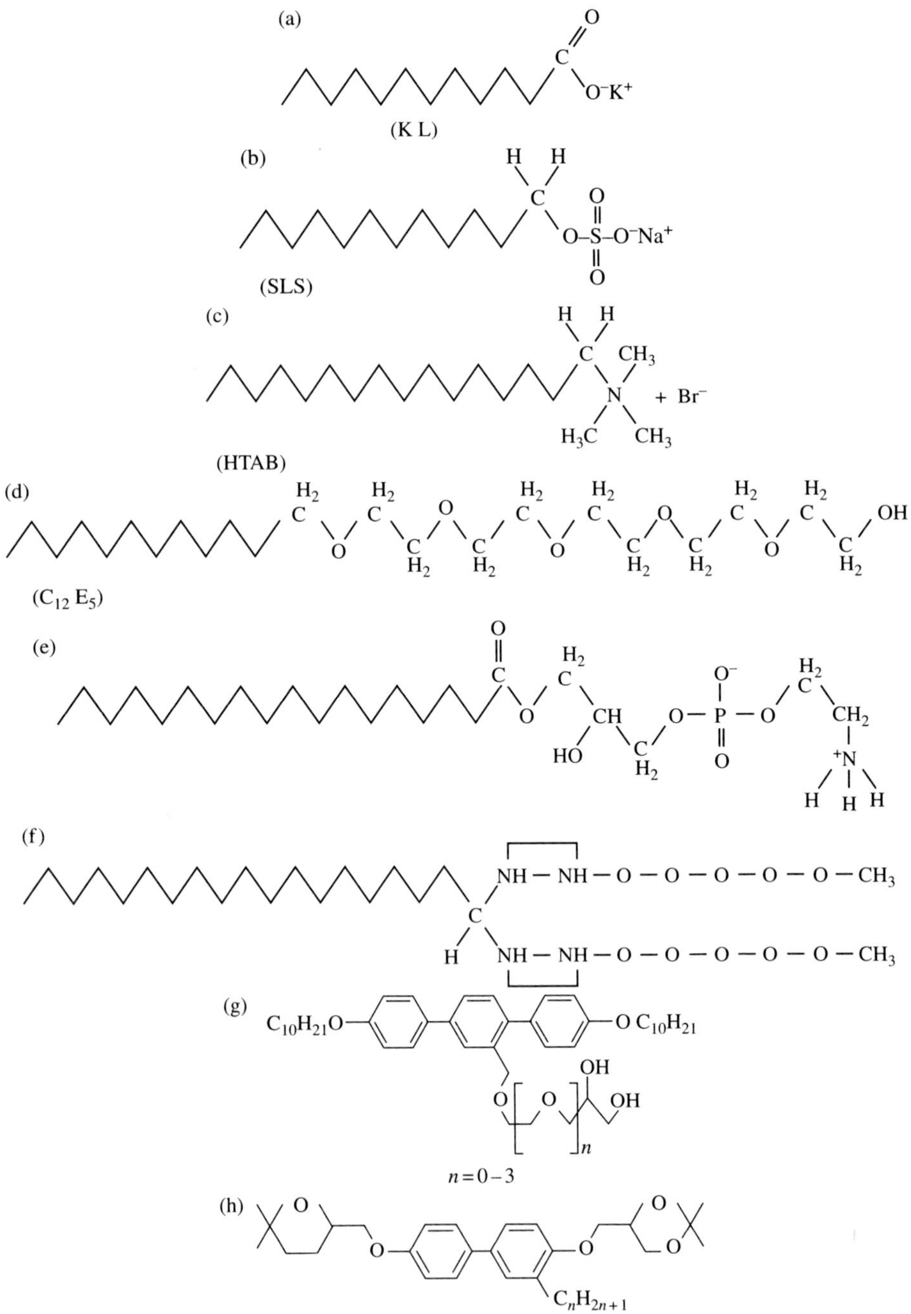

FIG. 1.2. Examples of different amphiphiles: (a) anionic, KL-potassium laurate; (b) detergent, SLS-sodium laurylsulfate; (c) cationic, HTAB or CTAB-hexadecyl trimethylammonium bromide; (d) nonionic, pentaoxyethylene dodecyl ether; (e) zwitterionic; (f) anelydes; (g) facial amphiphile; (h) bolaamphiphile.

and the alkyl chain is perpendicularly attached to the core [12]. In the presence of polar and non-polar solvents, these molecules form lyotropic mesophases, with nanosegregation properties [12,13].

As a final remark, it is important to note that a polar group is not always required to be hydrophilic (nor is a non-polar group always hydrophobic). The topology of the molecule and its insertion into the water network is also important to characterize the solubility in water [14].

1.1.3 *Definition of a lyotropic mixture*

Under suitable conditions of temperature and relative concentrations, mixtures of amphiphilic molecules and solvents can give rise to a *lyotropic mesophase.* In this type of system, amphiphilic molecules form self-assembled super-structures of several shape anisotropies and sizes.

Let us firstly classify lyotropics into three big families:

(a) Micellar systems, with molecular aggregates, called *micelles*, of small shape anisotropy, as sketched in Fig. 1.3(a). These micelles are aggregates of amphiphilic molecules, with typical dimensions of about 10 nm and shape anisotropy of order 1 : 2 in linear dimensions.

(b) Systems with aggregates of large shape anisotropy, of typical order 1 : 100 in terms of linear dimensions. These aggregates are sometimes called *infinite*, but we do not use this nomenclature. In Fig. 3(b), we sketch a long cylindrical aggregate.

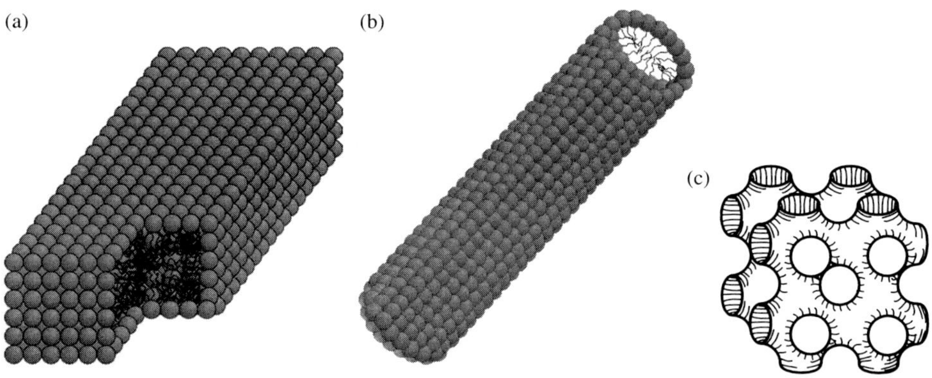

FIG. 1.3. Amphiphilic molecular aggregates. The polar head and the paraffinic chain of the molecules are represented by a sphere and a line, respectively: (a) sketch of an orthorhombic micelle. The cut in the right-down side shows the paraffinic chains in its inner part; (b) large anisotropic cylindrical aggregate; (c) sketch of a bicontinuous molecular aggregate with a cubic symmetry.

(c) Bicontinuous systems, in which the amphiphilic molecules self-assemble as a three-dimensional continuous structure at large scales (larger than 10^3 nm). Fig. 1.3(c) shows a sketch of a bicontinuous molecular aggregate with cubic symmetry.

1.1.4 *Self-assembled systems*

We now discuss the self-assembling properties of amphiphilic molecules in lyotropic mixtures.

In the case of mixtures of amphiphilic molecules and a solvent, one interesting concept is the *critical micellar concentration*, CMC, [6,15,16]. It is defined as the concentration of amphiphilic molecules above which they self-assemble into micelles. Let c be the concentration of amphiphilic molecules in the solution of amphiphiles and a solvent. For $c <$ CMC, the amphiphilic molecules remain isolated, without the formation of micelles. For $c >$ CMC, the fraction of isolated amphiphilic molecules remains almost constant, and the concentration of micelles increases with c (Fig. 1.4). The *hydrophobic/hydrophilic* effects are the most important mechanisms of micelle formation [6]. In water-based mixtures, the formation of micelles can also be understood in terms of the entropy of the structured water, since, for concentrations larger than CMC, the aggregation of amphiphilic molecules increases the water entropy [6,17].

Different theoretical approaches have been used for the understanding of the micellization process [6,14,18–22].

From the experimental point of view, some physico-chemical properties of these solutions, as detergency, equivalent conductivity, high-frequency conductivity, surface tension, osmotic pressure and interfacial tension, present remarkable behavior as c approaches CMC [15]. In actual mixtures, there is no well-defined concentration of amphiphiles at which all of these properties

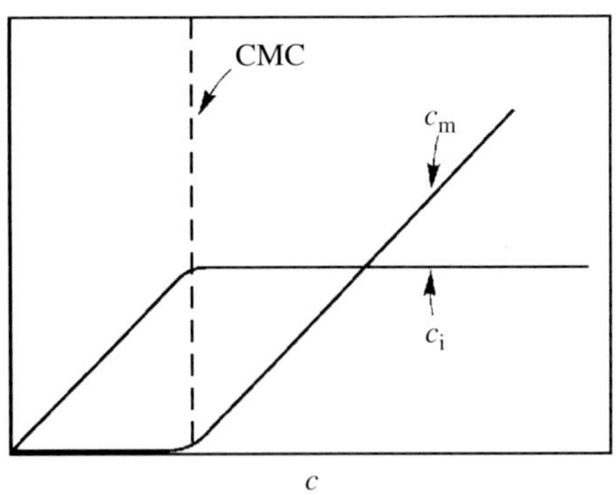

FIG. 1.4. Concentration of isolated amphiphilic molecules (c_i) and micelles (c_m) around CMC as a function of the concentration amphiphilic molecules (c) [14].

present a drastic modification in their behavior. It should be noted that CMC is a function of temperature [23] and that the lifetime of an amphiphile in a molecular aggregate is of the order of 10^{-5}–10^{-3} s [24]. At a given temperature, a good estimate of CMC, in terms of the chain length of the surfactant [25,26], is given by

$$\log_{10} \mathrm{CMC} = 1.6 - 0.3n_{\mathrm{c}} \tag{1.1}$$

where n_{c} is the number of carbon atoms in the chain.

Recently, nonlinear optical properties of amphiphilic solutions were investigated at amphiphile concentrations around CMC. Using a mixture of potassium laurate $[\mathrm{COOK(CH_2)_{10}CH_3}]$ and water, it has been shown [27] that the presence of micelles in the solution, in a concentration range up to $10^2 \times$ CMC, does not significantly modify the thermal conductivity, $\kappa \sim 0.3\ \mathrm{W\,K^{-1}\,m^{-1}}$, of this solution with respect to a solution of isolated amphiphilic molecules. On the other hand, the presence of micelles in the solution changes the behavior of the thermooptic coefficient, $\partial n/\partial T$, where n is the index of refraction of the solution and T is the temperature, as a function of amphiphile concentration. For $c \lesssim$ CMC, the thermooptic coefficient is almost constant and small (about $-2.5 \times 10^{-5}\ \mathrm{K^{-1}}$). Also, it is always negative in the domain of this investigation. Increasing c, the absolute value of $\partial n/\partial T$ increases almost linearly and reaches $\partial n/\partial T \sim -27 \times 10^{-5}\ \mathrm{K^{-1}}$ at $c \sim 0.8$ M. The behavior of the nonlinear index of refraction n_2 in terms of amphiphile concentration is strongly affected by the presence of the micelles in the solution. For $c \lesssim$ CMC, n_2 is negative and almost constant, $n_2 \sim -0.02 \times 10^{-7}$ esu. For $c >$ CMC, two tendencies were observed: from CMC until $10 \times$ CMC, there is a linear decrease of n_2 with c, which reaches about -0.12×10^{-7} esu; for larger values of c, n_2 tends to stabilize at $n_2 \sim -\,0.13 \times 10^{-7}$ esu, at concentrations of about $10^2 \times$ CMC. Comparing these results with the known dependence of the different physical parameters of amphiphilic solutions at concentrations around CMC [15,28], the absolute values of the thermooptic coefficient, $|\partial n/\partial T|$, and of the nonlinear refractive index, $|n_2|$, present the same qualitative behavior of the *high-frequency electric conductivity*, σ_{HF}, and the inverse of the *equivalent electric conductivity*, σ_{EQ}. The mobility of counterions in the double layer around the micelles seems to be strongly related to the nonlinear response of the medium to an electric field.

Besides the critical micellar concentration, CMC, the *critical micellar temperature*, CMT, is another concept that plays a similar role in the self-assembly of amphiphilic molecules [29,30]. The critical micellar temperature CMT is the lower temperature limit between the hydrated solid phase and the micellar phase. This temperature depends on the particular amphiphilic molecule and on the ionic strength of the mixture. As a working example, consider a mixture of sodium dodecylsulfate (SDS), NaCl and water [30]. At a 6.9×10^{-2} M concentration of SDS, CMT increases from about 15°C in a mixture without the salt to about 25°C in a sample with 0.6 M NaCl. Also, CMT was shown to present a small dependence on the SDS concentration, for a fixed salt concentration.

1.1.5 *Direct and inverted polymorphism*

Depending on temperature, type and concentration of the solvents, there may exist *direct* or *inverted* molecular aggregates in the lyotropic mesophases (see Fig. 1.5). Although commonly used in the field of colloidal systems, this nomenclature is obviously *ad hoc*. Geometrical parameters of the amphiphilic molecules, as the relation between the surface per polar head and the volume of the carbonic chain in the structure, affect the polymorphism in a lyotropic mixture, specially the direct and inverted forms [14].

Let us consider molecular aggregates, excluding the bicontinuous structures. In direct mesophases, the polar solvent is a continuous medium, in which the amphiphilic molecular aggregates are present. The paraffinic chains, as well as other non-polar solvents in the mixture, are confined inside the isolated aggregates (Fig. 1.5(a)). On the other hand, in the case of inverted mesophases, the polar solvent is confined in closed regions and the non-polar material is the continuous external medium.

In bicontinuous structures, the characterization of confined, polar or non-polar, material is not straightforward. Usually, in this case, the terminology of direct or inverted structures refers to the relative concentrations of polar and non-polar solvents with respect to the concentration of the principal amphiphile. In direct structures, the polar solvents have the largest concentration; in inverted structures, the largest concentration is of non-polar solvents. In Fig. 1.5(b), we sketch direct and inverted sponge phase structures.

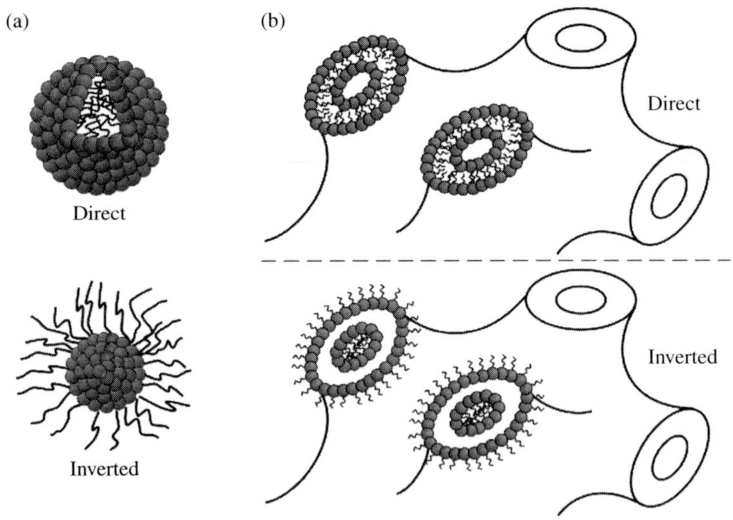

FIG. 1.5. Examples of direct and inverted structures: (a) micelles; in the sketch of the direct micelle, we draw a cut to show the paraffinic chains; (b) bicontinuous direct and inverted sponge phase structures.

1.1.6 *Lyotropic liquid crystalline phases*

In a micellar solution, there appears anisotropic liquid crystalline phases if we increase the concentration of amphiphilic molecules to values much larger than CMC. The typical concentration of amphiphilic molecules in a liquid crystalline mesophase is larger than $10^2 \times$ CMC. For example, in the case of the potassium laurate/water mixture, CMC = 0.008 M [31], and liquid crystalline phases are present for $c \gtrsim 2$ M [32], in a temperature range from approximately 20 to 350°C.

In a temperature versus amphiphile concentration phase diagram, the liquid crystalline region, at high temperatures, is limited by a domain with an isotropic solution of isolated molecules or even micelles. If micelles are present, this is called a micellar isotropic phase. At lower temperatures, it is limited by a crystalline-type region [32] (see Fig. 1.6). The Krafft line defines the function CMC(T) in the phase diagram.

If the temperature of the mixture is lowered, at a given amphiphile concentration, there may appear an intermediate gel phase [33–36], before the system reaches a solid crystalline state. This phase is stable but, if the temperature continues to be lowered, it becomes metastable and spontaneously transforms to a coagel and later to a crystalline phase [37].

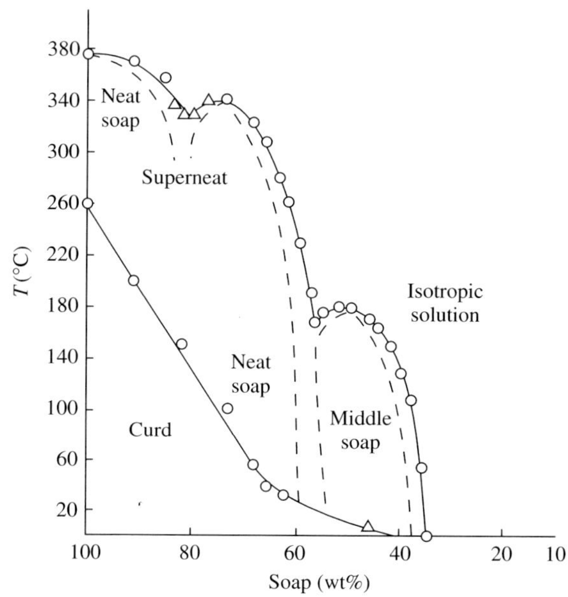

FIG. 1.6. Phase diagram of the potassium laurate/water mixture, in the temperature versus concentration plane (ref. [32]). The phases shown in the figure are discussed in the following sections.

Lyotropic liquid crystalline phases display long-range orientational order and, in some cases, long or medium-range positional order in one or two dimensions. There may even exist medium-ranged structures of three-dimensional character. Unlike the thermotropic mesophases, lyotropic nematics and cholesterics may also present short-range positional order among micelles, giving rise to a pseudo-lamellar structure.

In general, the paraffinic chains inside the molecular structures are in a liquid-like state, without positional order [33]. The order parameter of the different segments of the paraffinic chain, measured by nuclear magnetic resonance (NMR) technique with selectively deuterated samples [38], displays a decreasing profile from the polar head nearest-neighbor carbon towards the CH_3 end group (Fig. 1.7). Besides molecular diffusion, the paraffinic chains in the molecular aggregate describe several movements, as twist, bend, and rotations around particular axes [39].

The polymorphism in a mixture of amphiphilic molecules and solvents depends on different parameters of the amphiphile itself, as the ionic character of the polar head, the size and volume occupied by the head with respect to the parameters of the chain, the presence or absence of another surfactant (usually called *cosurfactant*) or of salt in the mixture, the pH and ionic strength of the solution, the purity of the compounds, and the temperature, among other factors. In some cases, these parameters are difficult to be controlled experimentally, which explains that reproducibility in some experiments in lyotropics is not easy to be achieved, specially if only temperature and concentrations of the different compounds of the mixture are taken into account. A salt, an alcohol, as well as other solvents, can be added to a binary lyotropic mixture in order to produce a reconstruction of the phase diagram, introducing new phases and modifying the topology. Another surfactant can also be added to binary or even ternary mixtures, already having an alcohol, in order to produce a reconstruction of the phase diagram.

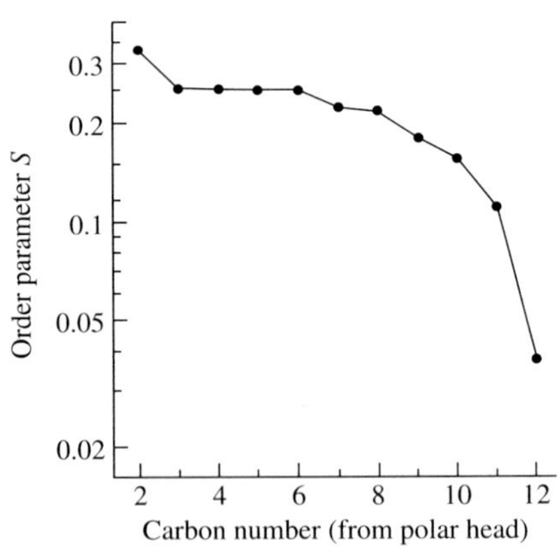

FIG. 1.7. Order parameter of the different CD (carbon-deuteron) bonds in the paraffinic chain inside a micelle (from DMR measurements of quadrupolar splittings [38]).

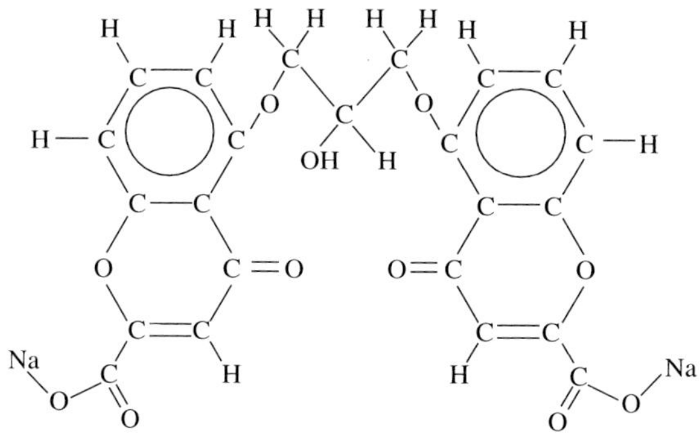

FIG. 1.8. Disodium chromoglycate (DSCG) molecule, which leads to a chromonic mixture.

Water, which is present in most of the lyotropic mixtures, plays a significant role in the stability of the different mesophases. Water molecules take part in ion–dipole and dipole–dipole interactions, and in hydrogen bonds, involving the hydrophilic groups of the amphiphilic molecules. We may say that there is always a certain amount of *bounded water* in the structure of amphiphilic molecules, giving rise to a hydration layer around them [32]. The lifetime of these bonds depends on the hydration number, defined as the number of water molecules orientationally bounded to an ion [14], and ranging from 1 to about 6. For example, in the case of commonly used materials in lyotropics (Na^+, K^+, and Cs^+), the exchange time between bounded and unbounded water molecules is about 10^{-10}–10^{-9} s [14].

Another type of lyotropic-like mixture is the *chromonic* [40], in which more complex molecules, as DSCG [$C_{23}H_{14}O_{11}Na_2$], as sketched in Fig. 1.8, are mixed with a solvent (water). This type of mixture presents a polymorphism that depends on the concentration of solvent. The structures in the phase diagram show some characteristics of the lyotropic phases and also of thermotropic columnar phases.

Depending on time and length scales, different experimental techniques can be used for studying lyotropic liquid crystals. Some of the most common of these techniques are NMR, for systems with ^{1}H and ^{2}H nuclei, and counterions as Li, Na and Cs [41–43], light scattering [44,45], neutron [46,47] and X-ray [33,37,48] scattering and diffraction, polarized light optical microscopy [49–51], conoscopy [52,53], and electric conductimetry [54,55].

1.1.7 *Structures and terminology*

The lyotropic liquid crystals provide perhaps the richest examples of polymorphism among complex fluids.

The *micellar isotropic phase* (labeled L_1 and L_2, for direct and inverted structures, respectively; see Fig. 1.9) can be found in different regions of the lyotropic phase diagrams (not only at higher temperatures, as it is usually expected).

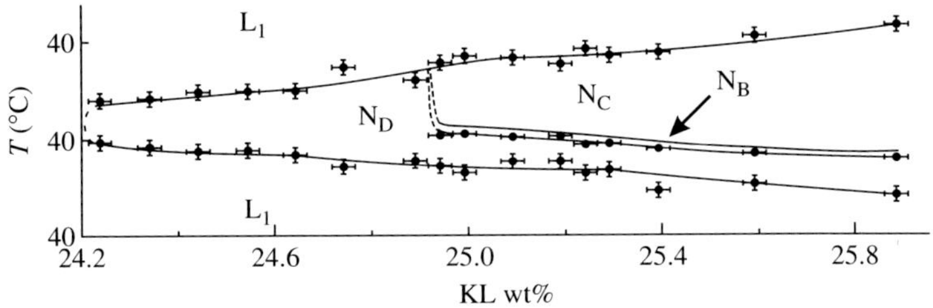

FIG. 1.9. Surface of the phase diagram of a mixture of potassium laurate, decanol (at 6.24 wt%) and water; L_1, N_C, N_B, and N_D represent the micellar isotropic, calamitic nematic, biaxial nematic, and discotic nematic phases, respectively (from ref. [82]).

This is due to the possibility of changing the shape anisotropy of micelles depending on temperature and relative concentrations of the compounds [56]. At low amphiphilic molecular concentrations ($c \gtrapprox$ CMC), micelles are mostly spherical in shape. At larger concentrations of amphiphiles (typically, of order $c \sim 10^2 \times$ CMC), although randomly oriented in space and in isotropic phases, micelles may have non-spherical shapes. In some particular lyotropic mixtures, isolated micelles have orthorhombic symmetry, and are piled up (locally) in small correlation volumes with a pseudo-lamellar structure, although these correlation volumes are randomly oriented in space [56]. This self-arrangement may also lead to an isotropic phase.

Three types of *lyotropic nematic phases* were identified, two of them of uniaxial character [57–61], N_C (calamitic nematic) and N_D (discotic nematic), and a third phase of biaxial character, N_B [61,62]. Figure 1.9 shows a particular section of the phase diagram of a mixture of potassium laurate, decanol and water. These mesophases are composed by micelles with short-range positional and long-range orientational order. The shape of the micelles depends on the particular mixture. Mixtures with only one amphiphile (e.g., decylammonium chloride/NH_4Cl/water and potassium laurate/KCl/water) form disk-like or cylinder-like micelles. A better picture of them could be an oblate (see Fig. 1.10(a)) or a prolate (see Fig. 1.10(b)) ellipsoid. These mixtures do not have the biaxial N_B phase. On the other hand, mixtures with more than one amphiphile (e.g. potassium laurate/decanol/water; sodium decylsulfate [$CH_3(CH_2)_9OSO_2ONa$]/decanol/water; potassium laurate/decylammonium chloride/water) display the three nematic phases. In these cases, micelles have an orthorhombic (brick-like) symmetry, as sketched in Fig. 1.3(a), or the shape of a flattened prolate ellipsoid [56].

Micelles are piled up in a pseudo-lamellar structure at short-range scales [61], and orientationally ordered depending on the particular nematic structure. In Fig. 1.11, we sketch the orientational fluctuations of brick-like micelles. The dots represent a particular surface of the micelles. In Fig. 1.11(a), the orientational

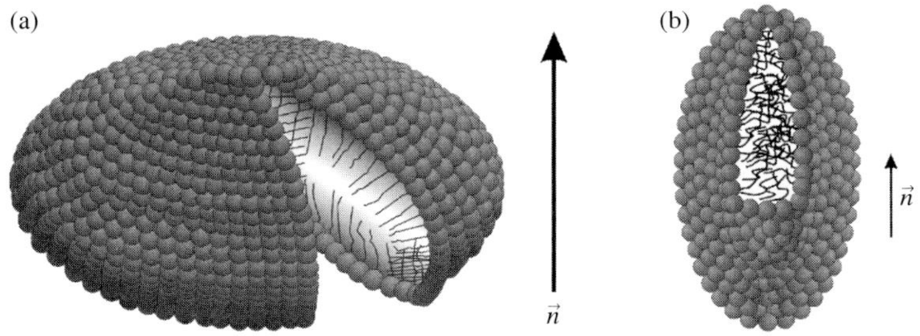

FIG. 1.10. Sketch of micelles in lyotropic mixtures with one amphiphile only; $\vec{n}$ is the director, or axis of symmetry, of the phase. The cuts show the paraffinic chains in inner part of micelles: (a) oblate ellipsoid or disk-shaped; (b) prolate ellipsoid or cylinder-shaped.

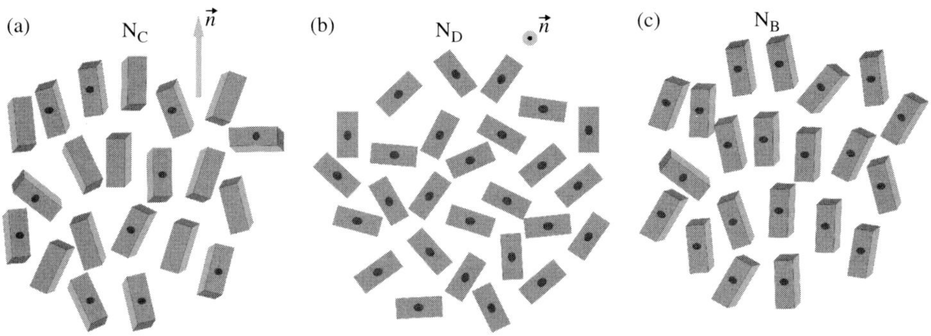

FIG. 1.11. Sketch of the micellar order in the context of the intrinsically biaxial (brick-like) micelles model. The dots represent a particular surface of the micelles and $\vec{n}$ is the director: (a) N_C phase; (b) N_D phase; (c) N_B phase.

fluctuations are full rotations around the long axes of the micelles, which gives rise to the N_C phase; in Fig. 1.11(b), the fluctuations are around the axes perpendicular to the largest surface of the micelles, originating the N_D phase; finally, in Fig. 1.11(c), there are small amplitude fluctuations around the three orthogonal axes of the micelles, which leads to the N_B phase.

A *cholesteric phase* can be formed if a nematic phase is doped with a chiral molecule or if a mixture contains a chiral amphiphilic molecule, as chiral soaps, detergents or alcohols [63–66]. In Fig. 1.12, we sketch an example of a cholesteric arrangement. As in the case of nematics, three types of lyotropic cholesterics were identified [66], a calamitic cholesteric phase, Ch_C, a discotic cholesteric phase, Ch_D, and a biaxial cholesteric phase, Ch_B. The typical concentration of chiral molecules in a cholesteric mesophase is 0.05 M%. The micelles spontaneously organize into a helical structure with a pitch that depends on different

FIG. 1.12. Sketch of a sequence of micelles in a cholesteric structure. The orientation of a prolate ellipsoidal micelle rotates around an axis perpendicular to the infinite-fold symmetry axes of the micelles (represented by the black arrow inside the micelles).

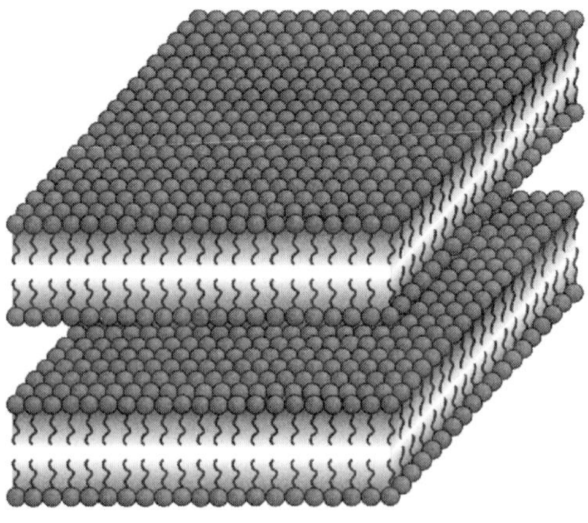

FIG. 1.13. Sketch of the lamellar L_α structure.

parameters, as temperature, pressure, concentration of chiral molecules [63], and the shape anisotropy of the micelles [67,68].

In the *lamellar phase*, amphiphilic molecules are organized as supermolecular aggregates, forming layers with a large shape anisotropy (see Fig. 1.13).

Comparing the thickness of the lamellae with any dimensions in the plane of the lamellae, the shape anisotropy is typically larger than $1:50$. Usually, the

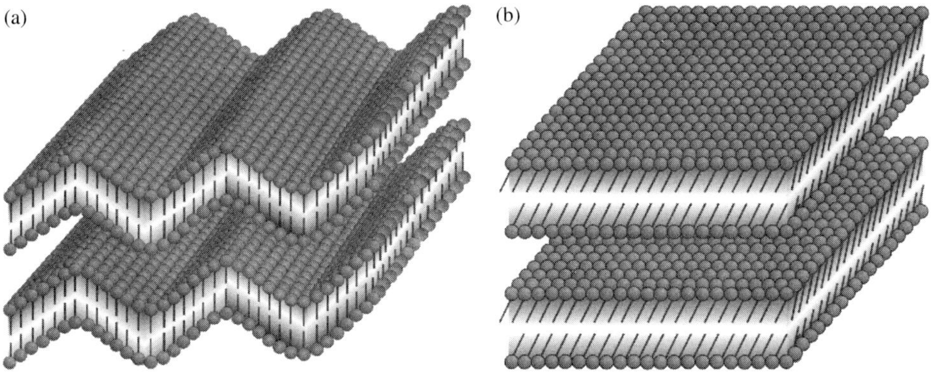

FIG. 1.14. Sketch of lamellar phase structures: (a) ripple phase $P_{\beta'}$; (b) $L_{\beta'}$ phase.

thickness of the layers is of the order of almost twice the length of the main amphiphilic molecule (in some particular mixtures, single-layered structures, called L_s, were observed). However, this length can increase due to the addition of a non-polar solvent. The thickness also depends on the temperature of the sample [33]. The length of a carbonic chain in a *trans* configuration (unfolded state) can be obtained from the empirical expression [69]

$$l = 0.15 + 0.127 n_c,$$

where l is given in nanometers, and n_c is the number of carbons in the chain, excluding the last carbon atom in the CH_3 ending group. If the layers are flat and the paraffinic chains are in a liquid-like state, the phase is called L_α (although the name "neat soap" has also been used [32]). On the other hand, if the chains are stiff and have positional order, different lamellar structures can be formed [70,71]. In the L_β phase, the chains are packed in a two-dimensional hexagonal ordering, with rotational disorder. Phase L_γ is formed by a sequence of layers with L_α and L_β structures. If the surface of the lamellae has an undulated topology [71], the phase is named *ripple-*$P_{\beta'}$ (see Fig. 1.14(a)); if there is an ordering of hexatic type, the phase is named $L_{\beta'}$ [72] (see Fig. 1.14(b)).

In the *hexagonal phase*, the amphiphilic molecules are packed as long cylinder-like aggregates, with a large shape anisotropy (see Fig. 1.15). The diameter of the cylinders is of the order of twice the length of the main amphiphilic molecule of the mixture, and the typical lengths are at least 50 times larger than the diameter. Parallel cylinders are packed on a two-dimensional hexagonal lattice, in the plane perpendicular to the axes. In direct structures (see Fig. 1.15(a)), the hydrocarbon chains inside the cylinders display a liquid-like ordering and the phase is labeled H_α. This phase was also called "middle soap" and labeled E or H_1 [32]. In the case of an inverted structure, which is present in mixtures with a large concentration of nonpolar solvents, the polar solvent is placed inside

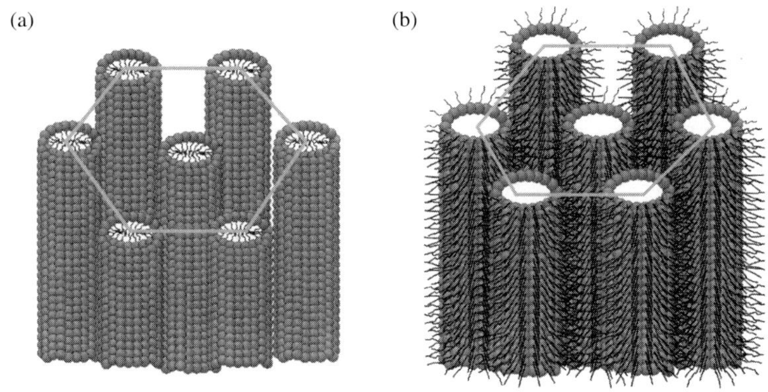

FIG. 1.15. Sketch of the hexagonal phase structure: (a) direct; (b) inverted.

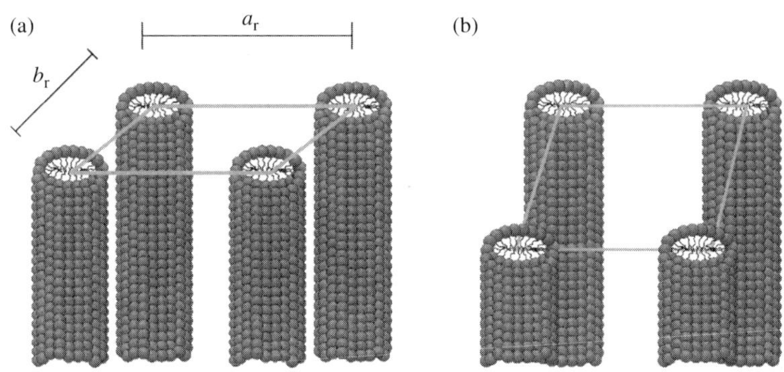

FIG. 1.16. Sketch of two-dimensional phase structures: (a) rectangular; (b) square.

the cylinders and the carbonic chains point outwards (see Fig. 1.15(b)). This phase was also labeled F or H_2 [32]. We will use the nomenclature H_α and H^i_α for direct and inverted phases, respectively. The subscript α indicates that the carbonic chains have a liquid-like ordering; the superscript i denotes the inverted configuration.

These large cylindrical molecular aggregates can also pack as a two-dimensional rectangular (Fig. 1.16(a)) or square (Fig. 1.16(b)) lattice, perpendicular to the axes of the cylinders. The *rectangular* and the *square phases* were labeled R and C (or K in the case of inverted aggregates), respectively [32].

Other mesophases displaying hexagonal lattices have been identified in lyotropics, such as the *"complex" hexagonal phase*, H_c [73], and the *R3m phase*, in which case X-ray diffraction patterns indicate the existence of a rhombohedral lattice, of space group $R3m$, and which will be called Rh.

Another example of polymorphism in lyotropics, as far as these lower-symmetry phases are concerned, is a system of molecular aggregates with the form of long unfolded ribbons [74]. These ribbons can be sketched as flattened cylindrical aggregates and packed on a two-dimensional lattice in the plane perpendicular to the axes of the ribbons. *Ribbon phases* usually appear in phase diagrams of mixtures with at least two amphiphiles (e.g. an alcohol and a soap).

Lyotropics may also display medium-ranged three-dimensionally ordered phases. One of them is the *cubic micellar phase*, Q_m [75,76], with micelles packed on a cubic (face-centered or body-centered) lattice (see Fig. 1.17). In the *hexagonal micellar phase*, H_m, micelles are packed on a hexagonal three-dimensional lattice (hcp structure) [77].

The polymorphism of lyotropics presents bicontinuous structures. In this case, amphiphilic molecules self-assemble in a three-dimensional structure. The bicontinuous character means that a molecule, with its head in the aggregate surface, can diffuse continuously through *all* of the structure without the need of going to the bulk in which the solvent is present. This characteristic feature of the bicontinuous structure is not present in phases with aggregates as micelles, cylinders or lamellae, in which case a molecule needs to go through the solvent in order to diffuse from one aggregate to another. *Bicontinuous cubic phases* (called Q_b; see, e.g., Figure 1.3(c)) were also identified in lyotropic mixtures [76]. Another bicontinuous structure found in lyotropics is the *sponge phase*, called L_3, as sketched in Fig. 1.18. Experimental observations of this phase indicate a microstructure with a surfactant bilayer of multiply connected topology separating two solvent domains over macroscopic distances [78,79].

Other lyotropic mesophases with lower symmetries were identified [32,80], as the so-called *"white phase"* or *square phase* (direct, C, or inverted, K), in which long cylinders with predominantly quadratic cross section pack in a two-dimensional tetragonal symmetry.

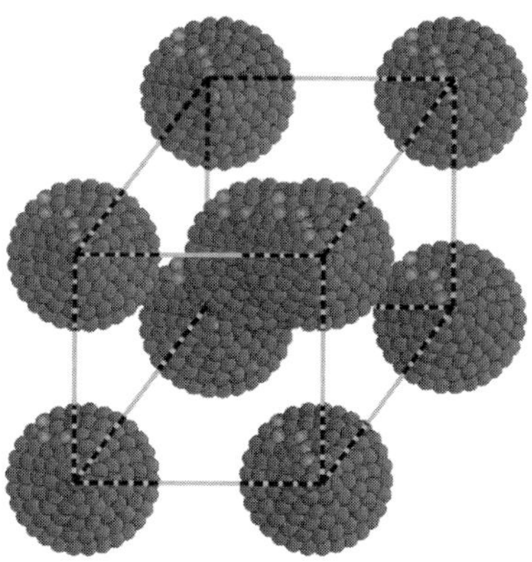

FIG. 1.17. Sketch of the direct micellar cubic (bcc) phase.

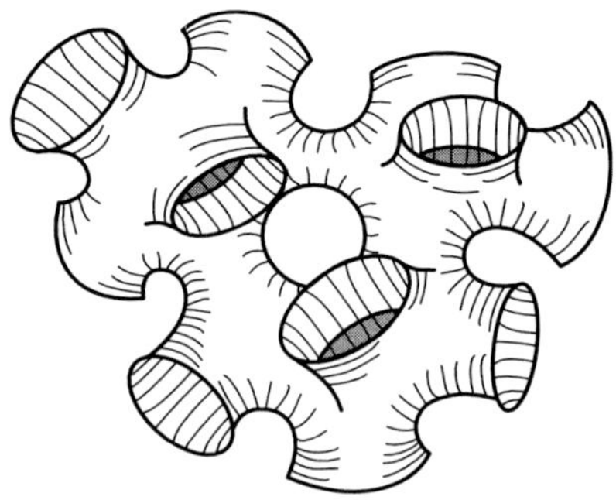

FIG. 1.18. Sketch of the sponge phase structure.

As a final remark about the polymorphism of lyotropics, we point out that there may be regions of phase coexistence, in some cases with more than two coexisting phases. Some of these coexisting phases can have a lower symmetry as compared with the cases discussed in this section. From the experimental point of view, mechanical procedures, as high-speed centrifugation, are commonly used to separate the different phases of the mixture.

1.2 An introductory example

1.2.1 *How to prepare a lyotropic mixture (specially for experimentalists)*

Each experimentalist develops his or her own method of preparation of the mixture, which obviously changes and improves with time. Some advice, however, should be relevant, in special for those who are beginning to work in this field.

The weighting procedure is very important. Attention should be paid to the precision of the equipment (a precision of at least 0.02 wt% is recommended) and to the weighting sequence of the different compounds of the mixture. It is recommended to prepare the mixture in a small tube with screw cap which leads to a very good closing. The volume of the tube depends on the quantity of the mixture to be prepared. However, as most of the experiments do not require large quantities of sample, a tube with a volume of about 10 ml can be used to prepare about 3 ml of the mixture. In order to guarantee the process of homogenization, it is important to let a large free space inside the tube.

The typical weight percentages of the lyotropic mixture sodium decylsulfate/decanol/water are 37.68/7.43/54.89, corresponding to the molar percentages 4.47/1.45/94.08.

We should begin by weighing the powder compounds. Then, we weigh the small liquid quantities, and the main solvent (just water, in this example). This solvent should be used to wash the walls of the tube, removing traces from the other components. After the weighting process, the tube is closed with its cap and sealed with paraffin in order to avoid any loss of mass.

The homogenization starts with the manual rotation of the tube around its axis (tilted with respect to the vertical direction by about 45°), for a few minutes, and after a sequence of shakes in a vortex and mechanical centrifugation. After a visual inspection indicates that the homogenization has been achieved, it is important to let the mixture rest, at about 25°C, for at least 24 h, before using it in the experiments.

1.2.2 *The potassium laurate (KL) lyotropic mixtures*

This soap is one of the most used compounds for the preparation of lyotropic mixtures. It is easily prepared from lauric acid, with a relatively cheap and straightforward synthesis. Binary, ternary and multicomponent mixtures with KL were shown to display lyotropic mesophases. As they lead to very rich phase diagrams, with many examples of lyotropic polymorphism, we pay special attention to these binary and ternary mixtures.

1.2.2.1 *The binary potassium laurate/water mixture* The potassium laurate molecule, with the carbonic chain in a *trans* (unfolded) geometry, is about 2.0 nm long and its surface per polar head is about 0.5 nm^2 at 100°C (with values depending on temperature and the phase under consideration [73]).

The phase diagram of this mixture [81], in terms of temperature T and the concentration c of amphiphiles, between the low-temperature crystallization line and the high-temperature isotropic phase transition, presents two lyotropic mesophases, H_α and L_α, and regions of phase coexistence (see Fig. 1.6). Changing the temperature at a fixed value of c, there is no transition between H_α and L_α. On the other hand, changing c at a given temperature, there may be a transition between H_α and L_α. The H_α region, also called "middle soap phase domain," is located at soap concentrations of about 50 wt% (where wt% means weight percent), in the middle of the phase diagram. The L_α phase, also called "neat soap phase," is located at a region of larger soap concentration of the phase diagram.

1.2.2.2 *The ternary potassium laurate/1-decanol/water mixture* A ternary mixture of potassium laurate, 1-decanol, and water is one of the most extensively studied lyotropic systems. We restrict the description to the neighborhoods of the regions of nematic phases. The polymorphism of this system is particularly rich. Also, it has led to the first identification of a biaxial nematic phase [62,82].

The topology of a section of the phase diagram, in terms of temperature and concentration of potassium laurate, at given concentration of 1-decanol, presents a low-temperature isotropic micellar phase, $L_1^{(l)}$, a high-temperature isotropic micellar phase, $L_1^{(h)}$, and three nematic phases, N_C, N_B, and N_D. The N_B phase is located between the two uniaxial phases (see Fig. 1.9).

The boundaries between the nematic–isotropic and nematic–nematic phase transitions are of first and second order, respectively [53]. The Landau point, at the convergence of a first-order and two second-order transition lines, is experimentally accessible in the case of this mixture [83].

From the analysis of early X-ray [59,60] and neutron [47] diffraction results, the micelles in the N_C and N_D nematic phases were assumed to be prolate and oblate ellipsoids, respectively. These conclusions were also based on the equivalence between the frame axes of the laboratory, where the diffraction patterns were actually obtained, and the frame axes attached to the micelles. It is important to note that in the time-scale of these diffraction experiments (at least minutes, for X-rays from synchrotron radiation) all the orientational and dynamical fluctuations of the micelles are averaged. The bare coherence length measured by Rayleigh scattering [84] is about 12 nm, larger than the typical dimensions of a micelle, of about 9 nm [61,85]. Therefore, no drastic modifications of the shape of the micelles are expected at the uniaxial–biaxial nematic phase transitions.

More recently, a different picture has been proposed on the basis of the analysis of detailed investigations of the profile of the X-ray diffraction bands for the three nematic phases [61,85]. All the available experimental results (diffraction, NMR, and light-scattering measurements, including the behavior in the presence of a magnetic field) can be explained if we assume the existence of similar (direct) micelles in the three nematic phases. These micelles are supposed to have a biaxial symmetry, as a flattened prolate ellipsoid, with three symmetry axes of order two, mutually orthogonal as in a "brick-like" geometry. The three nematic phases are a consequence of orientational fluctuations of these micelles (see Figs 1.11(a)–(c)).

The typical dimensions of a brick-like, orthorhombic, micelle are of the order $A' = 8.5$ nm, $B' = 5.5$ nm, and $C' = 2.6$ nm, where C' is associated with the potassium laurate bilayer, and A' and B' are dimensions in the plane perpendicular to the bilayer. The shape anisotropy of the micelles is about $3:2:1$ $(A':B':C')$. These dimensions undergo slight changes with temperature, but no drastic variations were detected at the nematic–nematic phase transitions. The orientational fluctuations which degenerate the symmetry axis (the *director* of the phase, which is represented by the unit vector $\mathbf{n}$) perpendicular to the largest surface of the micelles give rise to the N_D phase (see Fig. 1.11(b)). The orientational fluctuations which degenerate the symmetry axis in the plane of the largest surface of the micelle, along the largest axis of the flattened ellipsoid, give rise to the N_C phase (see Fig. 1.11(a)). Small amplitude orientational fluctuations along the three symmetry axes of the micelles give rise to the N_B phase (see Fig. 1.11(c)).

The model of intrinsically biaxial micelles in the three nematic phases was further confirmed by neutron diffraction measurements with contrast variation [86]. Neutron diffraction patterns of different samples (one of them with protonated KL and 1-decanol, and another one with perdeuterated KL and protonated

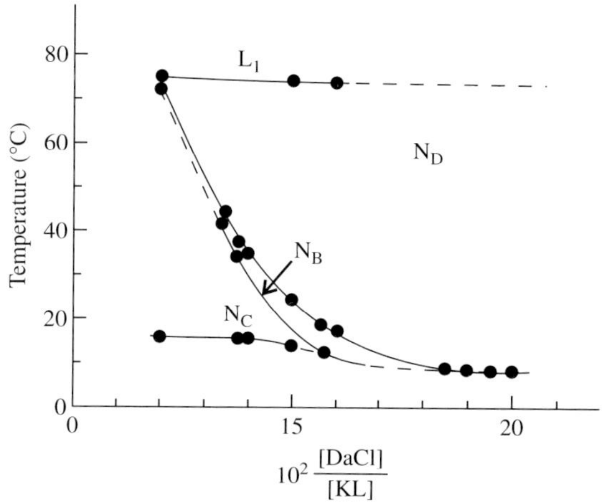

FIG. 1.19. Phase diagram of the potassium laurate/decylammonium chloride (DaCl)/water mixture (from ref. [90]).

1-decanol) in the N_C phase, with **n** parallel to the neutron incident beam, indicated a local biaxial ordering in this phase, in agreement with the model of intrinsically biaxial micelles.

The modification of the shape anisotropy of the micelles as a function of temperature seems to be the mechanism responsible for the presence of the low-temperature isotropic phase in the phase diagrams [87,88]. As the temperature decreases, the probability of *cis* configurations in the carbonic chains decreases and the bilayer thickness tends to increase, reducing the micellar shape anisotropy and favoring the appearance of an isotropic phase.

This particular polymorphism of the phase diagram, with three nematic phases, was only encountered in mixtures with at least two amphiphiles. Mixtures with only one amphiphile have only uniaxial nematic phases (discotic or calamitic, depending on the system); in these mixtures, there are no observations of transitions with an intermediate biaxial phase.

Other ternary lyotropic mixtures presenting a similar polymorphism are sodium decylsulfate (SdS or NadS), 1-decanol and water [89], and potassium laurate, decylammonium chloride (DaCl) and water [90], as sketched in Fig. 1.19.

1.3 The lyotropic mesophases

We now discuss some features of several lyotropic mesophases. In particular, we present a number of experimental results obtained with complementary techniques, which lead to a better characterization of phase structures and phase transitions.

1.3.1 *Micellar isotropic phases*

Micellar isotropic phases are optically isotropic, with a single index of refraction. In a polarizing light microscope (crossed polarizers), a sample holder filled with a lyotropic mixture in this micellar phase displays a black homogeneous and isotropic texture. The characteristic X-ray diffraction pattern of this phase presents a small-angle scattering due to the individual micelles at a typical value $s \sim 2 \times 10^{-2}$ nm^{-1} (s is the modulus of the scattering vector, given by $s = (2 \sin \theta)/\lambda_{\mathrm{X}}$, where 2θ is the scattering angle and λ_x is the X-ray wavelength [91]). An isotropic band can be present at $s \sim 2.0 \times 10^{-1}$ nm^{-1}, due to the inter-micellar correlations; at high-angles, due to the carbonic chains intercorrelations, there is an isotropic band at $s \sim 2.2$ nm^{-1}.

Depending on the particular region of the phase diagram, micelles can present different shapes and shape anisotropies. First, consider the case of direct micelles. In binary mixtures, at low concentrations of amphiphiles, just above CMC, direct micelles are spherical in shape, with typical diameter of about twice the length (carbonic chains in *trans* conformation) of the amphiphile. At larger concentrations of amphiphiles, particularly in the neighborhood of the domain of nematic phases, micelles can keep the same shape of the nematic phase (intrinsically biaxial, for mixtures with more than one amphiphile, oblate or prolate ellipsoids, for binary mixtures). In mixtures with more than one amphiphile, at small angles, intrinsically biaxial micelles present local ordering (pseudo-lamellar structure), which is also present in the nematic phases [85]. In this case of anisotropic micelles, the correlation volumes with about three micelles each, forming the so-called pseudo-lamellar structure, do not present the typical long-range orientational order of the neighboring nematics. In some cases, these anisometric micelles (or correlation volumes) can be oriented by means of a velocity gradient imposed to the sample [92]. The typical time-scale of the relaxation process of the shear-induced birefringence in the mixture of potassium laurate/1-decanol/water in the isotropic phase, in the neighborhood of the domain of the nematic phases, is of the order of 15 ms.

Under favorable conditions, the NMR technique provides a direct measurement of the magnetic surroundings of the counterion nuclei in the electric double-layer around the micelles. The main information is provided by the NMR line shape and its different rigid lattice contributions (spin–spin interactions and quadrupolar effects). The various microscopic environments of the nuclei give essentially two types of line broadening, homogeneous and inhomogeneous. In the first case, all contributions to the resonance line are centered around the Larmor frequency of a particular resonant nucleus, which is usually observed in systems with an isotropic distribution of local fields, as in the case of liquids. In the second case, systems show an inhomogeneous line broadening, and the resonance profile is composed by a distribution of lines around the Larmor frequency, but not centered on it [93]. Therefore, if it is possible to identify an inhomogeneous resonance line, the line shape is formed by the overlapping of individual peaks, such that the spectrum becomes a sort of histogram of distributions of local electric and magnetic fields in each one of the counterion nuclei. Consider, e.g., the micellar phase of a mixture of sodium lauryl sulfate (SLS) and water. This phase is characterized by strong dipolar-coupled ^{23}Na nuclei with an anisotropically broadened resonance line. Such anisotropy is characteristic of isotropic micellar structures. It also causes the appearance of rotational echoes [94]. The weak peak in the spectrum is identified as arising from quadrupolar-coupled ^{23}Na with a non-vanishing *electric field gradient* (EFG) at the resonant sites. Consequently, the motions of the SLS molecules are not fast enough to completely average out the EFG at the sites of sodium nuclei.

Cubic micellar phases are also optically isotropic and will be described in Section 1.3.6.

1.3.2 *Nematic phases*

The first classification of lyotropic uniaxial nematics was proposed about thirty years ago [58]. Based upon NMR experiments [95], measurements of the optical birefringence [49], and on the sign of the anisotropy of the diamagnetic susceptibility ($\Delta\chi = \chi_{\parallel} - \chi_{\perp}$, where $\chi_{\parallel}$ and $\chi_{\perp}$ are the susceptibilities parallel and perpendicular to the director of the phases), the lyotropic nematics were classified as type I and type II (no biaxial phases had yet been observed at that time). Type I mesophases have $\Delta\chi > 0$, negative optical anisotropy ($n_{\parallel} - n_{\perp} < 0$, where $n_{\parallel}$ and $n_{\perp}$ are the indices of refraction parallel and perpendicular to the director, respectively) and, in the presence of a strong enough magnetic field ($B \gtrsim 10$ kG), the director **n** aligns parallel to the field. Note that **n**, which is also called "optical axis," is the symmetry axis of the phase; in the case of uniaxial nematics, it is an infinite-fold axis. Type II mesophases have $\Delta\chi < 0$, positive optical anisotropy and, in the presence of a strong magnetic field, **n** aligns perpendicular to **B**. It is important to note that this classification cannot be directly applied to all lyotropic mixtures, since there are type II systems with $\Delta\chi > 0$ [55]. Perfluorated amphiphiles were shown to present lyomesophases where the signs of $\Delta\chi$ are inverted with respect to those of the carbonated amphiphiles [96].

Taking into account that the long axis of the paraffinic chains of the amphiphilic molecules tend to align perpendicular to the applied magnetic field **B**, and considering the macroscopic symmetry of the phases, it was proposed [97] that the micelles are cylinders and disks in type I and type II phases, respectively. The X-ray diffraction patterns of these phases [59] lead to the determination of the reciprocal space structures of uniaxial nematics. They can be depicted as a torus with the major axis parallel to **n**, and as an elongated hollow circular cylinder with an axis parallel to **n**, in type I and type II phases, respectively. This interpretation is consistent with the picture of cylindric and discotic micelles, but these micellar shapes are not the only configurations that could lead to the reciprocal space images obtained in the X-ray experiments. Hendrikx and Charvolin [60] proposed to label type I and type II mesophases as N_C (*calamitic* nematic phase) and N_D (*discotic* nematic phase), respectively.

When the biaxial nematic phase was identified [62], there was a question referring to what happens at the micellar level: Do the micelles change to a biaxial shape? Is there a mixture of disks and cylinders in the biaxial phase? Both possibilities account for almost all of the experimental results accumulated so far. Nevertheless, it is important to note that there is no direct experimental evidence of the presence of either cylinders or disks in lyotropic mixtures which display the three nematic phases. On the contrary, neutron scattering experiments [86] in N_C phases of ternary mixtures (with two amphiphiles) have clearly shown that there are no cylinders in this phase, supporting the model of *intrinsically biaxial micelles* discussed in Section 1.3.2. In binary mixtures, with only one (always uniaxial) nematic phase, symmetry reasons can be evoked to justify the existence of objects of higher symmetry, as disks and cylinders, but in mixtures with more than one amphiphile this argument cannot be directly applied.

A nematic phase of inverted micelles does not seem to have been characterized experimentally.

1.3.2.1 *The order parameter* The nematic phases of thermotropic liquid crystals are characterized [98,99] by a second-rank, traceless, symmetric tensor order parameter. In uniaxial (biaxial) phases, there are two (three) different eigenvalues. These eigenvalues are the symmetric invariants of the tensor. For example, the optical dielectric tensor $\overleftrightarrow{\epsilon}$, which has the same symmetry of the phases, can be chosen as the order parameter [1]. The anisotropic part of $\overleftrightarrow{\epsilon}$ can be written in a diagonal form,

$$\overleftrightarrow{\epsilon}_{\mathrm{a}} = \begin{pmatrix} \epsilon_{\mathrm{a}x} & 0 & 0 \\ 0 & \epsilon_{\mathrm{a}y} & 0 \\ 0 & 0 & \epsilon_{\mathrm{a}z} \end{pmatrix}, \tag{1.2}$$

where x, y, and z are the three orthogonal axes of the laboratory frame of reference, and the diagonal elements depend on the experimentally measured optical birefringences [53]. The birefringences are then proportional to the order parameters of the distinct phases. Optical birefringences are obtained as the difference between two of the refraction indices along the principal directions of the phase. In a uniaxial phase, there are only two different refraction indices. In a biaxial phase, however, there are three different refraction indices. Calling Δn and δn the birefringences that vanish in the uniaxial N_D and N_C phases, respectively, we have [53]

$$\begin{aligned} \epsilon_{\mathrm{a}x} &= \frac{4\langle n\rangle}{3}\left(\Delta n + \frac{\delta n}{2}\right), \\ \epsilon_{\mathrm{a}y} &= -\frac{2\langle n\rangle}{3}(\Delta n - \delta n), \\ \epsilon_{\mathrm{a}z} &= -\frac{4\langle n\rangle}{3}\left(\frac{\Delta n}{2} + \delta n\right), \end{aligned} \tag{1.3a}$$

where $\langle n\rangle$ is the average index of refraction of the mixture.

Since the birefringence of lyotropic nematics is of the order of 10^{-3}, thick samples usually have to be used to measure both Δn and δn. For example, using laser conoscopy [52,53], liquid crystalline films of about 2 mm have to be used to allow the observation of a reasonable number of conoscopic fringes (about six, three along each direction in the plane perpendicular to the laser beam), in order to measure birefringence with an accuracy of about 10^{-5}.

In the case of lyotropics, however, this simple definition of the order parameter does not seem enough, since the shape anisotropy of the micelles changes as a function of temperature and relative concentration of the compounds. This brings the need to introduce an additional non-critical order parameter for obtaining a complete description of the experimental nematic phase diagram [100]. This novel formulation will be introduced and developed in the forthcoming chapters.

1.3.2.2 *The calamitic nematic phase* N_C In mixtures based on amphiphiles with carbonic chains, the nematic phase N_C is characterized by $\Delta\chi > 0$, a negative optical anisotropy, and by a director **n** aligned parallel to the applied magnetic field **B**.

Usually, fields of about 10 kG are needed in order to orient samples 100 μm thick. However, an elegant method proposed by Brochard and de Gennes [101] allowed the orientation of samples in small (of the order of 100 G) magnetic fields. This procedure consists in introducing in the liquid crystal small magnetic grains (doping elements), which orient themselves in the magnetic field and, by a mechanical coupling with the director, turn out to orient the liquid crystal. In order to activate this mechanism, there should exist a minimum concentration of magnetic grains, which are supposed to promote a collective response of the liquid crystalline media to the small applied field [101,102]. Water based ferrofluids, also named *magnetic fluids*, which are colloidal suspensions of nanometric magnetic grains of typical dimensions of about 10 nm, dispersed in water, are the obvious candidates to be used. Liébert and Martinet reported the first experimental realization of a ferronematic lyotropic liquid crystal [103]. The typical concentration of magnetic grains for orienting liquid crystals, using samples about 200 μm thick, is of order 10^{12} grains/cm^3. At these concentrations, there are in the mixture about 10^6 micelles per ferrofluid grain, which corresponds, in a given direction of space, to about 100 micelles per grain. It has been observed that these concentrations of dopants do not lead to any modifications of the topology of the phase diagrams and of the values of the transition temperatures (at least, within 0.1°C). Nonlinear optical properties of the mixture, however, can be strongly affected by the doping. We will come back to this point in the sections of this book referring to nematic structures.

Textures and NMR identification A sample in the N_C phase, freshly prepared in a flat glass capillary (microslide), 200 μm thick, shows a typical schlieren texture in a polarizing light microscope (sample between crossed polarizers), as shown in Fig. 1.20. Applying a magnetic field $B = 8.4$ kG during 1 min, the texture presents inversion walls whose periodicity scales with the sample thickness [104].

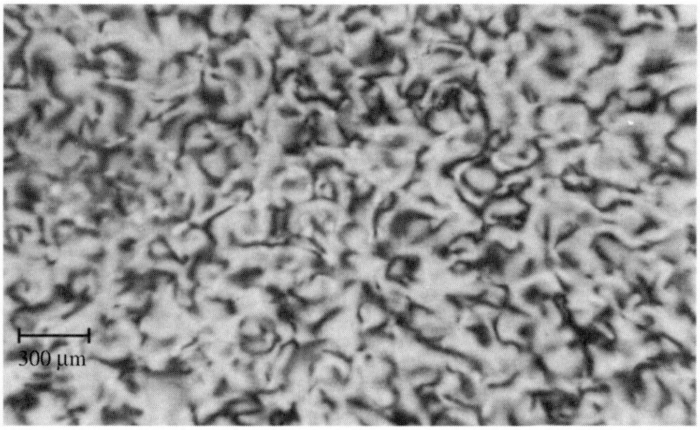

FIG. 1.20. Typical schlieren texture of a N_C phase (with a sample of thickness 200 μm of a potassium laurate/1-decanol/water mixture) in a polarizing light microscope, between crossed polarizer. The line in this figure corresponds to 300 μm.

This structure is unstable and, after some time in the field, there appear closed walls (ellipses), which relax to a final planar texture, with **n** parallel to **B** (see Fig. 1.21). However, since the process of alignment is slow, with a typical time scale of a few hours, it is possible to obtain the ratio between the bend elastic constant [105], k_{33}, and the anisotropy of diamagnetic susceptibility, $\Delta\chi$, by measuring the periodicity of the textures of the inversion walls [60–108]. Under some particular conditions, using ferrofluid doping, it is possible to obtain k_{33} and $\Delta\chi$ separately. For a potassium laurate/decanol/water mixture, $k_{33} \sim 2 \times 10^{-6}$ dyn and $\Delta\chi \sim 0.7 \times 10^{-8}$ cgs units.

Deuteron (^{2}H or D) NMR measurements of HDO (water molecules with H and ^{2}H), oriented in the N_C phase, show a typical quadrupolar splitting, which depends on temperature and the relative concentrations of the components of the mixture. For a ternary mixture of sodium decylsulfate, sodium sulfate and heavy water, this splitting is of the order of 600 Hz [109]. NMR measurements of some of the counterions of the mixture can also give information about the local ordering at the micellar length scale. If the sample with $\Delta\chi > 0$ spins around an axis perpendicular to the magnetic field, the NMR spectra is typical of a two-dimensional system, since the effect of the field is to spread the director in the plane perpendicular to the spinning axis.

Scattering and diffraction results At small angles, the typical X-ray diffraction pattern of the N_C phase, represented as a cut of the reciprocal space image of the phase, is shown in Fig. 1.22(b).

Let us define the laboratory frame of reference. The x axis is parallel to the applied magnetic field; the y axis is along the direction of the incident X-ray beam (the x and y axes define the horizontal plane). The z axis is parallel to the

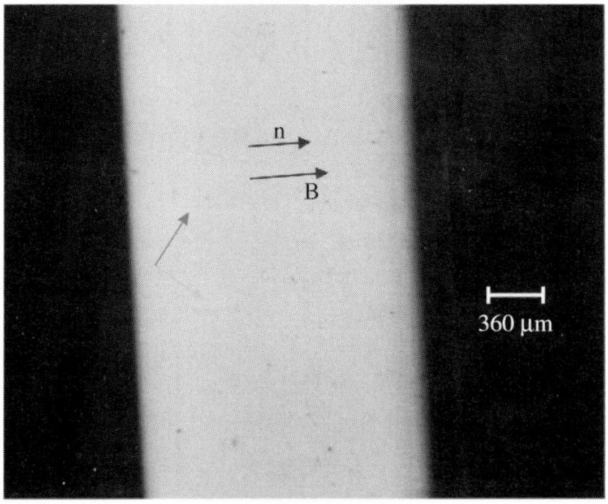

FIG. 1.21. Typical texture of a N_C phase in a polarizing light microscope, between crossed polarizer, about 1 h after the application of a magnetic field **B**, in a sample 100 μm thick. The arrows represent the directions of the polarizer and analyzer.

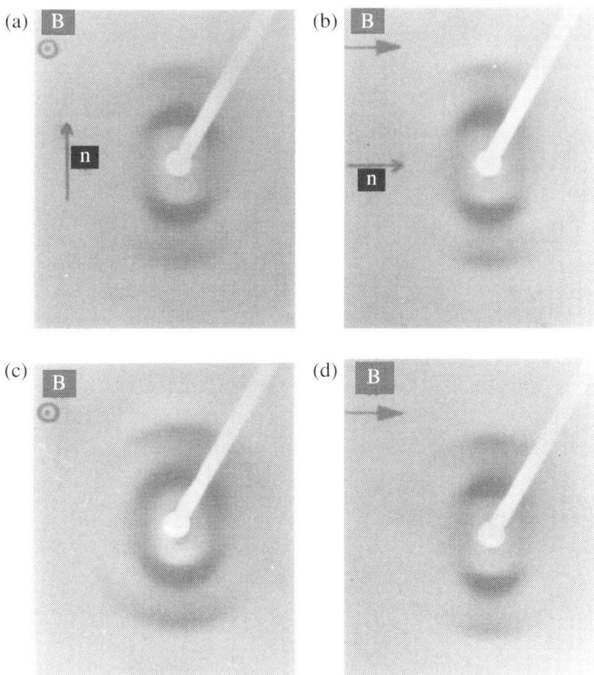

FIG. 1.22. Typical X-ray diffraction pattern of oriented sample (from ref. [61]): (a) N_D phase; (b) N_C phase; (c); and (d) N_B phase.

vertical direction. As discussed above, for samples with $\Delta\chi > 0$, the director $\mathbf{n}$ aligns parallel to the applied magnetic field $\mathbf{B}$ (consequently, along the x axis).

The reciprocal space image of this phase, obtained from a rotation of the pattern of Fig. 1.22(b) around $\mathbf{n}$, is a hollow cylindrical barrel, with the axis parallel to $\mathbf{n}$ and with more intense sides. A strong diffraction band is present along the direction perpendicular to $\mathbf{n}$ (that is, along the z axis). This band is due to the intermicellar positional correlation along this direction. For the potassium laurate/decanol/water mixture, this band gives a spacing $C \sim 4.9$ nm. An interesting feature in this pattern is the existence of a second-order band along the z axis, which was interpreted as originated from a pseudo-lamellar ordering [61] of the intrinsically biaxial micelles. In this structure, micelles pack along the direction perpendicular to the largest surface, with a typical mean-correlation distance, obtained from the width at half-height of the first-order diffraction band along the z axis, $\xi_C \sim 24$ nm, which corresponds to about 6 spacing distances for the potassium laurate mixture. In the direction parallel to the director (along the x axis), there is a broad and weaker band. This band has a structure that depends on temperature and on the particular nematic phase, and a maximum that moves in reciprocal space as a function of temperature [85]. This profile was used to describe other profiles in uniaxial and biaxial nematic phases. The diffraction band profiles of oriented N_D and N_B phases, in the x–y plane, were analyzed as a convolution product of the basic density profile of the band obtained for N_C with an ellipsoid of axes $1/A$, $1/B$, and $1/C$, where A, B, and C represent the orthogonal linear dimensions of the biaxial-like available volume per micelle. For the potassium laurate mixture, the broad band is associated with a characteristic distance $A \sim 10$ nm and a mean-correlation distance $\xi_A \sim 6$ nm (which corresponds to approximately $A/2$). This indicates a large polydispersity of the micellar dimensions in the direction parallel to $\mathbf{n}$. The difference between ξ_C and ξ_A can be understood taking into account that along the z axis a smaller polydispersity is expected since the length of the principal amphiphilic molecule defines a length scale for the structure, corresponding to the bilayer thickness. On the other hand, in the directions perpendicular to z (where the molecules pack side by side) there is no particular length scale, and the micelles can display a larger polydispersity. Besides the diffraction pattern in the N_C phase, we also need the patterns of all the three nematic phases in order to determine the complete set of distances of the available micellar volume in the context of the *intrinsically biaxial micelles model*.

At high angles, there is a broad band, almost isotropically oriented with respect to the y axis, at about 0.45 nm. It is interpreted as the mean distance between carbonic chains inside the micelles, due to the so-called "liquid-like" state of the chains.

1.3.2.3 *The discotic nematic phase* N_D In mixtures based on amphiphiles with carbonic chains, the discotic nematic phase is characterized by $\Delta\chi < 0$ and positive optical anisotropy. The director $\mathbf{n}$ aligns perpendicular to the applied magnetic field **B**. Usually, fields of about 10 kG are necessary for orienting samples of 100 μm of thickness, but ferrofluid doping reduces this field to about 100 G, as in the case of the N_C phase.

Different procedures can be used to obtain a well oriented N_D phase. One of them consists in rotating the sample around an axis perpendicular to the magnetic field (e.g. the z axis of the laboratory frame defined in the previous section), in order to have the director **n** oriented parallel to the z axis. Another possibility, mostly for thin samples (smaller than 100 μm), consists of using a combination of the magnetic field with a surface treatment which breaks the degeneracy of the field. A surface condition that orients **n** perpendicular to the surface, and a simultaneously application of **B** in this plane, will be enough to orient all the sample (bulk and surface layer).

Textures and NMR identification A sample in the N_D phase, inside a freshly filled flat glass capillary, between crossed polarizers, displays the texture shown in Fig. 1.23.

A sample with the director **n** along the z axis (which can be obtained by one of the procedures described before) presents a pseudo-isotropic texture under the polarizing microscope (observed along the z axis). This texture has an isotropic appearance, but this is due to the orientation of the director parallel to the direction of light propagation. This texture is typical of the homeotropic alignment of the director. If the sample orients with **n** in the x–y plane, the texture will show the usual planar alignment. A tip to distinguish between N_D, oriented in the homeotropic configuration, and a usual isotropic phase, in the case of samples placed in a flat glass microslide, is the observation of the borders of the flat

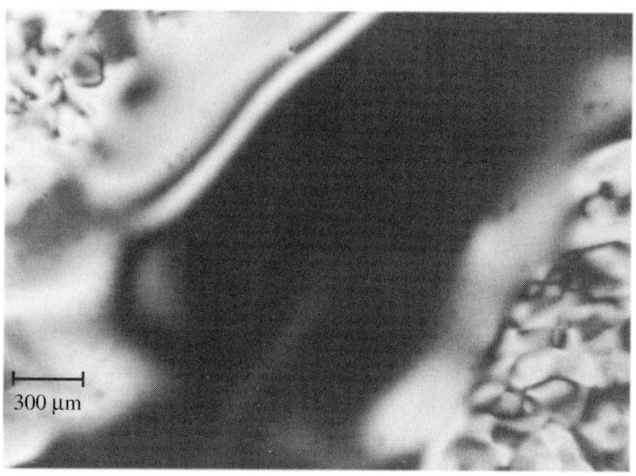

FIG. 1.23. Typical texture of an N_D phase in a 200 μm thick sample of a freshly prepared mixture of potassium laurate, 1-decanol, and water, in a polarizing light microscope, between crossed polarizers.

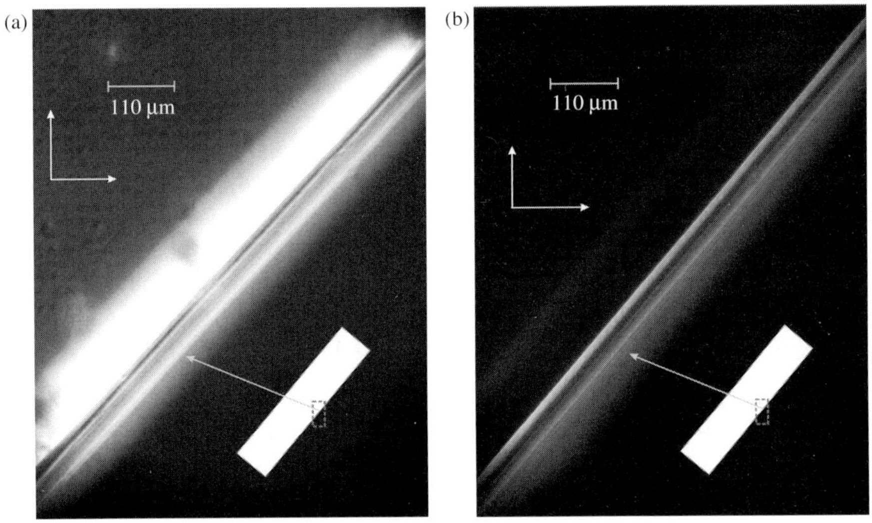

FIG. 1.24. Typical textures of N_D and L_1 phases in a polarizing light microscope, between crossed polarizers. Samples are placed in rectangular glass microslides sample-holders. The white arrows represent the directions of the polarizer and analyzer. (a) Pseudo-isotropic texture of a N_D phase in a homeotropic orientation of the director (see the bright borders of the sample holder). In the detail, in the right down corner of the figure, the microslide is sketched and the rectangle shows the corresponding region in the picture. (b) Micellar isotropic phase (L_1) in the same conditions of (a).

capillary under crossed polarizers, with the long axis of the capillary oriented at 45° of the light polarizing direction. In the N_D phase, these borders appear bright (Fig. 1.24(a)); in the isotropic phase, they remain black (Fig. 1.24(b)). This happens because it is usual that **n** tends to align perpendicular to the glass surfaces and, in the case of the borders of the flat capillary, the semi-cylindrical geometry forces **n** to turn 180° from the upper to the lower capillary surface (and assuming a planar orientation).

The ^2H magnetic resonance of oriented HDO in the N_D phase gives a typical quadrupolar splitting (e.g. of about 250 Hz in a mixture of decylammonium chloride, ammonium chloride and water [51]). In contrast to the N_C phase, if a sample in the N_D phase, with $\Delta\chi < 0$, spins around an axis perpendicular to the magnetic field, it aligns with **n** parallel to the spinning axis, leading to the typical spectra of an oriented sample.

Scattering and diffraction results At small angles, the typical X-ray diffraction pattern of the N_D phase, represented as a cut of the reciprocal space image of the phase, is shown in Fig. 1.22(a). As discussed before, for samples with $\Delta\chi < 0$, the director **n** aligns perpendicular to the applied magnetic field **B** (in the pattern shown in Fig. 1.22(a), **n** is parallel to the z axis). The reciprocal space image of this phase, obtained by rotating the pattern of Fig. 1.22(a) around **n**, is a hollow cylindrical barrel, with the axis parallel to **n** and with more intense caps. A strong diffraction band is present along the direction parallel to **n** (z axis). This band is due to the intermicellar positional correlation along this direction. For the potassium laurate/decanol/water mixture, this band is associated with a spacing distance $C \sim 4.9$ nm. As in the case of the N_C phase, a second-order band along the z axis is observed, which indicates the presence of the same pseudo-lamellar ordering [61]. In the context of the *intrinsically biaxial micelles model*, the micelles pack in a direction perpendicular to the bilayer plane, with a typical mean-correlation distance (obtained from the width at half-height of the first-order diffraction band along the z axis) $\xi_C \sim 30$ nm (which corresponds to about six spacing distances). In a direction perpendicular to the director (along the x axis), there is a broad and considerably weaker band, which also has some structure [85]. Its profile was described as discussed before. For the potassium laurate mixture, the broad band is associated with a characteristic distance $B \sim 8.5$ nm and a mean-correlation distance $\xi_B \sim 6$ nm, which indicates a large polydispersity of the sizes of the micelles in the plane perpendicular to **n**. According to the *intrinsically biaxial micelles model*, besides the diffraction pattern in the N_D phase, we also need the patterns in all of the three nematic phases in order to determine the complete set of distances of the available micellar volume. The complete description of the micelles in the three nematic phases will be presented in the next section.

At high angles, there is a broad band, almost isotropically oriented with respect to the y axis, at about 0.45 nm. It is interpreted as the mean distance between carbonic chains inside the micelles, due to the "liquid-like" state of these chains.

1.3.2.4 *The biaxial nematic phase* N_B The biaxial nematic phase N_B has three mutually orthogonal two-fold symmetry axes and two optical axes [52,53]. The experimental procedure to orient a sample inside a flat capillary, suitable for optical measurements, or a cylindrical capillary tube, for X-ray scattering and diffraction experiments, consists in applying a magnetic field, which is strong in usual samples and small in ferrofluid doped samples, and then, at typical time intervals of about 2 min, turning the samples, at alternate angles of $\pm 45°$, around an axis perpendicular to the field **B**, and returning again to the original position. For example, consider the case of a cylindrical glass capillary, with the axis parallel to the z direction, filled with a sample in the N_B phase. Initially, the sample is subjected to a magnetic field **B** parallel to the x axis. After some minutes in this configuration, the capillary is turned $+45°$ around the z axis and stays there for about 2 min. After these 2 min, the sample is turned back to the original orientation, and continuously turned to $-45°$, where it stays about 2 min also. After these operations, the sample turns back again, staying about 5 min in the initial orientation. This procedure is repeated until achieving a complete orientation of the sample. A good check of the quality of this alignment consists in observing the texture and the conoscopic figure. Symmetric conoscopic patterns and a homogeneous texture are signatures of a well-oriented sample. For the potassium laurate mixtures, the largest sides of the micelles orient parallel to the field **B**.

NMR spectroscopy experiments of ^{2}H, performed in a mixture of sodium dodecylsulfate, decanol and water, were explained with the assumption of two types of biaxial nematic phases, N_{Bx}^+ and N_{Bx}^-, in the partial isotherm of the phase diagram between the N_C and N_D phases [110]. In order to characterize the behavior in the presence of an external magnetic field, we can use the anisotropy of the diamagnetic susceptibility,

$$\Delta\chi = \chi_{33} - \frac{1}{2}(\chi_{11} - \chi_{22}), \tag{1.4}$$

where the subscripts refer to the three orthogonal directions in the orthorhombic basic unit. If $\Delta\chi > 0$, a biaxial phase aligns with the axis of the largest diamagnetic susceptibility parallel to the magnetic field (N_{Bx}^+). If $\Delta\chi < 0$, a biaxial phase aligns with the axis of the smallest diamagnetic susceptibility perpendicular to the magnetic field (N_{Bx}^-). According to the formulation of the NMR theory, the N_{Bx}^+ and N_{Bx}^- phases align with the largest residual *electric field gradient* component along and perpendicular to the applied magnetic field, respectively. On the basis of NMR quadrupole splitting measurements, Quist [110] concluded that the transitions between the nematic phases of this system are of first order. A similar study has been recently performed in another lyotropic mixture, tetradecyltrymethylammonium bromide/n-decanol/water [111].

Textures and NMR identification A sample in the N_B phase, placed in a flat and freshly filled glass capillary, presents a texture resembling the pattern of the N_C phase. An oriented sample, according to the techniques which have been described, displays a planar-type texture. In these conditions of orientation, both optical axes of the phase are oriented in a plane perpendicular to the x–y plane, which contains the z axis and the field **B**, with each axis forming an angle θ with respect to the z axis. This angle depends on temperature in the region of the N_B phase; it approaches 0° in the vicinity of the N_B–N_D phase transition, and 90° in the vicinity of the N_B–N_C phase transition. Only the measurements of the order parameters, proportional to Δn and δn, can identify the N_B phase unambiguously. As the conoscopic patterns of the N_C and the N_B phases are similar [53], measurements of birefringence become fundamental for establishing a distinction between these phases. However, the simple inspection of the conoscopic patterns of the N_B phase is not enough for a clear identification. Usually, the conoscopic fringes are thinner and more clearly defined in the N_B phase.

In a mixture of potassium laurate (KL), 1-decanol and D_2O, the temperature dependence of the ^{2}H magnetic resonance splittings of oriented samples in the nematic phases displays some anomalies (strong discontinuities of the first derivatives with respect to temperature) at the uniaxial–biaxial phase transitions [62,112]. The values of the splitting increase from the boundaries of the uniaxial–biaxial phase transitions towards the middle of the biaxial region. In the uniaxial phases, the splitting curves as a function of temperature present a smooth convex shape; in the N_B phase, this convex shape is more pronounced.

Scattering and diffraction results The small-angle X-ray diffraction patterns of the N_B phase, corresponding to two cuts of the reciprocal space image, are shown in Figs 1.22(c) and (d). For potassium laurate samples, the large axis of the micelles orients parallel to the field **B**. The reciprocal space image of this phase is a hollow barrel of elliptical cross section in a plane perpendicular to its long axis (i.e., a flattened ellipsoid), with more intense caps. Along the direction parallel the z axis, there is a strong diffraction band, due to intermicellar positional correlations. For a KL/decanol/water mixture, this band is associated with a spacing distance $C \sim 4.9$ nm. As in the case of the N_C and N_D phases, a second-order band along the z axis is observed, which indicates the presence of the same pseudo-lamellar ordering [61]. In the context of the *intrinsically biaxial micelles model*, the micelles pack in the direction perpendicular to their largest surface, with a typical mean-correlation distance, defined by the width at half-height of the first-order diffraction band along the z axis, $\xi_B \sim 30$ nm (which corresponds to about 6 spacing distances for the potassium laurate mixture). Along the x (parallel to **B**) and y axes, there appear broad and weaker structured bands, associated with different characteristic distances [85]. These profiles can be described as we have already done in a previous section. In the potassium laurate mixture, the broad bands are associated with the characteristic distances $B \sim 8.5$ nm and $A \sim 11.0$ nm (where A is along a direction parallel to **B**) and mean-correlation distances $\xi_{A,B} \sim 6$ nm, also indicating a large polydispersity of the sizes of the micelles in the plane perpendicular to the direction of the bilayer.

Now we have the complete set of distances of the available micellar volume in a potassium laurate mixture, $A \sim 11.0$ nm, $B \sim 8.5$ nm, and $C \sim 4.9$ nm, within the context of the *intrinsically biaxial micelles model*. In order to calculate micellar dimensions and the associated shape anisotropy, we assume that the micelle symmetry is the same of the available volume per micelle in the N_B phase and that water is equally covering all the surfaces of the micelles. For the potassium laurate mixture, this assumption [85] leads to micellar dimensions $A' \simeq 8.5$ nm, $B' \simeq 6.5$ nm, and $C' \simeq 2.6$ nm, from which we have the micellar shape anisotropy $3:2:1$.

At high angles, a broad band, almost isotropically oriented with respect to the y axis, is observed at about 0.45 nm. It is interpreted as the mean distance between carbonic chains inside the micelles, due to the "liquid-like" state of these chains.

It is interesting to make a comment about thermotropic biaxial nematics. In the published literature, despite the efforts of many investigators, there is no convincing result to clearly support the existence of a thermotropic biaxial phase. Comparing micelles with the molecules that are usually employed in the search for a thermotropic biaxial phase [113], the shape anisotropy of these molecules is always much larger than the values obtained for the micelles. This could be one of the reasons for the failure to find a thermotropic biaxial nematic phase. Molecules with a much smaller shape anisotropy should be necessary to provide reliable tests. On the other hand, micelles of the lyotropic mixtures are capable of undergoing continuous changes of shape anisotropy as a function of concentrations and temperature. These changes are a sort of tuning of the shape anisotropy required to break the symmetry of one of the uniaxial phases. Indeed, micelles undergo orientational fluctuations which can degenerate the axis of the director; if only small amplitude orientational fluctuations are allowed, the biaxial phase is stabilized. This property, however, has not been observed in molecules of thermotropic liquid crystals.

In a recent publication [114], Acharya and coworkers presented some experimental evidence, on the basis of X-ray diffraction measurements, supporting the existence of a biaxial nematic phase in bent-core thermotropic mesogens.

1.3.2.5 *Nematic–isotropic phase transition* The nematic–isotropic phase transition is of first order, associated with a jump in the birefringence (order parameter) measurements. In the potassium laurate mixture, the jumps from the N_C and the N_D phases to the isotropic phase are given by 10^{-3} and 0.4×10^{-3}, respectively [53]. Other mixtures (e.g. DaCl and SdS) present the same behavior and typical values for the jumps. In the phase diagrams, at a first-order transition, there are regions of coexistence of phases. The range of temperatures of this coexistence depends on the relative concentrations and the type of the components of the mixture. A typical texture of this coexistence region is shown in Fig. 1.25.

The existence of a high-temperature nematic–isotropic phase transition is expected for entropic reasons. However, some lyotropic mixtures also present a low-temperature transition to an isotropic phase [62]. This transition is attributed to the changes of the shape anisotropy of the micelles as a function of temperature. In particular, the parameter C' of the micelles is known to increase for decreasing temperatures [83,87].

The experimental behavior of the nonlinear refractive indices and birefringences in the neighborhood of the nematic–isotropic phase transition has been recently investigated [115]. The nonlinear optical birefringence shows a discontinuity at the transition temperature, in agreement with the linear birefringence measurements, which further supports the first-order nature of this phase transition.

1.3.2.6 *Nematic–nematic phase transition* This is one of the most interesting phase transitions in lyotropic mixtures, specially due to the presence of a biaxial phase between the two uniaxial phases. The temperature range of the N_B phase depends on the relative concentrations and type of the components of

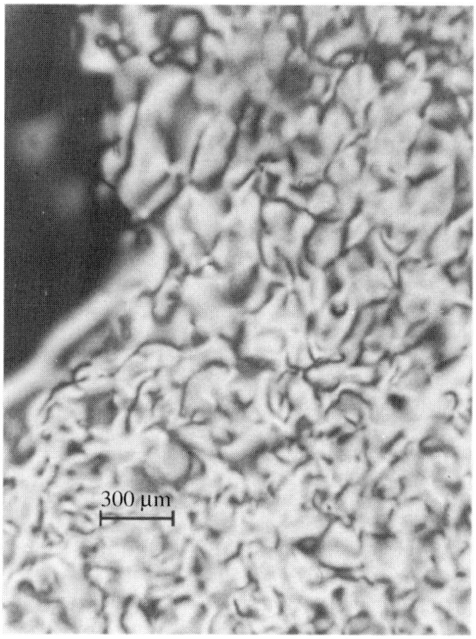

FIG. 1.25. Typical texture of coexisting nematic and micellar isotropic phases (for a 200 µm thick sample of a potassium laurate/1-decanol/water mixture) in a polarizing light microscope, between crossed polarizers. The isotropic phase domain is located at the upper left side of the figure.

the mixture. From the experimental point of view, at least two amphiphiles are necessary in order to give rise to a biaxial phase.

Mean-field approach In a temperature range of about $\Delta T/T_c \gtrsim 10^{-3}$, where $\Delta T = T - T_c$, and T_c is the transition temperature, the nematic–nematic phase transition was shown to be behave smoothly, as a second-order transition [53]. As it will be shown in Chapter 2, the invariants of a tensor order parameter, I_0, I_1, and I_2, can be written in terms of the diagonal elements of the dielectric tensor [98],

$$\begin{aligned} I_0 &= \epsilon_{ax} + \epsilon_{ay} + \epsilon_{az} = 0, \\ I_1 &= \frac{2}{3}(\epsilon_{ax}^2 + \epsilon_{ay}^2 + \epsilon_{az}^2), \\ I_2 &= 4\epsilon_{ax}\epsilon_{ay}\epsilon_{az}, \end{aligned} \tag{1.5}$$

where the first invariant is zero, since we are considering a traceless tensor. It should be noted, however, that this assumption does not account for changes in the micellar shape anisotropy, and thus may not be enough for the description of the lyotropic mixtures.

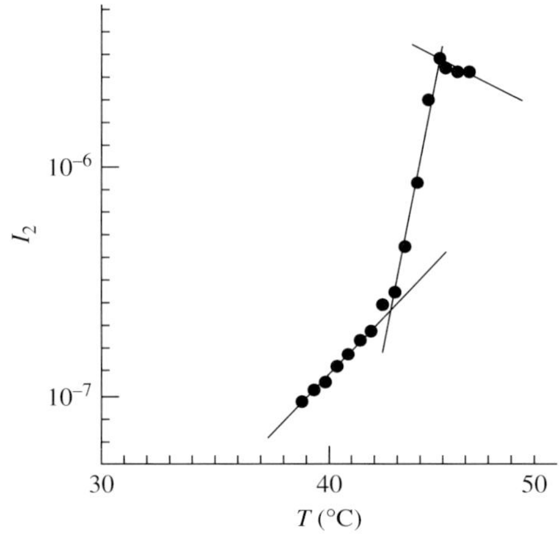

FIG. 1.26. Invariant I_2 of the dielectric tensor as a function of temperature, at the uniaxial–biaxial phase transition in a potassium laurate/decylammonium chloride (DaCl)/water mixture (from ref. [90]).

From measurements of optical birefringences, Galerne and Marcerou [53] have shown that these tensor invariants display a linear behavior in the uniaxial and biaxial nematic regions, in the vicinity of the phase transitions, with a precision of about 0.02°C between successive measurements. In Fig. 1.26, we show the I_2 invariant as a function of temperature for the potassium laurate/decylammonium chloride (DaCl)/water mixture [90].

Saupe and coworkers [83,112] used a mean-field approach in order to account for the available experimental results for these transitions in lyotropic mixtures. The measurements of one of the birefringences of the N_B and N_C phases, with a temperature precision of about 1 mK, already indicated small deviations of the uniaxial–biaxial transition from the predictions of this mean-field approach.

As in the case of the nematic–isotropic phase transition, the behavior of the nonlinear refractive indices and birefringences was also investigated in the neighborhood of the nematic–nematic phase transition [115]. In the vicinity of the transition temperature T_c, the nonlinear birefringence was shown to behave as $|T - T_c|^{-0.5}$ in both uniaxial and biaxial regions. The symmetry invariants of the order-parameter were shown to be linear functions of temperature in the uniaxial nematic domain, in good agreement with the mean-field prediction.

Critical behavior The uniaxial–biaxial phase transition was investigated from both theoretical [98,116] and experimental [117–119] points of view. The critical properties are supposed to belong to the universality class of the three-dimensional XY model [120]. Deviations from the mean-field behavior were observed [118,119] in measurements of the optical birefringence in the temperature range $\Delta T/T_c \sim 10^{-4}$ from the transition temperature of the uniaxial–biaxial transition. The critical exponents associated with the order parameter, β, and the susceptibility, γ, of the potassium laurate mixture were found to be $\beta = 0.38 \pm 0.03$ and $\gamma = 1.29 \pm 0.06$ (as compared with the theoretical predictions for the XY model [120], $\beta = 0.38$ and $\gamma = 1.35$).

1.3.3 *Cholesteric phases*

Cholesteric lyotropic phases, also named lyocholesterics, can be obtained by adding a chiral molecule to a mixture that displays a nematic phase. Maybe we should say doping, instead of adding, since it is usual that small amounts of a dopant molecule are already enough to produce a cholesteric mesophase. This procedure can be implemented in two ways: (a) by adding to the mixture an amphiphilic chiral molecule, which gives rise to an *intrinsic* cholesteric phase, and (b) by adding a non-amphiphilic chiral molecule, which gives rise to an *extrinsic* cholesteric.

Three types of lyotropic cholesterics were identified [63,65,66], $\mathrm{Ch_C}$, $\mathrm{Ch_D}$, and $\mathrm{Ch_B}$. The labels C, D, and B refer to the original nematic phases (calamitic, discotic, and biaxial), which were cholesterized by the addition of the chiral dopant. As in the case of nematics, mixtures with only one amphiphile give rise to only one of the cholesteric phases, $\mathrm{Ch_C}$ or $\mathrm{Ch_D}$. Mixtures with at least two amphiphiles can give rise to the three cholesteric phases. In this case (e.g. in a mixture of sodium decylsulfate, 1-decanol, water and brucine sulfate heptahydrate, $C_{46}H_{68}N_4O_{19}S$, [66]), it has been observed that micelles spontaneously pack in a helicoidal structure, keeping however their short-range pseudo-lamellar ordering.

Let us consider mixtures in which the principal amphiphile has carbon chains, that is, molecules which tend to have tails aligned perpendicular to the applied magnetic field. In the presence of a strong enough magnetic field (or of small fields, if the mixtures are doped with ferrofluids), the helical axes of $\mathrm{Ch_D}$ and $\mathrm{Ch_B}$ biaxial phases align parallel to the magnetic field. On the other hand, under the same conditions, $\mathrm{Ch_C}$ phases unwind in the presence of a field, giving rise to a planar alignment of the director. The director of the former nematic phase (which was the origin of the cholesteric phase) orients perpendicular to the helicoidal axis in both cases of the $\mathrm{Ch_C}$ and $\mathrm{Ch_D}$ phases. In the $\mathrm{Ch_B}$ phase, the largest dimension of the micelles orients along the helicoidal axis.

The pitch $P = 2\pi/q$ of the helicoidal structure, where q is the wave number, depends on different physico-chemical parameters [1,67,121], as temperature, concentration of the chiral dopant, c_m, and micellar shape anisotropy, S_a. In the context of the *intrinsically biaxial micelles model*, we write

$$P^{-1} \propto c_\mathrm{m}\, S_\mathrm{a}, \tag{1.6}$$

where $S_\mathrm{a} = A'/C'$. From this equation, we see that, at fixed micellar shape anisotropy, the concentration of chiral molecules is inversely proportional to the pitch. On the other hand, at fixed concentration of chiral molecules, the pitch decreases with increasing micellar shape anisotropy. In actual cholesterics, the typical length scale of the pitch is of the order of micrometers. A generalized Landau–Ginzburg–de Gennes theory, introducing the basic features of lyotropics, as the changing of the micellar shape anisotropy in terms of temperature and relative concentrations, has been developed to account for the dependence of P on the geometrical parameters [68].

Some features of the topology of the phase diagrams of the lyotropic cholesteric mixtures are fundamentally different from the corresponding lyotropic nematics [122–124]. No Landau points have been clearly observed, and large polyphasic regions surround the cholesteric domains (see Fig. 1.27).

1.3.3.1 *Intrinsic cholesterics* Intrinsic cholesterics are obtained by adding a chiral amphiphilic molecule to a nematic lyotropic liquid crystal. These molecules take part in the micellar structure as a non-chiral amphiphile. Examples of chiral molecules used to this purpose are l-n-lauroyl potassium alaninate [124], 2-sodium decylsulfate [125], and L,D-octanol.

1.3.3.2 *Extrinsic cholesterics* Extrinsic cholesterics are obtained by adding a chiral non-amphiphilic molecule to a lyotropic nematics. Depending on the electrostatic characteristics of the molecule (polar or nonpolar), it can be accommodated either in the inner or in the outer part of the micelle. Examples of chiral molecules used to this purpose are brucine sulfate heptahydrate, tartaric acid [63], cholesterol [63], L-sorbose, diacetone-sorbose, and diacetone-2-ceto potassium gulonate.

1.3.3.3 *Texture and NMR identification* Under the polarizing microscope (between crossed polarizers), a film of lyotropic cholesterics presents fingerprint,

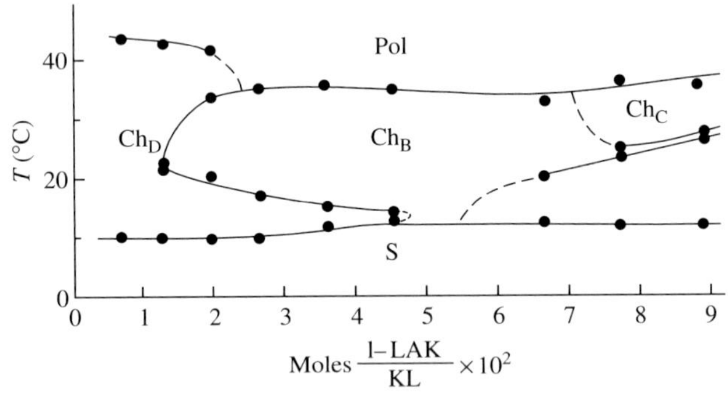

FIG. 1.27. Surface of the phase diagram of the lyotropic cholesteric mixture of potassium laurate (KL)/1-decanol/water/l-N-lauroyl potassium alaninate (l-LAK) [124]; POL and S refer to the polyphasic and gel regions, respectively.

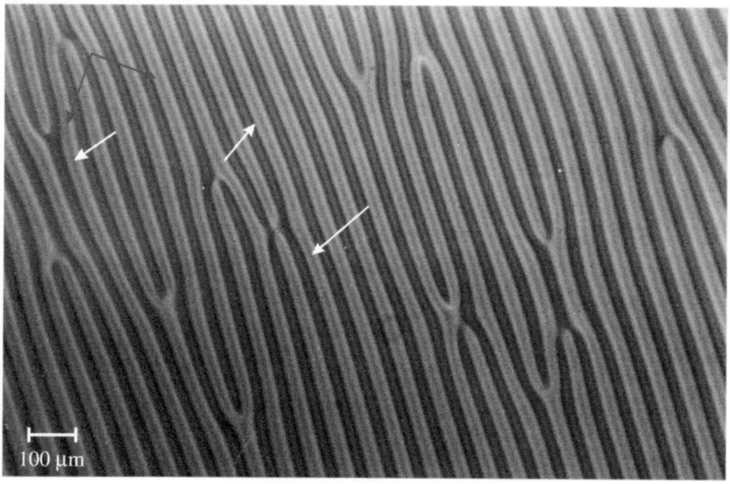

FIG. 1.28. Typical texture of an oriented cholesteric Ch_D lyomesophase (for a 200 μm thick sample of a sodium decylsulfate/1-decanol/water/brucine sulfate heptahydrate mixture) in a polarizing light microscope, between crossed polarizers. Black arrows represent the directions of the polarizer and analyzer. White arrows show some homeotropic regions in the center of fingerprint stripes.

chevron, focal conic or Grandjean textures [123], as shown in Fig. 1.28. A sequence of parallel stripes, with different colors (in the case of high birefringent samples and white light illumination) or different gray tonalities (in the case of samples with a small birefringence) can be identified in the texture.

Ch_C and Ch_D mesophases show, in the middle of the stripes of the (helicoidal) fingerprint texture, a pseudo-isotropic region (see Fig. 1.28), where the director is oriented in a homeotropic configuration. In the Ch_B type of cholesteric, however, this region is not present [66] (see Fig. 1.29).

The deuteron NMR spectra of the HDO mixture oriented in the Ch_C phase (non-spinning sample) shows an evolution of the quadrupolar splitting as a function of time due to the unwinding of the helix in the presence of the magnetic field. This happens, e.g., in a mixture based on potassium l-N-lauroylserinate, which has $\Delta\chi > 0$ [126]. Initially, a powder-type spectrum is observed and, after a few minutes in the presence of the magnetic field of the spectrometer, the intensity of the wings of the spectrum start to increase. After about 30 min in the spectrometer, a doublet with a quadrupole splitting becomes well defined and the splitting remains constant. Turning the sample by 90° in the spectrometer, the splitting is reduced to 1/2 of its value in the previous configuration. This is an indication that the cholesteric helix was unwound and the director is now oriented parallel to the magnetic field. In Ch_D samples, after some minutes in the magnetic field, the NMR spectra of ^2H show the typical doublet associated

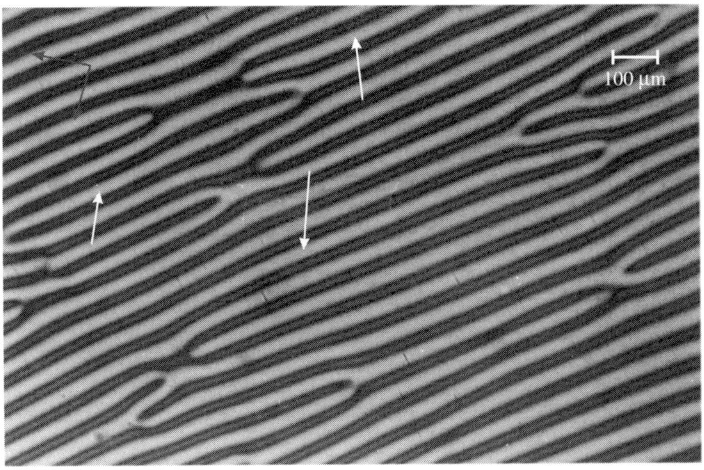

FIG. 1.29. Typical texture of an oriented cholesteric Ch_B lyomesophase (for a 200 μm thick sample of sodium decylsulfate/1-decanol/water/brucine sulfate heptahydrate mixture) in a polarizing light microscope, between crossed polarizers. Black arrows represent the directions of the polarizer and analyzer. White arrows show some planar regions in the center of fingerprint stripes.

with the quadrupolar splitting of an oriented sample (for samples with $\Delta\chi > 0$, the cholesteric helix orients parallel to the magnetic field) [127]. Turning the sample by 90° in the spectrometer, the spectra show two-dimensional characteristics immediately after the rotation [127]. A powder-type spectrum is observed if the samples spin around an axis perpendicular to the magnetic field.

1.3.3.4 *X-ray diffraction results* The structure and local ordering of the micelles in lyocholesterics, at length scales of nanometers (the typical length scale of P is micrometers), can be investigated by X-ray diffraction. It is interesting to ask how the chiral molecules modify the local order of the nematics to form a long-range cholesteric structure. In particular, it is interesting to investigate how the pseudo-lamellar order observed in some lyotropic nematics is affected by the presence of chiral molecules.

Two mixtures were studied with respect to their local order and structure in nanometric length scales [128]: sodium decylsulfate, 1-decanol, water, and brucine sulfate heptahydrate, called BS; and potassium laurate, 1-decanol, water, and D-2-octanol. The phase diagram of these mixtures display a Ch_C phase that unwinds the cholesteric helix in the presence of sufficiently strong magnetic fields. Comparing the diffraction patterns of field-unwind Ch_C mesophases with those of the corresponding N_C phases, it was possible to check that the pseudo-lamellar order observed in nematics is still present in cholesterics.

Although the general features of the diffraction patterns in nematics and cholesterics remain essentially unchanged, in the particular case of the extrinsic cholesteric mixture with BS the shape of the strong first-order band is slightly different with respect to the corresponding nematics (the edges of the diffraction band are bent towards small scattering vector intensities [128]). These results were interpreted as an indication of micellar deformations, with the corresponding swelling of the pseudo-lamellar order near their edges, due to the presence of large BS molecules (the available parallelepiped volume per BS molecule being about $1 \times 1.6 \times 0.7$ nm^3). Within the resolution of the experiments, this modification of the band shape has not been detected in the intrinsic cholesteric mixtures, which is an indication that the incorporated chiral molecules do not lead to a swelling of the pseudo-lamellar structure.

1.3.3.5 *The cholesteric–cholesteric phase transition* From the experimental point of view, the inspection of the texture evolution of a lyocholesteric sample under crossed polarizers as a function of temperature indicates that the Ch–Ch phase transitions are continuous [119,123]. From the theoretical point of view, there are a number of mean-field calculations to describe these transitions [68,129,130].

The chiral field Cholesteric phases provide an interesting topic of research since the micelles at the helicoidal structures are subjected to an elastic field. In contrast to the nematic phase in a magnetic field **B**, which couples to the (long-range) collective orientational modes of the director [1], the elastic field acts at the (much smaller) micellar scale. In cholesterics, this elastic field is the *chiral field*. The effect of this field in the Ch–Ch phase transition is a chirality-induced biaxiality [119,131], qualitatively equivalent to the induced biaxiality at the uniaxial–biaxial transition in nematics, but with a smaller amplitude.

In the mean-field approach, the induced biaxial order parameter ζ in the $\mathrm{Ch_{C,D}}$ phases is proportional to the square of the wave number, $q = 2\pi/P$, and inversely proportional to the reduced temperature,

$$\zeta \propto \frac{q^2}{T - T_c}. \tag{1.7}$$

Measurements of the optical birefringences (δn and Δn) lead to the determination of ζ (since $\zeta \propto \delta n/\Delta n$). Interferometric measurements performed in the $\mathrm{Ch_D}$–$\mathrm{Ch_B}$ transition for different cholesteric mixtures [119,131] show that the order of magnitude of ζ^2 at T_c is 10^{-2}, at a concentration of chiral molecules of about 10^{-2} M% (see Fig. 1.30). The corresponding susceptibility coefficient, which is essentially a ratio between the Landau coefficients of the free-energy expansion [131], was shown to be of the order of 10^{-10} cm^2 K, and the bare correlation length calculated from this estimate for the susceptibility is $\xi_0 \sim 10$ nm.

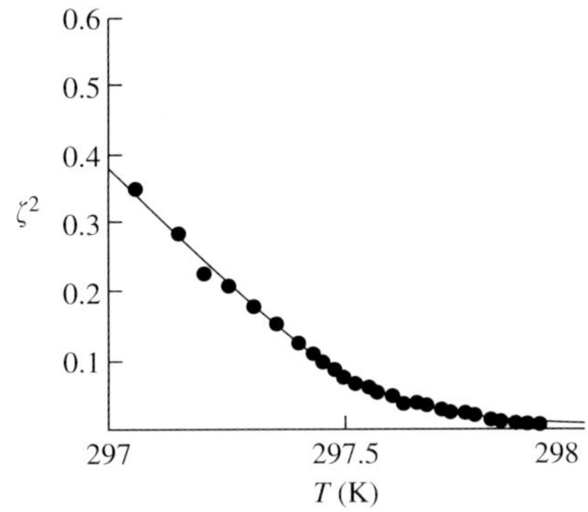

FIG. 1.30. Square of the induced biaxial order parameter (ζ) as a function of temperature in the neighborhood of the Ch_D–Ch_B phase transition, for a potassium laurate/decylammonium chloride/water/brucine sulfate heptahydrated mixture [119]. The line is a fitting from the minimization of the characteristic free-energy density.

These values are consistent with Rayleigh scattering measurements for nematics [84]. The experimental results of ζ versus T in the neighborhood of T_c show that the Ch–Ch transition becomes continuous (second-order) in a chiral field.

Critical behavior If we assume the picture of an elastic chiral field as an external field applied to a nematic medium, the uniaxial–biaxial nematic phase transition can be studied by using cholesterics with different concentrations of chiral molecules. The variation of the concentration of chiral dopants in a set of different mixtures simulates the application of fields of different strengths. In analogy with the Cotton–Mouton coefficient [1,118], the *cholesteric susceptibility* [119] C_M can be defined by the relation $C_M \propto \zeta q^{-2}$. The experimental critical exponent of the chiral susceptibility, in the temperature range $\Delta T/T_c \sim 10^{-4}$, in the neighborhood of the Ch–Ch phase transition in a mixture of potassium laurate, decylammonium chloride, water and BS, is $\gamma = 1.35$, in good agreement with theoretical results for the three-dimensional XY model (see Fig. 1.31).

1.3.4 *Lamellar phases*

Different types of lamellar phases, L_α, L_β, L_γ, $L_{\beta'}$, L_δ, $P_{\beta'}$, and L_s, have been observed in lyotropic mixtures [70,71]. Some examples of this polymorphism can be found in the lyotropic mixtures of phosphatidylcholine (usually called lecithin) and water [70,71], dimiristoyl-phosphatidylcholine (DMPC) and water [132], and dipalmitoyl-phosphatidylcholine (DPPC) and water [71]. Figure 1.32 presents the phase diagram of the mixture DMPC/water [133]. Note the presence of the phase sequence $L_\alpha \rightarrow P_{\beta'} \rightarrow L_{\beta'}$ as the temperature decreases, at a fixed water concentration. Also, note that the phases are separated by regions of phase coexistence.

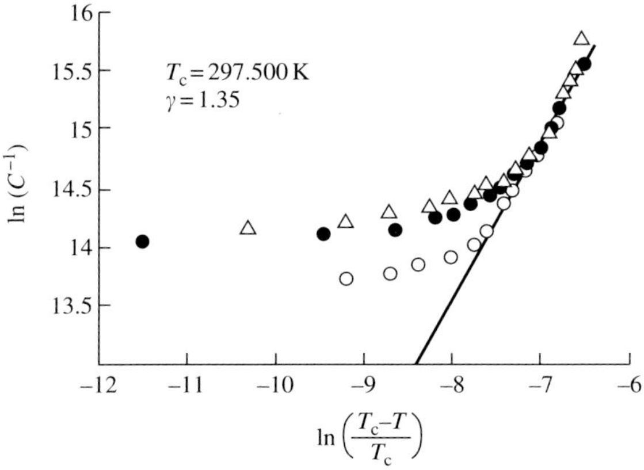

FIG. 1.31. Double logarithmic plot of the inverse of the cholesteric susceptibility, $C^{-1} \propto \zeta^{-1}q^2$, versus reduced temperature, $(T_c - T)/T_c$ [119]. The symbols $\triangle$, $\bullet$, and $\circ$ represent different concentrations of chiral molecules, from the smallest ($\circ$) to the largest ($\triangle$). The chiral doping represents the elastic field. The line is a linear fitting.

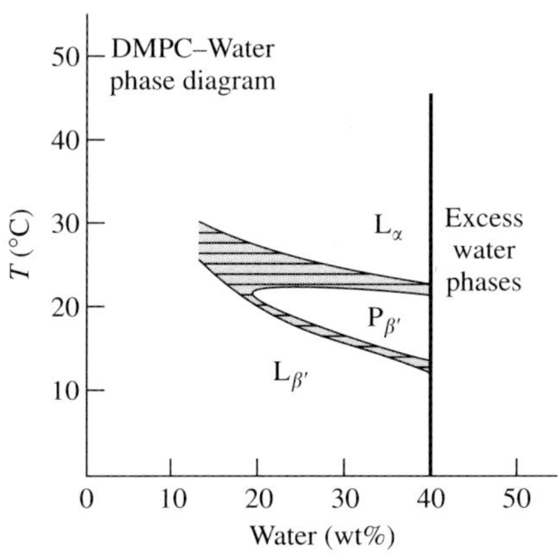

FIG. 1.32. Phase diagram of the mixture DMPC/water (ref. [133]).

The L_α phase is characterized by a one-dimensional periodicity. It is sometimes called *neat soap*, being usually found in regions of the phase diagrams with a high concentration of amphiphiles [32]. The carbonic chains are in a liquid-like state. The L_α phase is present in many lyotropic mixtures, e.g., potassium laurate/decanol/water, sodium caprylate/nonanol/water, potassium laurate/KCl/water, sodium octylsulfate/decanol/water, potassium laurate/water, and many others. At a given temperature, in ternary mixtures, the L_α phase is located almost in the middle of a triangular phase diagram (in which the vertices of the base, at left and right, represent 100 wt% of the solvent and of the principal amphiphile, respectively). In other words, the lamellar phase has about 30 wt% of the solvent and 60 wt% of the principal amphiphile [32]. In binary mixtures, these proportions are also observed. The cosurfactant (e.g. an alcohol in a ternary mixture), with a carbonic chain of smaller length than the chain of the principal amphiphile, acts to increase the radius of curvature of the surface of the supermolecular aggregate, which favors flatter interfaces. As we will see in the following sections, the domain of the lamellar phase in the triangular phase diagrams is located in regions with increasing amounts of the co-surfactant with respect to the location of the hexagonal phase. The L_α phase has the smallest surface per polar head with respect to other lyotropic structures. For the potassium laurate/water mixture, this parameter is a function of the temperature of the sample, reaching values of the order of 0.4 nm^2 [39]. DMR (deuterium magnetic resonance) studies performed with the lyotropic binary mixture of potassium laurate and D_2O show that the longitudinal and transverse relaxation times are also small as compared with those of the cubic and hexagonal phases. Different regimes of time were identified in this experiment with the L_α phase: a fast molecular reorientation of the O–D bonds, with a time scale of 10^{-11} s; a slow motion, with a time scale of 10^{-6} s; and short- and long-range diffusion, with time scales of 10^{-9} s and 10^{-7}–10^{-8} s, respectively [134]. These results suggest an important correlation between the paraffinic and the aqueous media around the (direct) amphiphilic aggregates.

The L_β phase differs from the L_α phase by the state of the carbonic chains. These chains remain stiff, perpendicular to the plane of the lamellae, with the axes organized on a two-dimensional hexagonal lattice. There is orientational disorder along the chain axes. The L_γ phase is formed by a sequence of layers with L_α and L_β structures.

The $L_{\beta'}$ phase (see Fig. 1.14(b)) differs from the L_β phase by the tilt of the axes of the carbonic chains with respect to the normal to the lamellae. Tilt angles from about 15° to 40° were observed in lipid–water mixtures [71].

In the L_δ phase, the carbonic chains remain stiff, perpendicular to the plane of the lamellae, with their axes organized as a two-dimensional square lattice. The chains, as in the case of the L_β phase, present orientational disorder with respect to their long axes.

The $P_{\beta'}$ phase presents rippled lamellae with the characteristics of the $L_{\beta'}$ phase (see Fig. 1.14(a)). The surface undulations of the lamellae, suggested by X-ray diffraction experiments, were observed by transmission electron microscopy (TEM) in a freeze-fracture replica of a sample of DMPC/water mixture [135]. The amplitude and wavelength of the observed undulations are about 4.5 and 16 nm, respectively. This wavelength presents a weak dependence on temperature and water content in the sample [136]. Symmetric and asymmetric undulations were observed in lyotropic mixtures with chiral amphiphiles, regardless of the chirality of the lipid molecules forming the bilayer [137,138].

In contrast to the L_α phase, NMR experiments in $L_{\beta'}$ and $P_{\beta'}$ phases of lyotropic mixtures indicate the same ordering of the carbonic chains in both phases [139].

Phase sequences involving lamellar–cubic–hexagonal sequences of structures were observed in some lyotropic mixtures (e.g. in mixtures of dodecyltrimethylammonium chloride and water, and hexaethylene glycol mono-n-dodecyl ether and water [140]). Sequences involving lamellar–rectangular–hexagonal phases were also identified in binary (Na laurylsulfate and water) and ternary (Na decylsulfate, decanol and water [60]) lyotropic mixtures.

Single-layered L_s phases were also observed in some lyotropic mixtures, e.g., Na caprylate/octan-1,8-diol/water [32], almost in the center of the triangular phase diagram. In this phase, the thickness of the lamellae and the length of the principal amphiphile are of the same order.

Another lamellar-type phase found in ternary lyotropic mixtures, in the water-rich region of the phase diagram, is the so-called *mucous woven phase*. It is made of a double-layered structure, as in the usual L_α phase, but with a smaller optical anisotropy. An example of lyotropic mixture with this phase is given by sodium caprylate/caprylic acid/water, at relative concentrations of about 10/10/80 wt% [141].

1.3.4.1 *Textures* The observation of a thin (~100 μm) non-oriented sample in the L_α phase shows a typical mosaic-like texture (see Fig. 1.33), with many defects and birefringence colors. In contrast to the case of nematics, there are sharp boundaries between different colors. Under particular conditions of the surfaces of the sample holder, it is possible to orient the lamellae parallel to these surfaces, and produce a pseudo-isotropic texture.

Pseudo-isotropic textures with oily streaks and terrace edges [142,143], *bâtonnets* [33], and Maltese crosses [144], can also be found in lamellar L_α lyomesophases. The texture of the $P_{\beta'}$ phase confined in a flat glass capillary presents typical oil streak patterns [145], similar to the L_α phase.

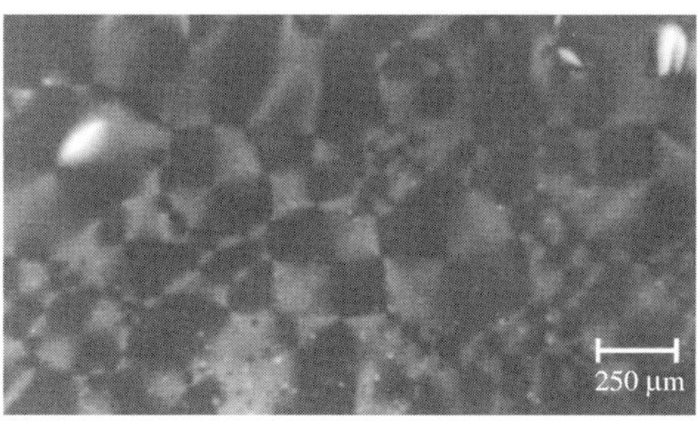

FIG. 1.33. Typical mosaic-like texture of a lamellar phase for a 100 μm thick sample of a potassium laurate/water mixture in a polarizing light microscope, between crossed polarizers.

1.3.4.2 *Scattering and diffraction results* At small angles, X-ray diffraction patterns of oriented lamellar L_α phases display a typical one-dimensional periodic structure, characterized by the presence of Bragg spots along the direction perpendicular to the lamellae. These spots have a half-height width which is essentially due to the experimental resolution of the diffraction setup, indicating that the order of the structure (correlation length) spreads over distances larger than 100 nm. The characteristic distances associated with the spots in the reciprocal space keep the ratios 1 : 2 : 3 : 4.... The repeating distance in the direct space, associated with the first-order peak, scales with the length of the principal amphiphilic molecule of the mixture and the amount of solvent between layers. For a mixture of approximately 80 wt% of potassium laurate and water, this repeating distance is about 3 nm. However, in order to sketch the structure in direct space, it is necessary to have additional information about the mixture (e.g., relative concentrations, volume occupied by the solvent, arrangement of the amphiphiles in monolayers or bilayers, etc). In this example, the thickness of the layers is about 2 nm [73]. At high angles, the diffraction pattern shows the typical band due to the mean distances between carbonic chains (~0.5 nm) in a liquid-like state.

In L_β and L_δ phases, at high angles, X-ray diffraction patterns indicate the organization of the carbonic chains as two-dimensional hexagonal (with typical lattice parameter of about 0.42 nm) and square (with typical lattice parameter of about 0.47 nm) lattices, respectively.

X-ray diffraction experiments performed in freely suspended hydrated DMPC multimembrane films [72] show that the commonly observed $L_{\beta'}$ phase is, in fact, composed by three different regions, with distinct phases, $L_{\beta F}$, $L_{\beta L}$, and $L_{\beta I}$ (see

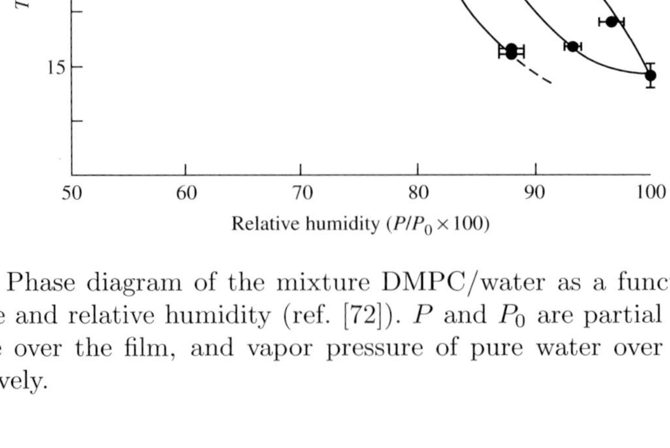

FIG. 1.34. Phase diagram of the mixture DMPC/water as a function of temperature and relative humidity (ref. [72]). P and P_0 are partial water-vapor pressure over the film, and vapor pressure of pure water over pure water, respectively.

Fig. 1.34). These three phases have tilted distorted hexagonal two-dimensional lattices, differing among them with respect to the values of the angle ϕ_a between the molecular tilt direction and the nearest-neighbor bond direction. $L_{\beta F}$ and $L_{\beta I}$ phases are associated with $\phi_a = 0°$ and $\phi_a = 30°$, respectively. In the $L_{\beta L}$ phase, ϕ_a increases continuously from 0° to 30°.

X-ray scattering and diffraction experiments were performed to investigate the behavior of the inter-layer distance (lattice parameter) of the lamellar phase as a function of the concentration of amphiphiles. This process is usually called *swelling* of the lamellar phase, although this terminology is also used to describe the introduction of nonpolar solvents in the lamellar structure. If ϕ is the volume fraction of amphiphiles in the mixture, the experimental small-angle X-ray data yield $q_{\mathrm{m}} = 2\pi\phi/\delta$, where q_{m} is the position of the first-order maxima of the scattering curve, and δ is the thickness of the bilayer [146].

1.3.4.3 *The* N_D–L_α *phase transition* In some phase diagrams of lyotropic mixtures, as cesium perfluorooctanoate (CsPFO) and water [147], and decylammonium chloride (DaCl), NH_4Cl and water [148], a nematic region separates the lamellar from the isotropic phases. Measurements [149] of the splay (k_{11}) and bend (k_{33}) elastic constants [105], and of the rotational viscosity (γ_1) in the N_D phase, in the vicinity of the N_D–L_α phase transition, show that k_{33} and γ_1 diverge at the transition, as the temperature approaches the lamellar phase from the nematic region, and that k_{11} is smooth and well behaved as a function of temperature. In the mixture DaCl/NH_4Cl/water, the critical exponent associated with the divergence of k_{33} depends on the relative concentrations of the components. The reported experimental values [150], in a range from 0.87 to 1.07, do not agree with either mean-field predictions (1/2) or calculations based on the analogy proposed by de Gennes [1] between liquid crystals and superfluid helium (2/3). For the CsPFO/water mixture, however, in a phase diagram in terms of temperature and relative concentration of amphiphiles, the N_D–L_α transition changes from second to first order at a tricritical point [96,147].

Electric conductivity (ac field) measurements [55,151] performed in the DaCl/NH_4Cl/water mixture indicate that, as a function of temperature, there is no discontinuity across the N_D–L_α transition of the values of the conductivity anisotropy, defined as $(\kappa_{\parallel} - \kappa_{\perp})/(\kappa_{\parallel} + \kappa_{\perp})$, where $\kappa_{\parallel}$ and $\kappa_{\perp}$ are the conductivities parallel and perpendicular to $\mathbf{n}$ in the N_D phase. This result is consistent with the picture that the L_α structure has many holes or channel-like defects, estimated as occupying about 30% of the total area of the lamellae in some cases [152].

In the vicinity of the phase transition, small-angle X-ray and neutron scattering results for a mixture of pentadecafluorooctanoate and water [153] indicate the occurrence of a transformation from disk-like to elongated-flattened micelles, which is a precursor of the *defect lamellar structure*. The L_α–N_D transition, as a function of temperature or relative concentrations of the compounds, should be regarded as a process of continuous growth of defects and holes, until there appear independent micelles, and the nematic structure is finally established.

In contrast to this picture, measurements of conductivity and density in a CsPFO/water mixture across the L_α–N_D transition [154] seem to indicate that there is no relevant modifications of the basic units (molecular aggregates) in both phases. This transition just involves the onset of translational order of the micelles, being equivalent to the nematic–smectic phase transitions in thermotropics [1].

An interesting effect is observed if a polymer is solubilized in a water-based lyotropic mixture presenting a L_α–N_D transition. Above a critical concentration, depending on the polymer molecular weight, the replacement of water by the polymer leads to a suppression of the lamellar phase [155]. The polymer network acts to disrupt the lamellar order, which favors the appearance of the nematic structure.

1.3.5 *Hexagonal and other two-dimensional ordered phases*

These phases are characterized by the two-dimensional periodicity of the high anisometric molecular aggregates. These long cylindrical-like aggregates are sometimes called *infinite aggregates*, since one of their dimensions can be larger than 100 nm. The monoclinic M phase will be discussed in Chapter 7.

1.3.5.1 *The hexagonal phase* Three types of lyotropic hexagonal phases were identified, direct (H_α), inverted (H^i_α), and complex (H_c) hexagonal phases. The viscosity and fluidity of the hexagonal phase are larger than in the lamellar phases.

The H_α phase is present in many lyotropic mixtures, as potassium laurate/decanol/water, sodium caprylate/*p*-xylene/water, sodium octylsulfate/decanol/water, potassium caprate/ocatanol/water, sodium myristate/water, potassium laurate/water, among several others. In ternary mixtures, at a given temperature, the H_α phase is located almost in the middle, near the base, of a triangular phase diagram [32], at about 50 wt% of solvent and about 45 wt% of the principal amphiphile (recall that the vertices of the base of the triangle, at left and at right, represent 100 wt% of the solvent and of the main amphiphile, respectively). In binary mixtures, these approximate concentrations are also observed.

In some lyotropic mixtures, there is a hexagonal phase in the neighborhood of a calamitic nematic phase. In the potassium laurate/KCl/water mixture, the N_C–H_α transition has been shown to be of first order, since there is phase coexistence as a function of temperature [156]. The H_α phase has a larger surface per polar head than the lamellar phase. For the potassium laurate/water mixture, this parameter is a function of temperature, reaching values of the order of $0.55\ nm^2$ [39].

DMR studies for the lyotropic binary mixture of potassium laurate/D_2O show that the longitudinal and transverse (with respect to the symmetry axis of the structure, the optical axis) relaxation times are also larger as compared with those of the cubic and lamellar phases. Different time regimes were also identified in experiments with the H_α phase, with the same order of magnitude of the corresponding regimes in the lamellar phase [134].

On the other hand, the H^i_α phase has been observed in a more restrict set of lyotropic mixtures, as octyl ammonium chloride, *p*-xylene and water, aerosol OT, *p*-xylene and water, aerosol OT, decanol and water, and aerosol OT, caprylic acid and water. In ternary mixtures, at a given temperature, the H^i_α phase is located at the right corner of the triangular phase diagram, near the base, at about 10 wt% of the polar solvent and 80 wt% of the main amphiphile [32].

The H_c phase was identified in some binary aqueous lyotropic mixtures with amphiphiles of large carbonic chains, as Na and K palmitates and stearates, in a small region between H_α and L_α phases [32]. The amphiphilic molecular aggregates seem to be made of a hollow cylinder formed by a bend double layer configuration, as an open vesicle, with water both inside and outside of the hollow cylinder. The lattice parameter is larger than in the H_α phase. Inverted H_c structures were also reported.

The shape and geometry of the supermolecular aggregates in the hexagonal phases depend on the relative concentrations of the principal amphiphile and cosurfactant and of the structure of the amphiphile itself. In particular, there is a dependence on the relation between the surface area per polar head and the volume occupied by the non-polar part of the molecule.

Textures A non-oriented thin (~100 μm) sample observed between crossed polarizer shows a typical fan-like texture (see Fig. 1.35). As in the textures of lamellar phases, there are sharp and well-defined boundaries between regions

FIG. 1.35. Typical fan-like texture of a hexagonal phase for a 100 μm thick sample of a potassium laurate/decanol/water mixture, in a polarizing light microscope, between crossed polarizers.

with different birefringence colors. Focal conics and non-geometric textures are also observed in some mixtures [142,143].

Diffraction results Small-angle X-ray diffraction patterns of an oriented H_α phase present Bragg peaks, with the characteristic distance ratios $1 : \sqrt{3} : \sqrt{4} : \ldots$, of a hexagonal two-dimensional structure in direct space. At high angles, the diffraction pattern shows a typical band due to the mean distances between carbonic chains ($\sim$0.5 nm) in a liquid-like structure.

As in the previous cases, additional information about the composition of the mixture is also necessary in order to allow a calculation of the dimensions of the molecular aggregates. For direct mesophases, we usually assume that the diameter of the cylinder is twice the principal amphiphilic molecular length. This assumption has to be checked self-consistently, with the data for the relative concentrations of the compounds of the mixture. In direct space, the image of the (direct) H_α phase is a two-dimensional hexagonal array of long parallel cylinders, each of them with the aliphatic chains in the inner part. In binary mixtures, the cylinders have a circular cross section. For a mixture of approximately 50 wt% of potassium laurate and water, the lattice parameter of the hexagonal array is about 4 nm and the cylinders have a diameter of about 3 nm [73].

If a cosurfactant is added to a binary lyotropic mixture, the cross section of the amphiphilic molecular aggregates, in a direction perpendicular to the long cylindrical axis, does not display a circular symmetry. The molecular aggregates loose their cylindrical shapes in order to form ribbon-like aggregates [74]. This result suggests that the addition to the original binary mixture of a cosurfactant with smaller chain favors the existence of flat surfaces in the amphiphilic aggregates.

The inverted hexagonal structure is formed by a hexagonal packing of cylindrical regions with the polar solvent inside the cores and the aliphatic chains pointing outwards of the cylinders. An eventual nonpolar solvent added to the mixture presenting this inverted phase will be inserted in the continuous medium of the chains, outside of the cylinders.

In the complex hexagonal phase H_c, the characteristic distances obtained from the Bragg peaks in the X-ray diffraction pattern keep the ratios [73] $1 : \sqrt{3} : \sqrt{4} : \sqrt{7} : \sqrt{9} : \sqrt{12}$. The structure of this phase can be modeled by a set of high anisometric (long) cylinders, arranged on a two-dimensional hexagonal lattice in the plane perpendicular to the cylinder axes. There is a bending of an amphiphilic double layer in order to close these cylinders, keeping the solvent inside and outside of them.

1.3.5.2 *The rectangular R phase* The rectangular phase is characterized by a two-dimensional orthorhombic array of high anisometric (long) parallel parallelepipeds of rectangular cross section on the plane perpendicular to the largest dimension of these aggregates (which are also called "ribbons") [157,158]. The R phase is located between lamellar and hexagonal phases. Examples of lyotropic mixtures showing this phase are KC_{16} (potassium palmitate, $CH_3(CH_2)_{14}COOK$)/water [32], NaC_{18} (sodium stearate, $CH_3(CH_2)_{16}COONa$)/water, KC_{18} (potassium stearate, $CH_3(CH_2)_{16}COOK$)/water, and Na decylsulfate/decanol/water [60]. In some lyotropic mixtures, the rectangular phase is in coexistence with the hexagonal phase.

Textures The rectangular phase is birefringent. Non-oriented samples of the R phase, between crossed polarizers, present a mosaic-like texture with many pseudo-isotropic regions and defects [32].

Diffraction results The array of long parallel parallelepipeds of the R phase forms a rectangular two-dimensional lattice on the plane perpendicular to the largest dimension of the molecular aggregate, with lattice parameters a_r and b_r. In the diffraction pattern, the positions of the Bragg diffraction peaks associated with the lattice parameters correspond to $a_r : b_r : 2a_r : 2b_r : 3a_r : 3b_r$. For the potassium laurate/water mixture, $(a_r/b_r) \sim 1.4$, and the diffraction data were explained in terms of long parallel rods of rectangular cross section, arranged with an orthorhombic symmetry [73].

1.3.5.3 *The square phase* This birefringent phase is characterized by a two-dimensional tetragonal array of highly anisometric parallel parallelepipeds of square cross section on the plane perpendicular to the largest dimension of the aggregate. The direct and inverted structures were labeled C and K, respectively. The C phase is found in the left corner of the triangular phase diagrams of some aqueous ternary lyotropic mixtures, as Na caprylate/nonan-1-ol/water, Na caprylate/octan-1-ol/water, Na caprylate/heptan-1-ol/water, and Na caprylate/hexan-1-ol/water [32]. The neighboring phases are L_1 and the lamellar phase. In many cases, the square phase is not found isolated, but in coexistence with L_1 and lamellar phases. Using NMR techniques, Tiddy [159] has shown the presence of two types of water molecules in this structure, essentially bounded to the amphiphilic aggregates.

Textures The textures observed in non-oriented samples in the square phase, placed between crossed polarizers, are of *bâtonnet* type. Many defects with well-defined borders and some regions with the pseudo-isotropic texture [32] are also present.

Diffraction results Long parallel parallelepipeds self-organize as a two-dimensional array in the plane perpendicular to the largest dimension of the molecular aggregates, with lattice parameter a_s. The characteristic distances associated with the Bragg peaks in the diffraction pattern keep the ratios $1 : 2 : 3 : 4$. In the particular case of the C phase, the thickness of the water layer between amphiphilic aggregates, calculated from diffraction results and taking into account the relative concentrations of the components of the mixtures, ranges from about 0.8 to 5 nm, for increasing water concentrations. Although the borders of the square-section aggregates are rounded in order to accommodate the polar heads at the surface, and minimize the contact of water with the inner parts of the aggregate, the scattering results lead to this particular square symmetry, different from the circular cross-sections of the aggregates.

1.3.6 *Three-dimensionally ordered phases*

There are two types of molecular aggregates, micelles and bicontinuous structures, with organized arrangements in three dimensions.

1.3.6.1 *Cubic phases* Cubic micellar phases are not so common in the area of lyotropic mixtures. They are usually located in narrow regions of the phase diagrams, between lamellar and hexagonal phases. Six cubic phases, two of them of micellar nature, have been identified so far [76,160]. All of them are optically isotropic but present shear-induced birefringence [140].

In terms of the location in the phase diagrams, cubic phases are found in the vicinity of micellar solutions, between lamellar and hexagonal phases. The viscosity and fluidity of these phases are larger than in lamellar and hexagonal phases.

Cubic micellar phase Two types of micellar cubic phases were identified in lyotropic mixtures: Q_m^{223}, with a primitive unit cell and a homogeneous lipid composition, and Q_m^{227}, with a fcc (face centered cubic) unit cell [76] of $Fd3m$ space group and heterogeneous lipid composition. Q_m^{223} was observed in chemically pure mixtures of lipids and solvents; Q_m^{227} requires mixtures of water-miscible and water-immiscible lipids and a solvent [76]. In the fcc structure, the characteristic distances associated with the Bragg peaks of the X-ray diffraction patterns keep the ratios $1 : \sqrt{4}/\sqrt{3} : \sqrt{8}/\sqrt{3} : \sqrt{11}/\sqrt{3}$.

Depending on the mixture, micelles in these phases can be spherical or anisometric (for instance, prolate ellipsoids). Examples of mixtures presenting micellar cubic phases are dodecyltrimethylammonium chloride (DTAC) and water, N, N, N-trimethylamino dodecanoimide and water, palmitoyllyso phosphatidylcholine (PLPC), and water, and dioleoylphosphatidylcholine, dioleoylglycerol and water.

The unit cell of the Q_m^{223} phase has six direct (disk-type) micelles of symmetry $\overline{4}2m$, and 2 spherical micelles of symmetry $m3$. The ratio between area and volume of these micelles is the same. The lattice parameters of the Q_m^{223} phase observed in DTAC/water and PLPC/water mixtures are 8.5 and 13.7 nm, respectively, always larger (about twice) than the lattice parameter of the neighboring hexagonal phase.

In the reverse cubic phase, Q_m^{227}, instead of micelles, spherical liquid globules composed by the solvent are packed in a cubic symmetry. This phase is observed, e.g., in the phase diagrams of a mixture of phosphatidylcholine, diacylglycerol and water, between the inverse hexagonal and inverse micellar isotropic phases, and coexists with another phase with excess water [161]. The unit cell of this phase is composed by two types of inverted micelles [161]: eight larger micelles in a tetrahedral diamond-like lattice and 16 smaller micelles in tetrahedral clusters at the four octants of the cell. The large and small micelles have different ratios between area and volume. The lattice parameter of the Q_m^{227} phase in PFL/water, where PFL is a lipid extracted from *Pseudomonas fluorescens* [162], is 14.6 nm, about three times larger than the lattice parameter of the neighboring hexagonal phase.

Bicontinuous cubic phases The structural description of the bicontinuous cubic phases can be discussed from two different points of view. We may consider either a structure of folded surfaces occupying the three-dimensional space or a network of rod-like aggregates.

The cubic bicontinuous structures identified in lyotropic mixtures are of the space groups $Ia3d$, $Im3m$, and $Pn3m$ [163]. The first structure is a bicontinuous structure of amphiphilic layer, which divides the space into two independent and unconnected solvent regions. The second structure is built by two independent networks of amphiphilic aggregates, without any connections, forming a labyrinth-type structure with cubic symmetry. The phases which were identified so far are Q^{230} ($Ia3d$) [164], Q^{224} ($Pn3m$) [165] and Q^{229} ($Im3m$) [166]. Another cubic phase, labeled Q^{212}, shows a complex intermediate character between the bicontinuous and the micellar phases, with a three-dimensional network of rods, enclosing micelles in the structure [166].

The Q^{230} phase was observed with both direct and inverted topology. This phase consists of interconnected rods, linked three-by-three, forming two three-dimensional networks intertwined and disconnected. It has been observed [167] in anhydrous fatty acid salts of divalent cations. It is also present in the water-rich region of the phase diagram of some lyotropic mixtures, as octaethylene glycol dodecyl ether ($C_{12}EO_8$) and water, and can be described in terms of a complex bicontinuous three-dimensional surface (the convoluted polar/non-polar interfaces) called *gyroid infinite periodic minimal surface*, G-IPMS, with space group $Ia\bar{3}d$ [168]. The microstructure of this phase can be modeled in terms of three-dimensional periodic *minimal surfaces*, free from self-intersections. These surfaces, generically called *triply periodic minimal surfaces*, TPMS, are characterized by a mean curvature that vanishes at every point [169]. Luzzati *et al.* [76] introduced the concept of *chaotic zones*, defined as the locus where the short-range disorder is maximal. The CH_3 end-groups of the hydrocarbon chains are located in the non-polar chaotic zones of the (direct) structure, which coincides with the G-IPMS.

The Q^{224} and Q^{229} phases, however, were observed in the inverted topology only. The Q^{224} phase consists of interconnected rods, tetrahedrally linked, forming two three-dimensional networks intertwined and disconnected. The Q^{229} phase seems to be formed by rods which are linked six-by-six and oriented along the sides of a cube. The convoluted surfaces (IPMS) of the Q^{224} and Q^{229} phases are the D-surface, with space group symmetry $Pn3m$, and the P-surface, with symmetry $Im3m$ [170].

The X-ray diffraction patterns of these phases present Bragg peaks which lead to the identification of the cubic symmetry of the structures in direct space. Examples of mixtures presenting bicontinuous cubic phases are [171] dodecyltrimethylammonium chloride/water and monoolein/oleic acid/water.

DMR studies carried out with the lyotropic binary mixture of potassium laurate and D_2O in the cubic phase show that the longitudinal relaxation time is intermediate with respect to the lamellar and hexagonal phases. On the other hand, the transverse relaxation time is larger than in the lamellar and hexagonal phases. Different regimes of time were also identified in this experiment: a fast molecular reorientation of the O–D bonds (with time scale of the order of 10^{-11} s), a slow motion (time scale of 10^{-6} s), and short- and long-range diffusion (with time scales of the order of 10^{-9} and 10^{-7}–10^{-8} s, respectively) [134]. The diffusion coefficient of the soap molecules in this cubic structure was found to be $D = 2 \times 10^{-6}\text{cm}^{2}/\text{s}$, at 90°C.

1.3.6.2 *Bicontinuous sponge phase (L_3)* In the L_3 phase, which is optically isotropic and does not display any long-range order, experimental observations indicate the existence of a microstructure in which a surfactant layer, of thickness δ, of multiply connected topology, separates two solvent regions over macroscopic distances, as illustrated in Figs 1.5(b) and 1.18 [172]. The bilayer presents local principal curvatures of opposite signs, which characterize a saddle-like topology. The structure of the sponge phase can be locally modeled with an IPMS [173]. This structural shape, which can be sketched as a melted cubic structure [174,175], has been confirmed by freeze-fracture electron microscopy observations [176]. From this point of view, an important difference between the L_3 and the cubic phase is the lack of space order and periodicity.

Structural aspects of the sponge phases have been investigated by a number of experimental techniques, including freeze-fracture electron microscopy, X-ray [146] and neutron [177] diffraction scattering, NMR [175], and light scattering [178], besides measurements of transport properties [179]. This phase is classified as symmetric or asymmetric, depending on the mean curvature at the center of the bilayer. Symmetric and asymmetric phases have zero and non-zero mean curvatures, respectively [180].

Sponge phases were observed in both water-rich and oil-rich mixtures, mostly in the vicinity of swollen lamellar phase domains. Some examples of mixtures presenting sponge phases are sodium bis-2-ethylhexyl sulfosuccinate (known as AOT), dodecanol and water [175], sodium dodecylsulfate (SDS), 1-pentanol, cyclohexane and water [181], AOT and brine (water and salt) [178], *n*-dodecylbetain, pentanol and water [178], and cetylpyridinium chloride, hexanol and brine [178].

Measurements of reduced electric conductivity (as compared with the conductivity of the same volume of water and ions) [179] in an oil-rich lyotropic mixture with L_1, L_3, and L_α phases lead to values about 10^{-2}, 0.65, and 1, respectively, where the conductivity has been measured in a direction parallel to the layers of the lamellar phases. These results are consistent with a picture of a labyrinth-like structure in the sponge phase.

X-ray and neutron scattering patterns of the sponge phase can be analyzed in different ranges of the intensity q of the scattering vector. For small values of q, patterns present a broad peak due to the correlation between pieces of the lamellae [146,175,179,182,183]. In q space, assuming a locally lamellar-type structure, the position of the peak is given by

$$q_m = \frac{2\pi}{d} = \frac{2\pi\phi}{\gamma_q \delta},$$

where d is the inter-layer repeating distance, ϕ is the volume fraction of amphiphile, and $\gamma_q \sim 1.5$ is a number larger than 1 [184]. In mixtures of SDS, pentanol, water and NaCl, and SDS, pentanol, dodecane and water [179], the values of the bilayer thickness obtained from SANS experiments are 2.9 and 3.1 nm, respectively. The thickness of the lamellae, α, in the neighboring lamellar phases are 1.9 and 2.1 nm, respectively, which gives $\gamma_q \sim 1.5$. In the lamellar phase, it has been shown that α depends on ϕ as $\alpha \propto \phi^{-1}$, although a more accurate analysis of the diffraction data suggests corrections to this dependence [180]. As in the case of the lamellar phase, this procedure for the L_3 phase is called *swelling* of the bilayer structure. In the range of high values of q ($q > q_m$), the patterns present the typical form factors of a flat bilayer. In mixtures with carbonic chain tails, the patterns also present the typical broad band due to the mean distance between the chains (at about 0.5 nm). In direct sponge phases, the layer thickness is about twice the length of the main amphiphile. In inverse water-based sponge phases, δ is the water layer which, as a rule, does not have a typical length scale and can vary according to the relative concentrations of the components of the mixture. A crossover from a behavior of the form $1/q^2$, for $q < 2\pi/d$, to $1/q^4$, for $q > 2\pi/d$, which are typically associated with the scattering of a flat bilayer with random orientation and a thin interface, respectively, is observed in the curves of the scattering intensity versus q.

The NMR technique is a powerful tool to check the bicontinuous character of the sponge structure [175]. For example, in the AOT/dodecanol/water mixture, the resonance spectrum of ^{23}Na shows a dispersion of the internuclear distances, which is not compatible with an isotropic micellar system, and thus becomes a signature of the L_3 phase.

The technique of fluorescence recovery after fringe pattern photobleaching was also used to investigate the topology of the sponge phase by studying the self-diffusion of guest fluorescent probes (amphiphilic molecules with fluorescent polar heads) embedded in the sponge structure host. In the particular case of a mixture of tetradecyldimethyl aminoxide (known as C_{14}DMAO), hexanol and water [146], $D_s/D_o \sim 0.65$, where D_s is the self-diffusion coefficient of the molecule in the structure, and D_o is the diffusion coefficient of the molecule in the lamellar phase. This ratio is almost constant in the dilute regime, decreasing with the volume fraction ϕ. This dependence with the volume fraction allows a comparison with predictions of different topological configurations for the host sponge structure.

One of the most spectacular properties of the L_3 phase is the flow-induced birefringence [178,185]. Under shear, there appears a velocity gradient and a transient flow birefringence, which relax with a characteristic time, depending on the system and relative concentrations. Given the shear rate $\dot{\gamma}$, this induced birefringence was shown to be proportional to $\dot{\gamma}\phi^{-2}$ [178]. Also, it has been shown that this birefringence becomes larger as the water content increases in the L_3 phases.

Let us look at some aspects of the phase transitions between the sponge and its neighboring phases, in particular the cubic and lamellar neighbors.

In the vicinity of the sponge–cubic transition, the NMR technique was used to measure the diffusion coefficient of the amphiphile (D_a) and water (D_w) in the AOT/NaCl/water mixture. The transition is of first order, with a large coexistence region of cubic and sponge phases. The ratio between D_w and the diffusion coefficient of water in NaCl brine of the same salinity (D_w^o) presents a linear dependence with $1/\phi$ [174] throughout the sponge and cubic regions. A single linear function can be fitted to the experimental data, without regard to the particular phase. In the sponge phase, $D_w/D_w^o \sim 0.5$ at $\phi \sim 0.2$; in the cubic phase, $D_w/D_w^o \sim 0.15$ at $\phi \sim 0.65$. The measurements of D_a versus ϕ in this lyotropic mixture also show a monotonic behavior, independently of the particular phase. In the sponge phase, $D_a \sim 2.8 \times 10^{-11}$ cm^2/s at $\phi \sim 0.1$; in the cubic phase, $D_a \sim 1 \times 10^{-11}$ cm^2/s at $\phi \sim 0.6$. These results suggest that there is a topological similarity between the structures of the cubic and the sponge phases.

The L_α–L_3 transition is also of first order, with coexistence regions in the partial isotherms of the phase diagrams [177,184]. SANS experiments show that locally both the (swollen) L_α and L_3 phases have similar morphology, with changes at large scales only. The structural transformation from one phase to the other seems to be triggered by modifications of the saddle and splay rigidity of the bilayer, induced by temperature or relative concentrations of the components [177].

1.3.7 *Lower-symmetry phases*

There are experimental evidences of the existence of lyotropic phases with lower symmetries, which are sometimes called "intermediate phases." Usually, they are observed in mixtures with long chain amphiphiles [186] or amphiphiles with restricted flexibility [187]. For example, a rhombohedral phase [188] was identified in the sodium dodecylsulfate/water mixture.

Another structure, called "mesh," in which the lamellae present pores or holes filled up with the solvent [189], has been identified in aqueous mixtures of long-chain nonionic surfactants, as poly-oxyethylene. These holes may or may not be correlated from one layer to the other. A mixture of nonaethylene glycol mono (11-oxa-14,18,22,26-tetramethylheptacosyl) ether (known as $C_{30}EO_9$) and

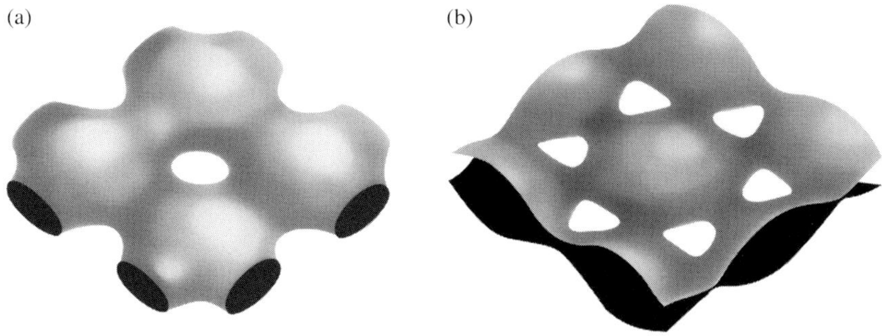

FIG. 1.36. Lower symmetry lyotropic phase structures: (a) sketch of one layer of the centered tetragonal mesh structure [190]; (b) sketch of one layer of the rhombohedral mesh structure [189].

water [189] presents intermediate phases between the lamellar and the hexagonal phases. The X-ray diffraction patterns of these phases present Bragg peaks that can be indexed in different ways, since the number of peaks is not enough to clearly identify the structure. Structural models of rods and boxes, with the hydrocarbon–water interface located in the surfaces of these elements, have been proposed to account for these structures [189]. For example, we have the following possibilities: (a) a centered tetragonal mesh (see Fig. 1.36(a)); (b) a bicontinuous centered tetragonal structure; (c) one layer of a *type a* hexagonal mesh from which a rhombohedral mesh and bicontinuous structures can be constructed; (d) the same model as in (c), but for a *type b* hexagonal structure. In the particular case of the $C_{30}EO_9$/water mixture, the structure of the intermediate phase seems to be a rhombohedral mesh of *type b* (see Fig. 1.36(b)).

The geometrical characteristics of the amphiphile are important parameters to define the topology of the phase diagram, particularly in the case of these intermediate structures. A common surfactant parameter is defined as $S_p = v/S_a l$, where v, S_a and l are the volume of the alkyl chain, the interfacial area of the chain, and the chain length perpendicular to the interface, respectively. The parameter S_p ranges from 1/2 to 1 [190], with $S_p = 1/2$ for a hexagonal structure, $1/2 < S_p < 2/3$ for a mesh intermediate phase, $2/3 < S_p < 1$ for cubic phases, and $S_p = 1$ for a lamellar phase. The physical concept underlying the definition of S_p is the idea of minimizing the interfacial curvature energy for a given structure, taking into account the characteristics of the amphiphiles, the temperature, and the relative concentrations of the components of the mixture.

1.4 Wetting of lyotropic phases

Wetting [191] with complex fluids, as liquid crystals and ferrofluids, is a rich field of research from both the fundamental point of view and the implications in technological aspects of the interaction between fluids and solids in interfaces.

Unlike the thermotropic liquid crystals, lyotropic mesophases seem to present some peculiarities, due to the presence of isolated amphiphilic molecules and the existence of molecular aggregates, which could lead to shape transformations at the interface. The selective adsorption of molecules of the mixture by the surface in contact with the fluid imposes a molecular concentration gradient from the interface to the bulk, which may lead to surface phase transitions with modifications of the topology of the molecular aggregates. In this section, we discuss two examples of wetting involving nematic and sponge phases in contact with solid substrates.

1.4.1 *Nematic phase*

The strength of the anchoring of thermotropic liquid crystals at a surface is measured by the *extrapolation length* b. If b is comparable to the molecular dimensions, the anchoring is strong; in the opposite case, if b is much larger than the molecular dimensions, the anchoring is weak [1]. In the absence of additional forces or torques, the director of the liquid crystalline phases is aligned along an "easy direction" associated with the surface. Therefore, the characterization of the anchoring refers to the behavior of the orientational angle of the director with respect to a given axis from the surface to the bulk of a certain phase.

In the treatment of nematic thermotropic liquid crystals, it is commonly assumed that the surface forces of the solid substrate are strong enough to impose a well-defined orientation of the phase director. In lyotropic mixtures, since the basic units are not isolated molecules, there is the possibility of a different behavior. Indeed, optical transmittance experiments [192] performed at the N_C phase of a mixture of decylammonium chloride, NH_4Cl and water in contact with amorphous non-treated glass surfaces were explained with the assumption of a gliding director of the N_C phase on the solid–liquid interface. Specifically, these experimental results were explained assuming the existence of an amphiphilic bilayer, that is, a kind of lamellar structure, which is stabilized by the glass surface. In contrast to the case of thermotropic nematic phases, there is no easy direction at the interface. Instead, there is only an easy plane, due to the azimuthal degeneracy at the cell walls (substrate) with respect to the bilayer symmetry of the lamellar-type structure stabilized at the interface. Experimental investigations of the N_C phase in contact with Langmuir–Blodgett (LB) films of lignin, cadmium stearate and behenic acid, with various numbers of layers, and with non treated glass surfaces, subjected to a magnetic field, have shown gliding movements of the orientation of the local director throughout the surface. These experiments were performed with samples at the N_C phase, with $\Delta\chi > 0$, oriented in a planar geometry, in applied magnetic fields along the plane of the interface. Regardless of the surface treatment, in the presence of a field, the director rotates around an axis perpendicular to the surface [192,193]. Also, the bulk orientation of the N_C director does not depend on the surface treatment. A possible explanation of these results is the stabilization on top of the glass surface, with or without any treatment, of a lamellar layer or bylayer of the amphiphilic molecules of the mixture. The micelles in the bulk *feel* essentially this layer and do *not* feel the bare glass surface covered by the LB film. This lamellar layer can be interpreted as a *self-surface-treatment*, created by the mixture itself, due to the role of adhesion properties [193]. It seems that this layer completely screens the effects of any possible LB film deposited onto the substrate.

1.4.2 *Sponge phase*

The wetting of lyotropic phases with bicontinuous structures includes new features which are not observed in the wetting of phases formed by simple micelles. Wetting a solid surface with a mixture of sodium dodecylsulfate (SDS), 1-pentanol, cyclohexane and water, in the L_3 phase, has been shown to induce an anisotropic structure at the interface [181]. A layer of the mixture, of about 10 nm from the surface, in contact with an anisotropic solid substrate, as a glass, with or without surface treatments, and beryllium sheets, undergoes a phase transition to an anisotropic phase. The experimental optical phase shift of this anisotropic phase, of the order of 0.1 rad for the SDS mixture, depends on the relative concentrations. This effect can be amplified by making a deposit of teflon over a glass surface and rubbing the surface along a certain direction. The uniaxial oriented surface in contact with a L_3 phase has been shown to induce a birefringent layer with an optical phase shift five times larger than in surfaces without this particular deposit [194]. The optical axis of the anisotropic phase at the interface, or the projection of this axis in the plane of the substrate, is oriented almost parallel to the direction of rubbing at the surface.

This wetting-induced birefringence has a static character, different from the usual shear-induced birefringence in complex fluids [1,195]. The self-screening process identified in the wetting of substrates with nematics is probably present in the case of the L_3 phase in contact with solid substrates.

Grazing X-ray scattering experiments [194,196], performed with synchrotron radiation and L_3 samples, in mixtures of SDS, 1-pentanol, cyclohexane (C_6H_{12}) and water, in contact with beryllium sheets, have shown the presence of a tetragonal phase with typical lattice parameters $a \sim 5$ nm and $5 \lesssim c \lesssim 13$ nm, in a layer at about 10 nm from the surface to the bulk. At high angles, the diffraction pattern shows an isotropic broad band with typical distances of 0.5 nm, due to the liquid state of the carbon chains. A L_3–tetragonal phase transition occurs at the surface, induced by what should be called a "surface field." The topology of the substrate defines the orientation of the symmetry axis of this induced surface layer.

1.5 Technological and industrial applications

Soaps and detergents, which are the basic elements of a lyotropic mixture, have many technological applications in industry, and are a common element of everyday life [197]. If someone has grease (a nonpolar material) in the hands, it is useless to use only water (a polar solvent) in order to remove the grease.

The addition of some drops of detergent (amphiphilic molecules) is necessary to clean up the hands (which should be well rubbed in the presence of the mixture between water and detergent). This process works because there appear amphiphilic molecular aggregates that are able to involve and imprison the grease molecules, so that the water finishes the job.

Some of the properties of micellar solutions make them very useful in practical applications [198]: (i) direct (inverted) micelles have a non-polar (polar) interior, in which non-polar (polar) groups can be solubilized; (ii) the available surface in these systems is very large (e.g. in a 0.05 M SDS/water mixture, it is about 1.5×10^8 cm^2 per liter of solution); (iii) dynamic processes are fast; (iv) transport processes, in particular diffusion of molecules in micellar systems, depend on the characteristics of surfactant.

In industry, the "neat phase" of surfactants (the lamellar phase, in our terminology) is the starting system for many different detergent products and processes, as spray-drying, chilling, and milling [142,143]. Lyotropic mixtures with hexagonal and isotropic phases have been used in the automobile industry and in the manufacture of textile soaps. The viscosity and fluidity of the lyotropic phases, which have an intrinsic connection with the structure of the phase, are important parameters of the transport processes for all sorts of pumping in industrial plants.

The knowledge of the physico-chemical properties of lyotropics is very useful in the technological applications of emulsions. An emulsion is a dispersed phase of droplets with typical sizes ranging from 0.1 to 10 μm [199]. An example of these applications is the use of oil in water, in food emulsions, and of oil in brine, in environmental accidents, as the spilling of oil. In this last case, the knowledge of properties of soaps mixed with water, salt and oil is fundamental in the cleaning processes.

Food industry also benefits from the accumulated research on lyotropic mixtures. This is particularly relevant in the case of lipids [200], including the new field of *molecular gastronomy* [201]. In a basically empirical fashion, the art of preparing sauces uses the knowledge of properties of amphiphiles in the presence of polar and nonpolar substances. An expert *chef* changes the relative amount of egg yolks, water, salt, milk and cream in the sauce, in order to control the fluidity and the "velvet texture" of the final products. The texture of the sauce strongly depends on the state of molecular aggregation and on the different length scales (from nanometers to millimeters).

Enhanced oil recovery, via micellar flooding of crude oil reservoirs, is one of the most important technological applications of lyotropic-like systems [202]. The commercial use of surfactants for this purpose started in the late 1920s with the patents of De Groot [203]. The introduction of amphiphiles in oil wells provokes the decreasing of the interfacial tension, which enhances oil recovery. In 1968, Gogarty and Tosch [204] suggested that the addition of cosurfactants and electrolytes, besides the commonly used surfactant, contributes to improving oil recovery processes.

The effectiveness in stabilizing a microemulsion under the particular conditions of an oil reservoir is a crucial criterion to choose a surfactant for this application [205]. Since the microemulsion is the most interesting type of phase for oil recovery, the presence of liquid crystalline phases in some lyotropic phase diagrams limits the application of some amphiphiles to this purpose [206]. Examples of lyotropic mixtures used in oil recovery are: SDS/1-butanol/water/NaCl/heptane, hexanol/hexadecane/potassium oleate/water, and pentanol/hexadecane/potassium oleate/water.

The industry of foams also benefits form the knowledge of properties of lyotropics and their basic components. In particular, it is relevant to know the conditions for the stability of thin films [207]. Studies of the lamellar–nematic transition and of surface instabilities in lamellar phases have direct consequences on the stability of thin films of foams.

The cosmetics industry is also an important activity using the knowledge of the physico-chemical properties of lyotropics, particularly in the case of surfactant emulsifying agents [208]. In cosmetics, these amphiphilic molecules are used as wetting, cleanser, foaming, solubilizer and modifier agents, which change the feelings of the human skin in contact with cosmetics. The thermodynamic stability of a molecular aggregate (in micellar isotropic phases or microemulsions) is a crucial problem for the preparation of new cosmetic products. The topology of the phase diagram of lyotropic mixtures, typically including water, a surfactant and a cosurfactant, at different temperatures, should also be known in order to design the final product.

Some of the parameters that play an important role in the process of formulating a cosmetic product, besides the stability of the system itself, are viscosity, "texture,"effectiveness to make foam, and temperature dependence [209]. In some of the applications, in particular if we need viscous fluids, lyotropic mixtures and microemulsions are used as a starting system in order to obtain the final emulsion state. Moreover, lyotropics and microemulsions are widely used in formulating cosmetic products, particularly for the solubilization of fragrances and flavor oils in aqueous systems. A crucial problem in this industrial application is the usual need of low viscosity products. If a new component is added to the lyotropic mixture (in order to reduce viscosity), the topology of the phase diagram may change drastically, with alterations of relevant mesomorphic properties.

Two types of lyotropic mixtures, water/ionic surfactant/alcohol, which is the basis of the preparation of emulsion and microemulsion, and water/nonionic surfactant/hydrocarbon, are the mostly used systems in the formulation of cosmetics. In the particular case of micellar systems, the property of leading to the solubilization of otherwise insoluble compounds characterizes the lyotropic mixtures as a good dispersion medium. Properties of emulsion systems can be strongly affected by the presence in the mixture of liquid crystalline structures [144]. For example, the stability of emulsion droplets against flocculation processes depends on the nature of the amphiphilic layer of separation. If the layers of the lamellar-type structure come into contact, depending on the strength of the interfacial tension, the onset of fluctuations may turn the system unstable, with the coalescence of droplets and the formation of micelles. From the empirical point of view, the stability of an emulsion with water and oil is improved (from minutes to years) upon the addition of a liquid crystalline material to the original emulsion [210].

Applications of lyotropic-type systems in the biological area are very promising. Drug delivery processes, using vesicles as the transporting medium, are an important example of these applications. In these processes, drugs are dispersed inside vesicles, which are protected from body enzymes. Due to the addition of molecules incorporated to the amphiphilic bilayer, there is a selective interaction with particular materials from the digestive or circulatory system, so that the drugs are liberated according to the prescription of a physician.

A new interesting example of a self-assembled structure of amphiphilic molecules is the *tubule*, designed for encapsulating solids and liquids. This structure consists of a hollow phospholipid bilayer cylinder with a typical length of tens or hundreds of microns and a diameter from 0.1 to about 0.7 mm. The wall thickness ranges from less than 10 nm to well over 50 nm. Numerous applications, as miniature microwave circuits and absorptive filters, are under evaluation from the commercial point of view. Tubules [211] with 0.5 μm of diameter and a characteristic length in a range from 50 to 200 μm are formed by some synthetic phospholipids [212] with diacetilenic moieties in the acyl chains. Circular dichroism experiments show that tubules have a chiral structure [213,214].

Amphiphilic metal complexes have been used to form templates for mesoporous silicates in cathalitic processes [215,216], which are also examples of applications of these molecules.

1.5.1 *Velocity gradient sensors*

We know at least three technological devices using the property of shear-induced birefringence of micellar isotropic and sponge lyomesophases: a vibration detector [217], a monitoring device for the blood pumps in hemodialysis [218], and a lyotropic stethoscope [219]. These devices are based on effect that optically isotropic, micellar and sponge, phases become birefringent under the action of a velocity gradient.

As these devices are based on the same working principle, we only discuss the details of the monitors for the blood pumps used in hemodialysis.

In an actual pump, the blood goes through pipes that are progressively and periodically squeezed by hard plastic arms (with the number and shape depending on the particular machine) against the inner metal body of the pump. With a peristaltic-type movement, the fluid is pumped through the artificial kidney to be cleaned and then sent back to the body. The time variation of the flux of fluid, before, during and after each squeezing cycle, is an important parameter of the pump.

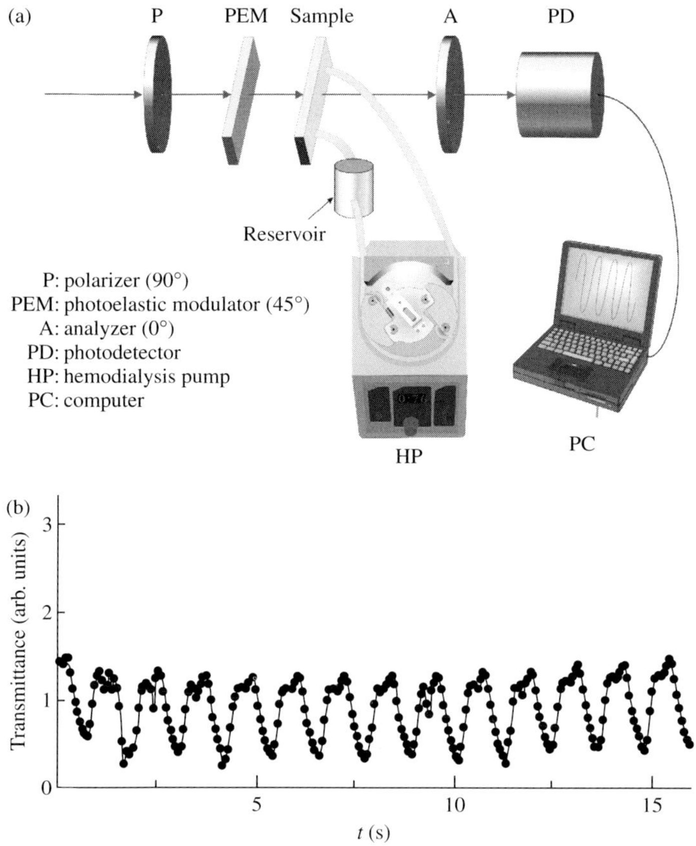

FIG. 1.37. (a) Sketch of the monitoring device for hemodialysis blood pumps using a lyotropic sponge mesophase. The angles in the figure are measured with respect to the vertical direction. (b) Optical transmittance as a function of time in the hemodialysis blood pump device.

In Fig. 1.37(a) we sketch a monitoring device. The liquid crystal, which simulates the blood in actual machines, is placed in the *reservoir* and fills the flexible pipes (which pass through the pump) and the glass cell. The glass cell is positioned along the trajectory of the polarized laser beam. The linearly polarized laser beam reaches a photoelastic modulator (PEM), at a fixed frequency, and then reaches the cell, the linear analyzer, with the optical axis perpendicular to the orientation of the polarized laser beam, and finally reaches the photodetector. The signal is detected by a lock-in amplifier, at the frequency of the PEM. The transmittance as a function of time is registered in a computer (Fig. 1.37(b)). In the absence of the velocity gradient, the sample is isotropic and no transmittance is measured by the photodiode. As the velocity gradient is generated by squeezing the pipe, the sample becomes birefringent, with the optical axis oriented along the direction of the liquid crystal flow, and a finite transmittance is measured by the photodiode. This happens because the direction of the flow is at 45° with respect to the optical axis of the linear analyzer. The relaxation time of the birefringence has to be smaller than the time between two successive squeezings. The shape of the time dependent transmittance function is related to the time variation characteristics of the liquid crystal flow.

Since the relaxation time of the induced birefringence depends on the particular mixture and phase (L_1 or L_3), it is interesting to develop different sensors, adapted to several frequencies. The typical relaxation times of the birefringence in L_1 and L_3 phases are of the order of 10^{-2} and 3 s, respectively.

1.6 Interfaces with biology

Amphiphilic and water molecules are important components of the human body. The structures formed by amphiphilic molecules in the human body, and the processes involving the connections with water, other polar and nonpolar molecules, are essential ingredients for the existence of life. As some of these processes are essential for the stability of lyotropic mixtures, there is a large interface between research in lyotropics and in biological sciences [220]. We now present some examples of this interface.

A lipid is an important structural component of the living tissues in plants, animals and microorganisms. Lipid molecules may be neutral, phosphatides, sphingolipids, glycolipds, and terpenoids [221]. Phospholipids have polar and nonpolar parts; under dispersion in an aqueous medium, they self-organize in micelles and more complex structures, depending on relative concentrations and temperature. In the presence of water, porphyrins, which are essential for life, also form liquid crystalline structures. In particular, phospholipids in the presence of increasing amounts of water, were shown to reduce their marked endothermic transition, usually observed in differential thermal analysis experiments [222], corresponding to the melt of the hydrocarbon chains. Mixtures of natural phospholipids, with highly unsaturated chains, and water display a transition between crystalline and liquid crystal phases at temperatures lower than in the biological environment [223]. At room temperature, the chains of these molecules have a high mobility, which improves the permeability of the membranes that are formed.

One of the most expressive similarities between the living systems and lyotropics, specially in the case of water-based mixtures, is the dominant presence of water which, in some mixtures, reaches concentrations larger than 96 M%. If water is mixed with suitable solvents, biological compounds or even molecules commonly found in the human body may present liquid crystalline phases. An interesting example of liquid crystalline polymorphism with biological lipids is a mixture of the mitochondrial lipid and water [224]; there were observations of three lamellar and one hexagonal phases, depending on temperature and relative concentrations of lipid and water. Polypeptides are systems containing two or more aminoacids and one or more peptide groups, as poly-γ-benzyl-L-glutamate (PBLG); if they are mixed with different solvents, polypeptides may present liquid crystalline phases [225]. Cholesterol and fatty acids form esters, which also have liquid crystalline properties. In particular, cholesterol is used to induce cholesteric mesophases in originally lyotropic nematic structures (Section 1.3.3).

Micellar solutions of amphiphiles also play an important role in biological processes [198]: (i) the digestion of fats by the human body requires solubilization by bile salt micelles, which act as the detergent of the body through the digestive tract; (ii) the micellar catalysis, which exploits the large surface available in micellar systems (an ester hydrolysis by an acid, e.g.). Particularly in the case of the fat solubilization problem, *in vitro* micellar solutions of lyotropics are used to study the temperature dependence of solubilization rate of fat acids in micelles. It was shown that in actual lyotropic mixtures this rate can increase by a factor of 4 at temperatures above the *penetration temperature* (at which the liquid crystalline phases appear).

Phospholipids and proteins are the basic structural blocks of the cell membrane [221]. About 50% of the mass of most animal cell plasma membranes are made of lipidic membranes. Lipids form a bimolecular layer (bilayer) which controls the traffic of substances to and from the inner part of the cell (Fig. 1.38). Despite the differences between the liquid crystalline structures and the living cell, lyotropics can be used as *models* to study some aspects of the cell membranes. These studies of cellular models started in the beginning of the twentieth century; the models refer to animal and vegetal membranes (as the example of soybean lecithin [226]).

Biological membranes inside cells serve to different purposes, as dividing compartments with different functions, and regulators of the flux of substances with different electric properties. These membranes usually have encrusted proteins.

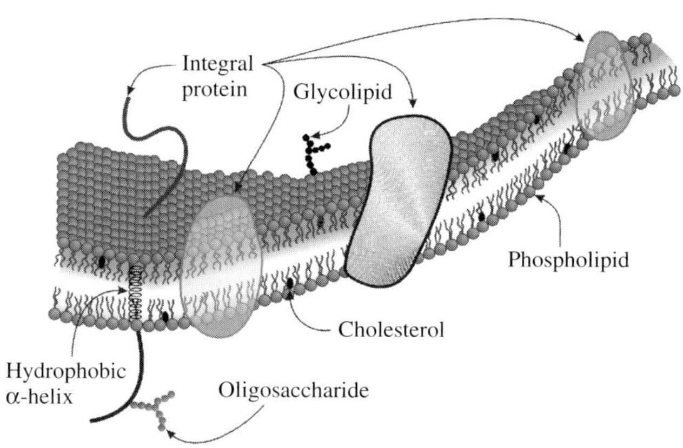

FIG. 1.38. Sketch of a cellular membrane of a living cell.

Investigations of the permeability of macromolecules through a lyotropic model membrane, in a lamellar-type phase, brings important insights into problems regarding the form (structure) versus functionality in actual membranes. Amphiphilic molecules of biological membranes usually have two hydrocarbon chains from glycerol derivatives. The membrane prevents transport from one compartment to the other, without a biological command. From this point of view, the equilibrium in terms of the concentration gradients of different substances is not passively achieved. In plants and bacteria, the dominant lipid membranes are glycolipids. In animals, the most common membranes are phospholipids. In the human body, molecular species with multiple positive charge are rather rare, and biological membranes have a net negative surface charge. The diffusion coefficients of molecules across the membrane and on its surface can be very different. As this behavior resembles the similar behavior of a lamellar liquid crystals, it is one of the reasons why lyotropics can be used as a model system to study diffusion processes in biological membranes. For example, the permeability to both strongly polar (water) and non-polar (hydrocarbons) molecules by the stratum corneum (the outer part of the human skin) can be studied by using lyotropic lamellar model systems [198].

One of the most interesting and important problems in living systems is the investigation of the interaction between lipids and proteins. According to a current view on this problem, proteins and cholesterol are incorporated into the bilayer structure, with its consequent disruption. Cholesterol molecules are interdigitated between lipid molecules of the bilayer, with the hydroxyl groups in the external water layer. The bilayer is conceived as a dispersion medium for the proteins and cholesterol. The stability of the structure and the permeability properties if the bilayer is in contact with hosts, as proteins and cholesterol, can also be investigated in lyotropic model membranes and in cholesteric phases (in the particular case of chiral molecules incorporated into micelles or even in bicontinuous structures). The presence of cholesterol in lipid bilayers (Fig. 1.38) was shown to lower the transition temperature between gel and liquid crystalline phases [227]. It disturbs the crystalline arrangement of the hydrocarbon chains in the gel phase structure. This process can be studied *in vitro* by using a lipid/cholesterol/water mixture.

Another aspect of the similarity between lyotropics and living systems is related to the process of aging of the cell. As the living system becomes older, it is known that the cell membrane becomes stiffer, with a drift towards crystallinity [221]. In a typical binary lyotropic water-based mixture, at fixed temperature and for increasing concentration of amphiphiles, there is a phase sequence from micellar isotropic to hexagonal, and then lamellar, and finally to a crystalline phase [32]. The aging of a cellular membrane keeps some similarities with the concentration-driven phase transition sequences of lyotropic systems. In lyotropics, the loss of water is the way to crystallinity. In human cell membranes, the aging process is certainly more complex than a simple loss of water, so that many other aspects have to be taken into account. However, the role of the water molecules and their interactions with the membrane structure can certainly be one of the main aspects of this problem. For example, the investigation of the role of the hydration layers around different amphiphilic molecular aggregates in lyotropics, from micelles to the lamellae, can provide relevant insights on the time dependence of the characteristics of membranes of living systems [6].

Shape transformations in biological lyotropic-like systems play a key role in many processes in the human body. For example, in the propagation of nerve impulses over the synaptic clef separating nerve cells by vesicles [198]. Vesicles, which are a blister-type bylayered structure ranging in diameter from micrometers to millimeters, undergo a transformation and participate in this propagation under the action of an electric excitation. This process is reversible and, due to the hydrolysis by an appropriate enzyme, the system is ready to receive and deliver new signals.

Amphiphilic biocompatible molecular aggregates are used as drug vectors. Medicines can be encapsuled inside micelles or vesicles and injected in the living system. These closed structures are used, e.g., as simple containers to transport drugs within the blood system.

Cell membranes in living systems are usually in contact with excess water. Therefore, due to topology and stability in the presence of excess water, the bicontinuous cubic phases Q^{224} and Q_m^{227} are also good systems in order to investigate typical processes in living cells. There are some proposals about biological similarities between the Q^{224} phase and the plasma membranes of thermoacidophilic archae-bacteria [228], in particular with respect to the digestion of fats [166]. The Q_m^{227} phase was observed in mixtures with lipids commonly present in biological membranes, as phosphatidylcholine and phosphatidyl ethanolamine. The stability of the structure in the presence of fatty acids can lead to interesting information about the leaking processes (disruption of the membrane) in living systems [76]. Cubic phases can also be used as templates for synthesizing nanoporous materials and nanocomposites. An important property of these phases is the large bilayer surface area (about 100 m^2/g). Biosensors should be developed by employing the immobilization of proteins by covalent attachment to the head group or by incorporation into the bilayer.

A promising field of research in the interface between lyotropics and biology, which may give relevant information in both areas, is the investigation of radiation-induced phase transitions. Recently [229], it was observed that lyotropic liquid crystalline samples (KL/1-decanol/water mixture), originally in the N_C nematic phase, subjected to irradiation with protons (doses of the order of 10 Gy), undergo a phase transition to a hexagonal phase. Fan-like textures (Fig. 1.39) were observed after the irradiation. At smaller doses ($\sim$1 Gy) no phase transition was observed. The microscopic process responsible for this phenomenon is not yet known. The radiation seems to destabilize the micelles in the solution, favoring the presence of large molecular aggregates.

A deeper understanding of the biological processes in living systems will certainly reveal other similarities between apparently disjoint fields of research.

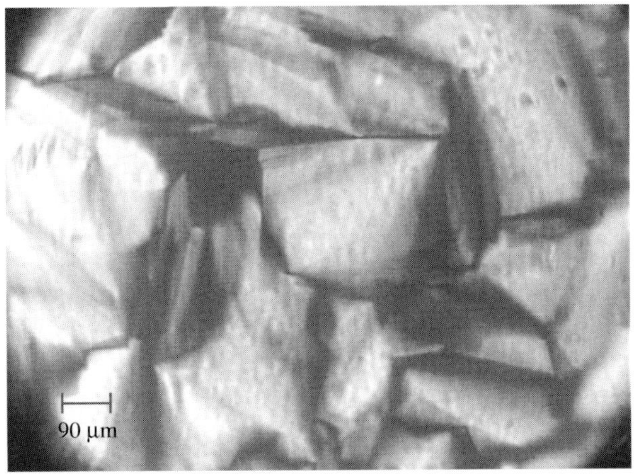

FIG. 1.39. Typical fan-like texture of a phase induced by radiation (protons of 40 Gy) in a sample of a previously nematic phase, in a polarizing light microscope, between crossed polarizers (for a 200 μm thick sample of a potassium laurate/decanol/water mixture).

Interfaces will become more frequent an much larger in this truly multidisciplinary subject. Research in lyotropic mixtures is expected to play a privileged role in this new context.

References

[1] P. G. de Gennes and J. Prost (1993). *The Physics of Liquid Crystals*, Clarendon Press, Oxford.

[2] T. P. Hoar and J. H. Schulman (1943). *Nature (London)* **152**, 102.

[3] L. M. Prince (Ed.) (1976). *Microemulsions. Theory and Pratice*, Academic Press, New York.

[4] I. Danielsson and B. Lindman (1981). *Colloid. Surf.* **3**, 391.

[5] C. J. F. Böttcher and P. Bordewijk (1978). *Theory of Electric Polarization*, Elsevier, Amsterdam.

[6] C. Tanford (1980). *The Hydrophobic Effect: Formation of Micelles and Biological Membranes*, Wiley, New York.

[7] D. Markovitsi, A. Mathis, J. Simon, J. C. Wittman, and J. Le Moigne (1980). *Mol. Cryst. Liq. Cryst.* **64**, 121.

[8] F. M. Menger and C. A. Littau (1991). *J. Am. Chem. Soc.* **113**, 1451.

[9] F. M. Menger and J. Ding (1996). *Angew. Chem.* **108**, 2266.

[10] F. Hildebrandt, J. A. Schröter, C. Tschierske, R. Festag, R. Kleppinger, and J. H. Wendorff (1995). *Angew. Chem., Int. Ed. Engl.* **34**, 1631.

[11] J. A. Schröter, C. Tschierske, M. Wittenberg, and J. H. Wendorff (1998). *J. Am. Chem. Soc.* **120**, 10669.

[12] J. H. Fuhrhop and D. Fritsch (1986). *J. Am. Chem. Soc.* **19**, 130.
[13] M. Kölbel, T. Beyersdorff, X. H. Cheng, C. Tschierske, J. Kain, and S. Diele (2001). *J. Am. Chem. Soc.* **123**, 6809.
[14] J. N. Israelachvili (1991). *Intermolecular and Surface Forces*, Academic Press, New York.
[15] W. C. Preston (1948). *J. Phys. Colloid Chem.* **52**, 84.
[16] A. W. Adamson (1960). *Physical Chemistry of Surfaces*, Interscience, New York.
[17] H. S. Frank and M. W. Evans (1945). *J. Chem. Phys.* **13**, 507.
[18] J. N. Israelachvili, D. J. Mitchell, and B. W. Ninham (1976). *J. Chem. Soc. Faraday Trans. 2* **72**, 1525.
[19] L. Leibler, H. Orland, and J. C. Wheeler (1983). *J. Chem. Phys.* **79**, 3550.
[20] W. E. McMullen, W. M. Gelbart, and A. Ben-Shaul (1984). *J. Phys. Chem.* **88**, 6649.
[21] R. G. Larson *et al.* (1985). *J. Chem. Phys.* **83**, 2411.
[22] A. T. Bernardes, V. B. Henriques, and P. M. Bisch (1994). *J. Chem. Phys.* **101**, 645.
[23] J. E. Adderson (1971). *J. Pharm. Pharmac.* **23**, 311.
[24] R. J. Hunter (1993). *Foundations of Colloid Science*, vol. 1, 5th edn, Oxford Science Publications, New York.
[25] K. Shinoda (1963). *Colloidal Surfactants*, Academic Press, New York.
[26] G. Gunnarsson, B. Jönsson, and H. Wennerström (1980). *J. Phys. Chem.* **84**, 3114.
[27] F. L. S. Cuppo (2002). PhD Thesis, Institute of Physics, University of São Paulo, Brazil.
[28] O. R. Howell and H. G. B. Robinson (1936). *Proc. Roy. Soc. (London)* **A155**, 386.
[29] M. C. Carey and D. M. Small (1970). *Am. J. Med.* **49**, 590.
[30] N. A. Mazer, G. B. Benedek, and M. C. Carey (1976). *J. Phys. Chem.* **80**, 1075.
[31] W. U. Malik and A. K. Jain (1967). *J. Inorg. Nucl. Chem.* **29**, 2825.
[32] P. Ekwall (1975). In *Advances in Liquid Crystals*, Ed. G. H. Brown, Academic Press, New York, vol. 1.
[33] V. Luzzati, H. Mustacchi, A. Skoulios, and F. Husson (1960). *Acta Crystallogr.* **13**, 660.
[34] J. M. Vincent and A. E. Skoulios (1966). *Acta Crystallogr.* **20**, 432.
[35] J. M. Vincent and A. E. Skoulios (1966). *Acta Crystallogr.* **20**, 441.
[36] J. M. Vincent and A. E. Skoulios (1966). *Acta Crystallogr.* **20**, 447.
[37] V. Luzzati (1968). In *Biological Membranes*, Ed. D. Chapman, Academic Press, New York.
[38] B. Mely, J. Charvolin, and P. Keller (1975). *Chem. Phys. Lipids*, **15**, 161.
[39] J. Charvolin and P. Rigny (1973). *J. Chem. Phys.* **58**, 3999.
[40] N. H. Hartshorne and G. D. Woodard (1973). *Mol. Cryst. Liq. Cryst.* **23**, 343.

[41] C. L. Khetrapal, A. C. Kunwar, A. S. Tracey, and P. Diehl (1975). In *NMR Basic Principles and Progress*, Eds P. Diehl, E. Fluck, and R. Kosfeld, Springer-Verlag, Berlin.
[42] G. Lindblom and G. Orädd (1994). *Prog. NMR Spectrosc.* **26**, 483.
[43] B. Halle and I. Furó (1998). In *Phase Transitions in Complex Fluids*, Eds P. Tolédano and A. M. Figueiredo Neto, World Scientific, Singapore.
[44] K. A. Stacey (1956). *Light Scattering in Physical Chemistry*, Academic Press, New York.
[45] V. Degiorgio and R. Piazza (1998). In *Phase Transitions in Complex Fluids*, Eds P. Tolédano and A. M. Figueiredo Neto, World Scientific, Singapore.
[46] B. Jacrot (1976). *Rep. Prog. Phys.* **39**, 911.
[47] Y. Hendrikx, J. Charvolin, M. Rawiso, L. Liébert, and M. C. Holmes (1983). *J. Phys. Chem.* **87**, 3991.
[48] V. Luzzati, H. Mustachiand, and A. Skoulios (1957). *Nature* **180**, 600.
[49] A. Saupe (1977). *J. Colloid Interfac. Sci.* **58**, 549.
[50] D. Demus and L. Richter (1980). *Textures of Liquid Crystals*, VEB Deutscher Verlag für Grundstoffindustrie, Leipzig.
[51] T. Haven, K. Radley, and A. Saupe (1981). *Mol. Cryst. Liq. Cryst.* **75**, 87.
[52] M. Born and E. Wolf (1980). *The Principles of Optics*, Pergamon Press, Oxford.
[53] Y. Galerne and J. P. Marcerou (1983). *Phys. Rev. Lett.* **51**, 2109.
[54] N. Boden, S. A. Corne, and K. W. Jolley (1984). *Chem. Phys. Lett.* **105**, 99.
[55] P. J. Photinos and A. Saupe (1986). *J. Chem. Phys.* **84**, 517.
[56] A. M. Figueiredo Neto (1998). In *Phase Transitions in Complex Fluids*, Eds P. Tolédano and A. M. Figueiredo Neto, World Scientific, Singapore.
[57] K. D. Lawson and T. J. Flautt (1967). *J. Am. Chem. Soc.* **89**, 5489.
[58] K. Radley, L. W. Reeves, and A. S. Tracey (1976). *J. Phys. Chem.* **80**, 174.
[59] J. Charvolin, A. M. Levelut and E. T. Samulski (1979). *J. Phys. (Paris) Lett.* **40**, L-587.
[60] Y. Hendrikx and J. Charvolin (1981). *J. Phys. (Paris)* **42**, 1427.
[61] A. M. Figueiredo Neto, Y. Galerne, A. M. Levelut, and L. Liébert (1985). *J. Phys. (Paris) Lett.* **46**, L-499.
[62] L. J. Yu and A. Saupe (1980). *Phys. Rev. Lett.* **45**, 1000.
[63] K. Radley and A. Saupe (1978). *Mol. Phys.* **35**, 1405.
[64] L. J. Yu and A. Saupe (1980). *J. Am. Chem. Soc.* **102**, 4879.
[65] M. Acimis and L. W. Reeves (1980). *Can. J. Chem.* **58**, 1533.
[66] A. M. Figueiredo Neto, Y. Galerne, and L. Liébert (1985). *J. Phys. Chem.* **89**, 3939.
[67] M.C. Valente Lopes and A.M. Figueiredo (1988). *Phys. Rev. A* **38**, 1101.
[68] L. Longa, M. Bachteler, A. M. Figueiredo Neto, G. Cholewiak, and F. A. Oliveira (2000). *Liq. Cryst.* **27**, 1669.

[69] C. Tanford (1972). *J. Phys. Chem.* **76**, 3020.
[70] V. Luzzati, T. G. Krzywicki, and A. Tardieu (1968). *Nature* **218**, 1031.
[71] A. Tardieu, V. Luzzati, and F. C. Reman (1973). *J. Mol. Biol.* **75**, 711.
[72] G. S. Smith, E. B. Sirota, C. R. Safinya, and N. A. Clark (1988). *Phys. Rev. Lett.* **60**, 813.
[73] F. Husson, H. Mustacchi, and V. Luzzati (1960). *Acta Crystallogr.* **13**, 668.
[74] Y. Hendrikx and J. Charvolin (1988). *Liq. Cryst.* **3**, 265.
[75] K. Fontell (1990). *Colloid Polym. Sci.* **268**, 264.
[76] V. Luzzati, R. Vargas, P. Mariani, A. Gulik, and H. Delacroix (1993). *J. Mol. Biol.* **229**, 540.
[77] M. Clerc (1996). *J. Phys. (France) II* **6**, 961.
[78] D. Roux, C. Coulon, and M. E. Cates (1992). *J. Phys. Chem.* **96**, 4174.
[79] G. Porte (1992). *J. Phys. Condens. Matter*, **4**, 8649.
[80] G. H. Brown, J. W. Doane, and V. D. Neff (1971). *A Review of the Structures and Physical Properties of Liquid Crystals*, CRC Press, Cleveland, OH.
[81] J. W. McBain and W. C. Sierichs (1948). *J. Am. Oil Chem. Soc.* **25**, 221.
[82] A. M. Figueiredo Neto, L. Liébert, and Y. Galerne (1985). *J. Phys.Chem.* **89**, 3737.
[83] G. Melnik, P. Photinos, and A. Saupe (1989). *Phys. Rev. A* **39**, 1597.
[84] M. B. Lacerda Santos, Y. Galerne, and G. Durand (1984). *Phys. Rev. Lett.* **53**, 787.
[85] Y. Galerne, A. M. Figueiredo Neto, and L. Liébert (1987). *J. Chem. Phys.* **87**, 1851.
[86] Y. Hendrikx, J. Charvolin, and M. Rawiso (1986). *Phys. Rev. B* **33**, 3534.
[87] Y. Galerne, A. M. Figueiredo Neto, and L. Liébert (1985). *Phys. Rev. A* **31**, 4047.
[88] M. J. de Oliveira and A. M. Figueiredo Neto (1986). *Phys. Rev. A* **34**, 3481.
[89] R. Bartolino, T. Chiaranza, M. Meuti, and R. Compagnoni (1982). *Phys. Rev. A* **26**, 1116.
[90] E. A. Oliveira, L. Liébert, and A. M. Figueiredo Neto (1989). *Liq. Cryst.* **5**, 1669.
[91] L. V. Azároff (1968). *Elements of X-Ray Crystallography*, McGraw-Hill, New York.
[92] P. R. G. Fernandes and A. M. Figueiredo Neto (1995). *Phys. Rev. E* **51**, 567.
[93] A. M. Stoneham (1969). *Rev. Mod. Phys.* **41**, 82.
[94] M. Mehring (1983). *High Resolution NMR in Solids*, 2nd edn, Springer-Verlag, Berlin and New York.
[95] B. J. Forrest and L. W. Reeves (1981). *Chem. Rev.* **1**, 1.
[96] N. Boden, P. H. Jackson, K. McMullen, and M. C. Holmes (1979). *Chem. Phys. Lett.* **65**, 476.

[97] F. Fujiwara, L. W. Reeves, M. Suzuki, and J. A. Vanin (1979). In *Solution Chemistry of Surfactants*, Ed. K. L. Mittal, Plenum Press, New York, vol. 1.
[98] M. J. Freiser (1970). *Phys. Rev. Lett.* **24**, 1041.
[99] C. S. Shih and R. Alben (1972). *J. Chem. Phys.* **57**, 3055.
[100] P. Tolédano and A. M. Figueiredo Neto (1994). *Phys. Rev. Lett.* **73**, 2216.
[101] F. Brochard and P. G. de Gennes (1970). *J. Phys. (Paris)* **31**, 691.
[102] A. M. Figueiredo Neto and M. M. F. Saba (1986). *Phys. Rev. A* **34**, 3483.
[103] L. Liébert and A. Martinet (1979). *J. Phys. (Paris) Lett.* **40**, L-363.
[104] A. M. Figueiredo Neto, Ph. Martinot-Lagarde, and G. Durand (1984). *J. Phys. Lett.* **45**, L-793.
[105] E. B. Priestley (1979). In *Introduction to Liquid Crystals*, Eds E. B. Priestley, P. I. Wojtowicz, and P. Sheng, Plenum Press, New York.
[106] T. Kroin and A. M. Figueiredo Neto (1987). *Phys. Rev. A* **36**, 2987.
[107] T. Kroin, A. J. Palangana, and A. M. Figueiredo Neto (1989). *Phys. Rev. A* **39**, 5373.
[108] A. J. Palangana and A. M. Figueiredo Neto (1990). *Phys. Rev. A* **41**, 7053.
[109] F. Y. Fujiwara and L. W. Reeves (1980). *J. Phys. Chem.* **84**, 653.
[110] P. O. Quist (1995). *Liq. Cryst.* **18**, 623.
[111] A. A. Melo Filho, A. Laverde Jr., and F. Y. Fujiwara (2003). *Langmuir*, **19**, 1127.
[112] A. Saupe, P. Boonbrahm, and L. J. Yu (1983). *J. Chim. Phys. (Paris)* **80**, 7.
[113] K. Praefcke (2001). *Mol. Cryst. Liq. Cryst.* **364**, 15.
[114] B. R. Acharya, A. Primak, and S. Kumar (2004). *Phys. Rev. Lett.* **92**, 145506.
[115] F. Cuppo, S. Gomez, and A. M. Figueiredo Neto (2003). *Phys. Rev. E* **67**, 051711.
[116] R. Alben (1973). *Phys. Rev. Lett.* **30**, 778.
[117] P. Boonbrahm and A. Saupe (1984). *J. Chem. Phys.* **81**, 2076.
[118] G. Melnik, P. Photinos, and A. Saupe (1988). *J. Chem. Phys.* **88**, 4046.
[119] Z. A. de Sant'Ana and A. M. Figueiredo Neto (1992). *Phys. Rev. A* **46**, 7630.
[120] J. C. Le Guillou and J. Zinn-Justin (1977). *Phys. Rev. Lett.* **39**, 95.
[121] A. M. Figueiredo Neto, L. Liébert, and A. M. Levelut (1984). *J. Phys. (Paris)* **45**, 1505.
[122] A. M. Figueiredo Neto and M. E. Marcondes Helene (1987). *J. Phys. Chem.* **91**, 1466.
[123] G. Melnik and A. Saupe (1987). *Mol. Cryst. Liq. Cryst.* **145**, 95.
[124] M. E. Marcondes Helene and A. M. Figueiredo Neto (1988). *Mol. Cryst. Liq. Cryst.* **162B**, 127.
[125] P. Diehl and A. S. Tracey (1975). *FEBS Lett.* **59**, 131.
[126] M. R. Alcantara, M. V. M. C. Melo, V. R. Paoli, and J. A. Vanin (1984). *Mol. Cryst. Liq. Cryst.* **107**, 303.

[127] M. R. Alcantara, M. V. M. C. Melo, V. R. Paoli, and J. A. Vanin (1984). *Mol. Cryst. Liq. Cryst.* **107**, 359.

[128] A. M. Figueiredo Neto, A. M. Levelut, Y. Galerne, and L. Liébert (1988). *J. Phys. (Paris)* **49**, 1301.

[129] H. R. Brand and H. Pleiner (1985). *J. Phys. (Paris) Lett.* **46**, 711.

[130] P. Tolédano, A. M. Figueiredo Neto, V. Lorman, and V. Dmitriev (1995). *Phys. Rev. E* **52**, 5040.

[131] T. Kroin, A. M. Figueiredo Neto, L. Liébert, and Y. Galerne (1989). *Phys. Rev. A* **40**, 4647.

[132] G. S. Smith, E. B. Sirota, C. R. Safinya, R. J. Plano, and N. A. Clark (1990). *J. Chem. Phys.* **92**, 4519.

[133] G. S. Smith, C. R. Safinya, D. Roux, and N. A. Clark (1987). *Mol. Cryst. Liq. Cryst.* **144**, 235.

[134] J. Charvolin and P. Rigny (1973). *Chem. Phys. Lett.*, **18**, 515.

[135] J. A. N. Zasadzinski, J. Schneir, J. Gurley, V. Elings, and P. K. Hansma (1988). *Science* **239**, 1013.

[136] D. C. Wack and W. W. Webb (1988). *Phys. Rev. Lett.* **61**, 1210.

[137] K. Sengupta, V. A. Raghunathan, and J. Katsaras (1999). *Phys. Rev. E* **59**, 2455.

[138] J. Katsaras, S. Tristam-Nagle, Y. Liu, R. L. Headrick, E. Fontes, P. C. Mason, and J. F. Nagle (2000). *Phys. Rev. E* **61**, 5668.

[139] G. Kothe and C. Mayer (1994). In *The Molecular Dynamics of Liquid Crystals*, Eds. G. R. Luckhurst and C. A. Veracini, Kluwer Academic, Dordrecht.

[140] R. R. Balmbra, J. S. Clunie, and J. F. Goodman (1969). *Nature* **222**, 1159.

[141] P. Ekwall and L. Mandell (1969). *Kolloid-Z.* **233**, 938.

[142] F. B. Rosevear (1954). *J. Am. Oil Chem. Soc.* **31**, 628.

[143] F. B. Rosevear (1968). *J. Soc. Cosmetic Chem.* **19**, 581.

[144] S. E. Friberg and M. A. El-Nokaly (1985). In *Surfactants in Cosmetics*, Ed. M. M. Rieger, *Surfactant Science Series*, vol. 16, Marcel Dekker, New York.

[145] A. A. Hidalgo (2000). PhD Thesis, Institute of Physics, University of São Paulo, Brazil.

[146] A. Maldonado, W. Urbach, R. Ober, and D. Langevin (1996). *Phys. Rev. E* **54**, 1774.

[147] N. Boden and M. C. Holmes (1984). *Chem. Phys. Lett.* **109**, 76.

[148] Y. Lee and L. W. Reeves (1976). *Can. J. Chem.* **54**, 500.

[149] T. Haven, D. Armitage, and A. Saupe (1981). *J. Chem. Phys.* **75**, 352.

[150] E. Zhou, M. Stefanov, and A. Saupe (1988). *J. Chem. Phys.* **88**, 5137.

[151] P. J. Photinos, L. J. Yu, and A. Saupe (1981). *Mol. Cryst. Liq. Cryst.* **67**, 277.

[152] M. C. Holmes and J. Charvolin (1984). *J. Phys. Chem.* **88**, 810.

[153] M. C. Holmes, M. S. Leaver, and A. M. Smith (1995). *Langmuir* **11**, 356.

[154] N. Boden and K. W. Jolley (1992). *Phys. Rev. A* **45**, 8751.

[155] M. R. Kuzma, W. Wedler, A. Saupe, S. Shin, and S. Kumar (1992). *Phys. Rev. Lett.* **68**, 3436.
[156] A. M. Figueiredo Neto and L. Q. Amaral (1983). *Mol. Cryst. Liq. Cryst.* **95**, 129.
[157] A. Skoulios (1978). *Ann. Phys.* **3**, 421.
[158] S. Alpérine, Y. Hendrikx and J. Charvolin (1985). *J. Phys. (France) Lett.* **46**, L27.
[159] G. J. T. Tiddy (1972). *J. Chem. Soc. Faraday Trans.* **68**, 379.
[160] J. Charvolin and J. F. Sadoc (1988). *J. Phys. (France)* **49**, 521.
[161] J. M. Seddon, E. A. Bartle, and J. Mingins (1990). *J. Phys. Condens. Matter* **2**, SA285.
[162] P. Mariani, E. Rivas, V. Luzzati, and H. Delacroix (1990). *Biochemistry* **29**, 6799.
[163] J. M. Seddon and R. H. Templer (1993). *Philos. Trans. Roy. Lond. A* **344**, 377.
[164] V. Luzzati and P. A. Spegt (1967). *Nature* **215**, 701.
[165] A. Tardieu (1972). PhD Thesis, Université de Paris-Sud, Orsay, France.
[166] P. Mariani, V. Luzzati, and H. Delacroix (1988). *J. Mol. Biol.* **204**, 165.
[167] V. Luzzati, A. Tardieu, and T. Gulik-Krzywicki (1968). *Nature* **217**, 1028.
[168] M. Impéror-Clerc and A. M. Levelut (2001). *Eur. Phys. J. E* **4**, 209.
[169] A. Goetz (1970). *Introduction to Differential Geommetry*, Addison Wesley, Reading, MA.
[170] S. T. Hyde (1990). *J. Phys. (Paris) Colloque* **51**, 209.
[171] V. Luzzati, R. Vargas, A. Gulik, P. Mariani, J. M. Seddon, and E. Rivas (1992). *Biochemistry* **31**, 279.
[172] D. Roux, M. E. Cates, U. Olsson, R. C. Ball, F. Nallet, and A. M. Bellocq (1990). *Europhys. Lett.* **11**, 229.
[173] J. C. C. Nitsche (1989). *Lectures on Minimal Surfaces*, Cambridge University Press, Cambridge.
[174] B. Balinov, U. Olsson, and O. Söderman (1991). *J. Phys. Chem.* **95**, 5931.
[175] M. Magalhães, D. Pusiol, M. E. Ramia, and A. M. Figueiredo Neto (1998). *J. Chem. Phys.* **108**, 3835.
[176] R. Strey, W. Jahn, G. Porte, and P. Bassereau (1990). *Langmuir* **6**, 1635.
[177] G. Porte, J. Appell, P. Bassereau, and J. Marignan (1989). *J. Phys. (France)* **50**, 1335.
[178] G. Porte, M. Delsanti, I. Billard, M. Skouri, J. Appell, J. Marignan, and F. Debeauvais (1991). *J. Phys II (France)* **1**, 1101.
[179] D. Gazeau, A. M. Bellocq, D. Roux, and T. Zemb (1989). *Europhys. Lett.* **9**, 447.
[180] J. Engblom and S. T. Hyde (1995). *J. Phys. II (France)* **5**, 171.
[181] M. Magalhães, A. M. Figueiredo Neto, and P. Tolédano (2000). *Phys. Rev. E* **62**, 5847.
[182] M. Skouri, J. Marignam, J. Appell, and G. Porte (1991). *J. Phys. II (France)* **1**, 1121.

[183] M. Filali, J. Appell, and G. Porte (1995). *J. Phys. II (France)* **5**, 657.
[184] G. Porte, J. Marignan, P. Bassereau, and R. J. May (1988). *J. Phys. (France)* **49**, 511.
[185] L. Porcar, W. A. Hamilton, and P. D. Butler (2002). *Phys. Rev. Lett.* **89**, 168301.
[186] U. Henriksson, E. S. Blackmore, and G. J. T. Tiddy, and O. Söderman (1992). *J. Phys. Chem.* **96**, 3894.
[187] P. Kékicheff and G. J. T. Tiddy (1989). *J. Phys. Chem.* **93**, 2520.
[188] P. Kékicheff and B. Cabane (1988). *Acta Crystallogr. B* **44**, 395.
[189] J. Burgoyne, M. C. Holmes, and G. J. T. Tiddy (1995). *J. Phys. Chem.* **99**, 6054.
[190] S. T. Hyde (1994). *Pure Appl. Chem.* **64**, 1617.
[191] P. G. de Gennes (1985). *Rev. Mod. Phys.* **57**, 827.
[192] E. A. Oliveira, A. M. Figueiredo Neto, and G. Durand (1991). *Phys. Rev. A* **44**, R825.
[193] S. Fontanini, A. Strigazzi, G. Barbero, M.C. Salvadori, and A.M. Figueiredo Neto (1998). *Liq. Cryst.* **24**, 793.
[194] M. Magalhães (2002). PhD Thesis, Institute of Physics, University of São Paulo, Brazil.
[195] L. Landau and E. M. Lifchitz (1969). *Électrodynamique des Milieux Continus*, Éditions MIR, Moscow.
[196] M. Magalhães, A. M. Figueiredo Neto, G. Barbero, and A. C. Tromba (2003). *J. Phys. Chem. B* **107**, 13949.
[197] R. H. Ferguson (1932). *Oil & Soap* **9**, 5.
[198] D. F. Evans and H. Wennerström (1994). *The Colloidal Domain where Physics, Chemistry, Biology and Technology Meet*, VCH Publishers, New York.
[199] S. Friberg and K. Larsson (1976). In *Advances in Liquid Crystals*, Ed. G. H. Brown, Academic Press, New York, vol. 2.
[200] F. D. Gunstone and F. A. Norris (1983). *Lipids and Foods*, Pergamon Press, Oxford, UK.
[201] H. This (1996). PhD Thesis, University Paris VI, France.
[202] D. O. Shah and R. S. Schechter (Eds) (1977). *Improved Oil Recovery by Surfactant and Polymer Flooding*, Academic Press, New York.
[203] M. De Groot (1929). US Patent no. 1,823,439.
[204] W. B. Gogarty and W. C. Tosch (1968). *J. Petrol. Tech.* **20**, 1407.
[205] T. A. B. M. Bolsman (1982). In *Proc. 2nd European Symposium on Enhanced Oil Recovery*, Technip, Paris, p. 3.
[206] Y. Barakat, L. N. Fortney, R. S. Schechter, W. H. Wade, and S. W. Yiv (1982). In *Proc. 2nd European Symposium on Enhanced Oil Recovery*, Technip, Paris, p. 11.
[207] S. Friberg (1978). In *Advances in Liquid Crystals*, Ed. G. H. Brown, Academic Press, New York, vol. 3.

[208] M. M. Rieger (Ed.) (1985). *Surfactants in Cosmetics, Surfactant Science Series*, vol. 16, Marcel Dekker, New York.
[209] M. Podzimek and S. E. Friberg (1980). *J. Dispers. Sci. Technol.* **1**, 341.
[210] S. E. Friberg (1979). *J. Soc. Cosmet. Chem.* **30**, 309.
[211] M. S. Spector, J. V. Selinger, and J. M. Schnur (1997). *J. Am. Chem. Soc.* **119**, 8533.
[212] P. Yager and P. E. Schoen (1984). *Mol. Cryst. Liq. Cryst.* **106**, 371.
[213] J. M. Schnur, B. R. Ratna, J. V. Selinger, A. Singh, G. Jyothi, and K. R. K. Easwaran (1994). *Science* **264**, 945.
[214] M. S. Spector, A. Singh, P. B. Messersmith, and J. M. Schnur (2001). *Nano Lett.* **1**, 375.
[215] C. T. Kresge, M. E. Leonowicz, W. J. Roth, and J. C. Vartuli (1992). *Nature* **359**, 710.
[216] H. B. Jervis, M. E. Raimondi, R. Raja, T. Maschmeyer, J. M. Seddon, and D. W. Bruce (1999). *Chem. Commun.* **20**, 2031.
[217] INPI (1988). Brazil, PI.8.801.918, April 15.
[218] INPI (2000). Brazil, PI.9.903.473-5, August 13.
[219] INPI (1998). Brazil, PI.9.805.500, November 20.
[220] G. T. Stewart (1974). In *Liquid Crystals and Plastic Crystals*, Eds G. W. Gray and P. A. Winsor, *Ellis Horwood Series in Physical Chemistry*, John Wiley, New York.
[221] G. H. Brown and J. J. Wolken (1979). *Liquid Crystals and Biological Structures*, Academic Press, New York.
[222] D. Chapman and D. T. Collin (1965). *Nature* **206**, 189.
[223] D. Chapman (1974). In *Liquid Crystals and Plastic Crystals*, Eds G. W. Gray and P. A. Winsor, *Ellis Horwood Series in Physical Chemistry*, John Wiley, New York.
[224] T. Gulik-Krzywicki, E. Rivas, and V. Luzzati (1967). *J. Mol. Biol.* **27**, 8.
[225] C. Robinson (1956). *Trans. Faraday Soc.* **52**, 571.
[226] M. Antonietti, M. Neese, G. Blum, and F. Kremer (1996). Langmuir **12**, 4436.
[227] B. D. Ladbrooke, R. M. Williams, and D. Chapman (1968). *Biochim. Biophys. Acta* **150**, 333.
[228] V. Luzzati, A. Gulik, M. De Rosa, and A. Gambacorta (1987). *Chem. Scripta* **27B**, 211.
[229] A. M. Figueiredo Neto, E. Szanto de Toledo, and N. Carlin. In preparation.

2

BASIC CONCEPTS OF PHASE TRANSITIONS

2.1 Introduction

In the following sections, we discuss a number of examples of phase transitions and critical phenomena in simple magnetic and liquid crystalline systems. We take advantage of these systems, and of standard mean-field calculations and Landau expansions, in order to give an overview and some illustrations of the main concepts in this area. Although there are many reviews of the work in the area [1]–[3], we first describe second-order transitions, critical and multicritical behavior in magnetic model systems. We then discuss first-order nematic–isotropic transitions, which have been observed in both thermotropic and lyotropic liquid crystals, and the transitions to a biaxial phase, as observed in lyotropic systems [4]. Finally, we refer to non-ordering parameters and to the reconstruction of phase diagrams, which are useful concepts to understand the behavior of lyotropic mixtures in terms of temperature and concentration.

Although the Landau theory of phase transitions discards local fluctuations, and does not give an accurate picture of the critical behavior, it does give a unifying, overwhelmingly simple, and qualitative overview of phase transitions and phase diagrams of all sorts of physical systems. It is particularly useful to describe the rich structures in the temperature–concentration phase diagrams of lyotropic mixtures. It will be used throughout this book as the basic theoretical framework for understanding phase transitions in lyotropic liquid crystals.

2.2 Critical and tricritical behavior in simple uniaxial ferromagnetic systems

According to the Landau theory of "second-order" phase transitions, the free-energy density of a system can be written as a power series in terms of the invariants of a fundamental quantity known as the "order parameter."For example, in zero external field, the order–disorder transition of a simple uniaxial ferromagnet is described by the free-energy density

$$f = f_0(T) + \frac{1}{2}A(T)m^2 + \frac{1}{4}C(T)m^4 + \cdots , \qquad (2.1)$$

where the magnetization per ion m is the order parameter, and $f_0(T)$, $A(T)$, and $C(T)$, are smooth functions of temperature. Due to the up-down symmetry of the magnetization of this uniaxial system, the invariants of the order parameter are even powers of m. For positive coefficients, $A, C > 0$, the free energy is minimum at $m = 0$, which corresponds to the disordered (paramagnetic) phase. For $A < 0$, with $C > 0$, $m = 0$ becomes a maximum of the free energy, but there appears

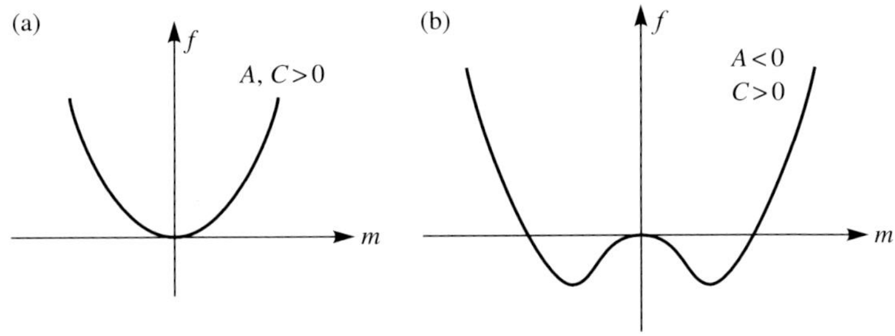

FIG. 2.1. Free-energy density f versus magnetization m for (a) $A, C > 0$, and (b) $A < 0$ and $C > 0$.

two symmetric, stable, minima, at $\pm m_0$ (see Fig. 2.1). The critical temperature is given by $A(T_c) = 0$.

If we drop the regular term $f_0(T)$, the neighborhood of this simple critical point is fully described by the truncated expansion,

$$f = \frac{1}{2} a_0 (T - T_c) m^2 + \frac{1}{4} m^4, \tag{2.2}$$

where a_0 is a positive constant, T_c is the critical temperature, and the coefficient $C(T_c)$ has been set to unit. For $T < T_c$, we have the order parameter

$$m_0 = \pm \sqrt{a_0} (T_c - T)^{1/2}, \tag{2.3}$$

whose (continuous) behavior near T_c leads to the definition of the (classical) critical exponent $\beta = 1/2$. This is an example of a continuous (second-order) transition, which is characterized by the smooth (continuous) variation of the order parameter with temperature.

If we include an external magnetic field, and consider the field (H) versus temperature (T) phase diagram, the Landau free-energy density may be written as

$$g = -mH + \frac{1}{2} a_0 (T - T_c) m^2 + \frac{1}{4} m^4, \tag{2.4}$$

with the external (applied) field in suitable units. In the H–T space, there is a line of first-order transitions, at $H = 0$ and $T < T_c$, with the coexistence of two phases, characterized by opposite values of m_0 and the same value of the free energy (see Fig. 2.2). It is important to remark that the order of a phase transition depends on the choice of the thermodynamic path in the phase diagram.

We now give an example of a multicritical point (and of a phase diagram with lines of first- and second-order transitions). In zero field, keeping terms up to sixth order, and still considering a uniaxial ferromagnet, we write the truncated

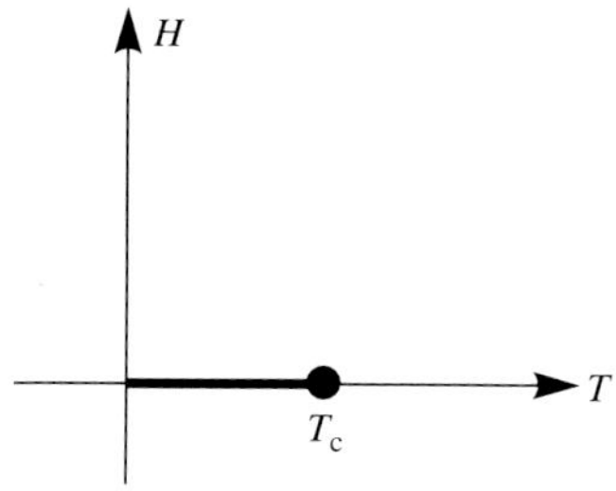

FIG. 2.2. Phase diagram, in terms of external field H and temperature T, of a simple uniaxial ferromagnet. The first-order boundary (solid line) ends at the critical temperature T_c.

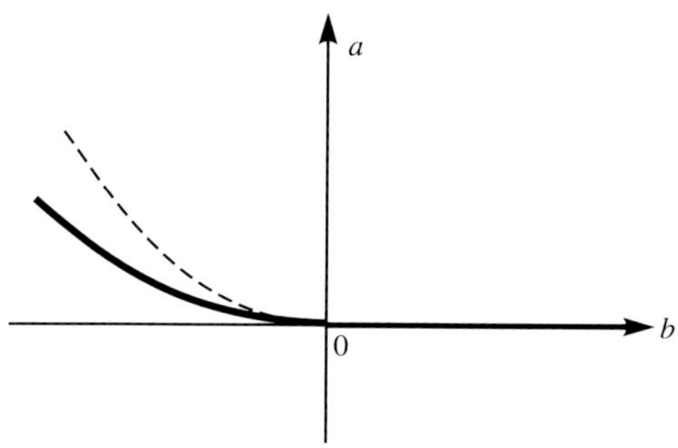

FIG. 2.3. Phase diagram in terms of the Landau coefficients a and b. The first-order boundary (heavy solid line; given by $a = 3b^2/16$ with $b < 0$) meets the critical line (given by $a = 0$ with $b > 0$), with zero slope, at a tricritical point ($a = b = 0$). The dashed line is the stability limit of the ordered solution.

expansion

$$f = \frac{1}{2}am^2 + \frac{1}{4}bm^4 + \frac{1}{6}m^6, \tag{2.5}$$

where a and b can change sign (depending on temperature and an extra field variable, as pressure or chemical potential, for example), but the coefficient of the sixth-order term is positive. It is straightforward to draw the $a - b$ phase diagram sketched in Fig. 2.3. We have to look for solutions of the equation

$$\partial f/\partial m = am + bm^3 + m^5 = 0. \tag{2.6}$$

Although there is a disordered solution, $m = 0$, for all values of the parameters, it is a stable minimum of the free energy for $a > 0$ only. The ordered phase ($m \neq 0$) is a stable minimum for: (i) $b > 0$ and $a < 0$ and (ii) $b < 0$ and $a < b^2/4$. These results lead to a line of continuous (second-order) transitions at $a = 0$, for $b > 0$, and a region of coexistence of two stable solutions, $m = 0$ and

$m \neq 0$, for $b < 0$, with $0 < a < b^2/4$. A comparison between the values of the free energy associated with each one of these solutions leads to the location of a line of coexistence of phases (first-order transitions) at $a = 3b^2/16$, with $b < 0$ (see Fig. 2.3). Both the second-order and the first-order boundaries meet, with the same (zero) slope, at the so-called tricritical point [5] (which is located at the origin in the a–b phase diagram of this simple example).

2.3 Phase diagrams with bicritical and tetracritical points

Phase diagrams with bicritical or tetracritical points are associated with systems with two distinct order parameters. A well-known example is given by the field versus temperature phase diagram of an anisotropic two-sublattice antiferromagnetic crystal [6] (with the external field applied along the easy axis of magnetization). A convenient order parameter is a staggered magnetization, in other words, a magnetization per site that points along a certain direction in sublattice A, and along the opposite direction in sublattice B, giving rise to the characterization of the typical, staggered, antiferromagnetic alignment. Below the Néel transition temperature, and for sufficiently weak fields, the ordered antiferromagnetic phase is associated with a non-zero staggered magnetization along the easy axis (and no perpendicular component of the staggered magnetization). As the field is increased, there may be a (first-order) transition to a spin-flop phase, with a net magnetization along the crystal axis, and a component of the staggered magnetization along a perpendicular direction (see Fig. 2.4). Let us call m_1 and m_2 the distinct components of the staggered magnetization along and perpendicular to the easy axis, respectively. The Néel phase, at smaller fields, is characterized by $m_1 \neq 0$ and $m_2 = 0$, and the spin-flop phase is associated with $m_1 = 0$ and $m_2 \neq 0$. At the disordered paramagnetic phase, we have $m_1 = 0$ and

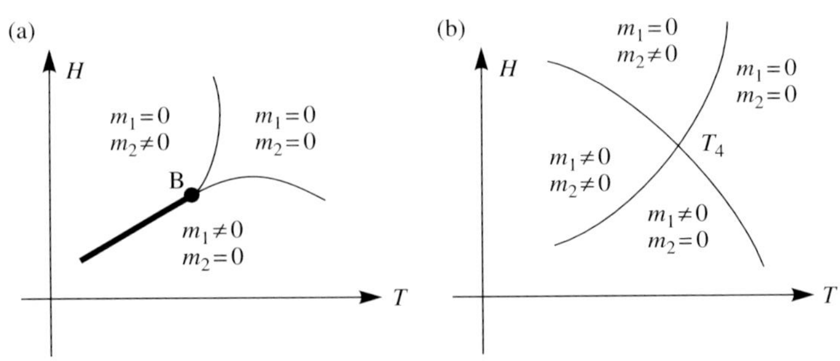

FIG. 2.4. Phase diagram of a two-sublattice anisotropic antiferromagnet in terms of the external field along the easy direction, H, and the temperature T: (a) the heavy solid line indicates a first-order boundary that ends at a bicritical point B; the light solid lines represent second-order transitions; (b) there are four lines of second-order transitions meeting at the tetracritical point T_4.

$m_2 = 0$. The bicritical point is the terminus of the first-order boundary (and the meeting point of the second-order lines between the ordered and the disordered regions of the phase diagram). In the case of some antiferromagnetic crystals, however, instead of a first-order transition, there may be an intermediate phase, with $m_1 \neq 0$ and $m_2 \neq 0$, and a tetracritical point of confluence of four lines of second-order transitions.

The Landau free energy for a magnetic system with two scalar order parameters, m_1 and m_2, is written as

$$f = f_0 + \frac{1}{2}am_1^2 + \frac{1}{2}bm_2^2 + \frac{1}{4}\left[cm_1^4 + 2dm_1^2m_2^2 + em_2^4\right] + \cdots , \tag{2.7}$$

where a and b may change sign, but c, d, and e are positive parameters. Taking into account up to fourth-order terms, we have to look for solutions of the equations

$$\frac{\partial f}{\partial m_1} = am_1 + cm_1^3 + dm_1m_2^2 = 0, \tag{2.8}$$

and

$$\frac{\partial f}{\partial m_2} = bm_2 + em_2^3 + dm_1^2m_2 = 0. \tag{2.9}$$

In the a–b phase diagram, there are four possibilities:

(i) There is always a disordered solution, $m_1 = 0$ and $m_2 = 0$, which is a minimum of the free energy (i.e. a stable solution) for $a > 0$ and $b > 0$ only (see Fig. 2.5).

(ii) The Néel ordered phase,

$$m_1^2 = -\frac{a}{c} > 0 \quad \text{and} \quad m_2 = 0, \tag{2.10}$$

which is stable for $a < 0$ and $a < b(c/d)$ (see Fig. 2.5).

(iii) The spin-flop phase,

$$m_1 = 0 \quad \text{and} \quad m_2^2 = -\frac{b}{e} > 0, \tag{2.11}$$

which is stable for $b < 0$ and $a > b(d/e)$ (see Fig. 2.5).

(iv) The mixed phase, given by

$$m_1^2 = \frac{-ae + bd}{ce - d^2} > 0 \quad \text{and} \quad m_2^2 = \frac{-cb + ad}{ce - d^2} > 0, \tag{2.12}$$

which requires a more detailed analysis of stability.

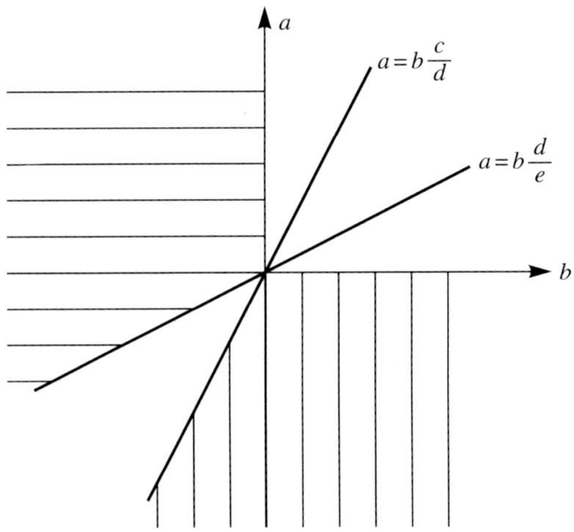

FIG. 2.5. Regions of stability of the Néel ordered phase [$a < 0$, and $a < b(c/d)$; vertical hatches] and of the spin-flop phase [$b < 0$, and $a > b(d/e)$; horizontal hatches].

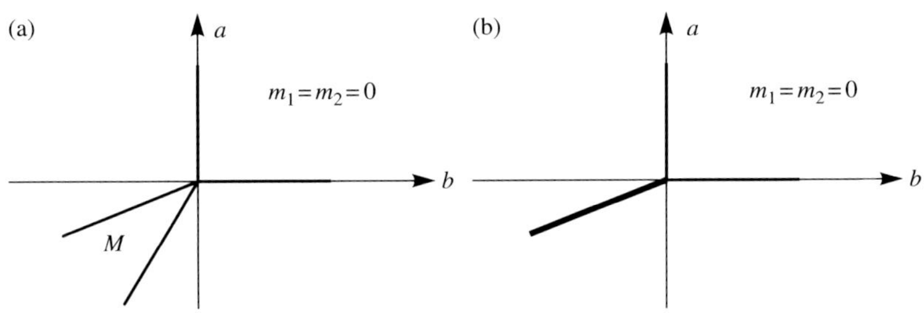

FIG. 2.6. (a) Phase diagram for $ce - d^2 > 0$. There is a mixed phase (M) and four lines of second-order transitions meeting at the origin (tetracritical point); (b) Phase diagram for $ce - d^2 < 0$. There is no mixed phase. The first-order line, drawn as a heavy solid line, given by $a = b(c/e)^{1/2}$, meets the second-order lines ($a = 0$ and $b = 0$) at the origin (which is now a bicritical point).

The regions of existence and stability of the mixed solution depend on the sign of the denominator of Eqs. (2.12):

(a) If $ce - d^2 > 0$, the mixed solutions exist and are stable for $a < b(d/e)$ and $a < b(c/d)$. In Fig. 2.6(a), we see that there are common limits of stability and no coexistence of phases. All transition lines are second-order and $a = b = 0$ is a tetracritical point.

(b) If $ce - d^2 < 0$, the mixed solutions exist for $a > b(d/e)$ and $a < b(c/d)$, which also corresponds to the region of coexistence of the Néel and spin-flop phases. However, if we look at the second derivatives, it is easy to see that these solutions are not stable. We then compare the free energies associated with the Néel and spin-flop solutions in order to show that there is a first-order boundary at $a = b(c/e)^{1/2}$ (see Fig. 2.6(b)).

2.4 Modulated phases and Lifshitz multicritical points

Modulated structures are experimentally found in a large number of quite different systems, ranging from helimagnets to cholesteric liquid crystals. The description of these structures requires the introduction of a space-dependent order parameter and its associated free-energy density.

Consider a simple layered magnetic system, with uniform behavior on the x–y planes and spacial dependence of the magnetization along the z direction. The Landau free energy, $F = F_{\mathrm{b}} + F_{\mathrm{grad}}$, may be written [7] as the sum of a bulk term,

$$F_b = \int \mathrm{d}z \{am_z^2 + m_z^4 + \cdots\}, \tag{2.13}$$

and a gradient term,

$$F_{\mathrm{grad}} = \int \mathrm{d}z \left\{ b\left(\frac{\mathrm{d}m_z}{\mathrm{d}z}\right)^2 + \left(\frac{\mathrm{d}^2 m_z}{\mathrm{d}z^2}\right)^2 + \cdots \right\}, \tag{2.14}$$

where m_z is the magnetization per ion along the z direction, a and b can change sign, but we are setting to unit the coefficients of the second derivative and the quartic terms. The description of cholesteric liquid crystals also involves the splitting of the free energy into bulk and gradient contributions, which are however considerably more complicated. We now transform to the Fourier representation,

$$m_z = \sum_q m_q \exp(\mathrm{i}qz), \tag{2.15}$$

where the sum is over the first Brillouin zone. We then have

$$F = N\sum_q \Lambda(q) m_q m_{-q} + N \sum_{q_1,q_2,q_3} m_{q_1} m_{q_2} m_{q_3} m_{-q_1-q_2-q_3} + \cdots, \tag{2.16}$$

where N is the number of sites along the z direction, and

$$\Lambda(q) = a + bq^2 + q^4. \tag{2.17}$$

As the coefficient of the quartic term is positive, the analysis of the (second-order) transition depends on the behavior of $\Lambda(q)$. There are two possibilities:

(i) If $b > 0, \Lambda(q)$ as a function of the wave number q has a minimum at $q = 0$. If this minimum, $\Lambda(0) = a$, is positive there is just a disordered phase ($m_q = 0$ for all q). For $a = 0$, however, there is a phase transition between a disordered ($m_q = 0$ for all q) and a uniformly ordered phase ($m_0 \neq 0$, and $m_q = 0$ for $q \neq 0$).

(ii) If $b < 0$, the coefficient $\Lambda(q)$ is a minimum at

$$q_c^2 = -\frac{1}{2}b. \tag{2.18}$$

Thus, if

$$\Lambda(q_c) = a - \frac{1}{4}b^2 > 0, \tag{2.19}$$

there is just a disordered phase. At $a = b^2/4$, there is a (second-order) phase transition to a modulated structure (with a main harmonic component associated with the wave number q_c).

In Fig. 2.7 we draw the a–b phase diagram. Note that $q_c = 0$ along the uniform-disordered critical line ($a = 0$ and $b > 0$), and that $q_c = (-b/2)^{1/2}$ along the modulated-disordered critical line ($a = b^2/4$ with $b < 0$). The special point $a = b = 0$ is known as the Lifshitz multicritical point [8].

We now turn to the analysis of the ordered region of this a–b phase diagram. In the neighborhood of the Lifshitz point, it should be enough to consider the main harmonic components of the magnetization. Thus, instead of using the complete Fourier representation, we write

$$m_z = M_1 \cos(q_c z) + M_3 \cos(3 q_c z) + \cdots, \tag{2.20}$$

where the coefficients M_1, M_3, ... are supposed to be real, symmetry arguments are invoked to discard even harmonic components, and we are also discarding an arbitrary phase factor. Inserting this expression into the Landau free energy,

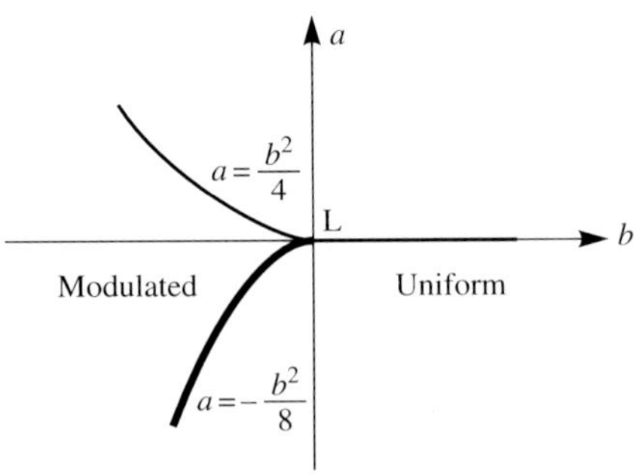

FIG. 2.7. Phase diagram with disordered, modulated, and uniform phases. At the Lifshitz point L, the first-order boundary (heavy solid line), meets smoothly with the second-order lines.

we have

$$f = \frac{1}{2}\Lambda(q_c)M_1^2 + \frac{1}{2}\Lambda(3q_c)M_3^2 + \frac{3}{8}(M_1^4 + M_3^4) + \frac{3}{2}M_1^2M_3^2 + \cdots . \tag{2.21}$$

As $\Lambda(3q_c) > \Lambda(q_c)$, the third harmonic component plays the role of a higher-order correction, and can be dropped in the immediate neighborhood of the transition. Thus, close to the Lifshitz point we can write the truncated form

$$f = \frac{1}{2}\Lambda(q_c)M_1^2 + \frac{3}{8}M_1^4. \tag{2.22}$$

In the modulated phase, we have $q_c = (-b/2)^{1/2} \neq 0$, and

$$M_1^2 = -\frac{2}{3}\Lambda(q_c). \tag{2.23}$$

In the uniform phase, we have

$$M_1^2 = -\frac{2}{3}\Lambda(0) = -\frac{2}{3}a. \tag{2.24}$$

Comparing the associated expressions for the free energy, we calculate the asymptotic form of the first-order boundary in the immediate neighborhood of the Lifshitz point,

$$a = -\frac{1}{8}b^2, \tag{2.25}$$

for $b < 0$, as drawn in Fig. 2.7. We note that, close to the transition, higher-order harmonic components can generally be neglected.

2.5 The nematic–isotropic phase transition and the Maier–Saupe model

A simple order parameter with up-down symmetry, as the magnetization per ion of a uniaxial ferromagnet, or the staggered magnetization of the antiferromagnets, is no longer adequate for describing the long-range order in uniaxial nematic liquid crystals [4, 9]. In fact, rigid rod molecules tend to align with respect to a preferred direction (of a director axis). A possible candidate for the status of an order parameter is the dipole moment $m = \cos\theta$, where θ is the angle between the rods and a director axis (which will be chosen along the z direction). However, as there is no arrow associated with the rigid rods, there should be no distinctions between positive and negative values of $\cos\theta$, and the dipole m cannot work as an order parameter. It turns out that a suitable order parameter is the quadrupole moment,

$$S = P_2(\cos\theta) = \frac{1}{2}(3\cos^2\theta - 1), \tag{2.26}$$

where P_2 stands for the second Legendre polynomial. At high temperatures, there should be no preferred direction, so that $\langle\cos^2\theta\rangle = 1/3$, and $S = 0$. At low temperatures, $\theta = 0$, so that $S = 1$, according to the anticipated behavior of an order parameter.

We now review the pioneering approach of Maier and Saupe for the uniaxial nematic–isotropic first-order transitions [10]. It was published about 50 years ago, and can be regarded as the counterpart of the Curie–Weiss mean-field formulation for the phase transition in ferromagnetic systems.

Consider a pair of rigid rods, at positions i and j, characterized by the angles θ_i and θ_j with respect to the z direction. We can always write a multipole expansion for the interaction energy. If we only keep the lowest order, cylindrically symmetric, part of the angular interaction, we may write the pair potential

$$u_{ij} = -AS_iS_j, \tag{2.27}$$

where $S_i = (3\cos^2\theta_i - 1)/2$, and A is a positive coefficient (which may depend on volume or concentration, for example). In a mean-field approximation, we replace this pair potential by a single-particle term,

$$u_i = -AS_i\langle S_j\rangle = -AS_iS, \tag{2.28}$$

where the thermal average $\langle S_i\rangle = S$ is given by the self-consistent condition

$$S = \langle S_i\rangle = \frac{\int \mathrm{d}\Omega_i S_i \exp(\beta AS_iS)}{\int \mathrm{d}\Omega_i \exp(\beta AS_iS)}, \tag{2.29}$$

and we are using the standard notation of statistical physics ($\mathrm{d}\Omega_i$ refers to the integration over the solid angle, and $1/\beta = k_\mathrm{B}T$, where k_B is the Boltzmann constant and T is the absolute temperature). It is easy to see that $S = 0$ (disordered phase) is a solution for all temperatures. Below a certain temperature, however, there appears a nematic, $S > 0$, solution, which is shown schematically in Fig. 2.8. A complete analysis of the stability of these solutions requires the knowledge of the free energy associated with the Maier–Saupe equation of state. Anyhow, the graph in Fig. 2.8 indicates the existence of three solutions below a certain temperature, so that there should be an intermediate temperature at which the ordered and disordered solutions correspond to the same free energy, which leads to a first-order transition (with coexistence of phases).

In the standard mean-field treatments, the Maier–Saupe equation of state (2.29) is used to write an expression for the free energy, which should then be minimized with respect to the order parameter S. In the following paragraphs we adopt an alternative, but equivalent, point of view, which consists in the definition of the exactly soluble Maier–Saupe model.

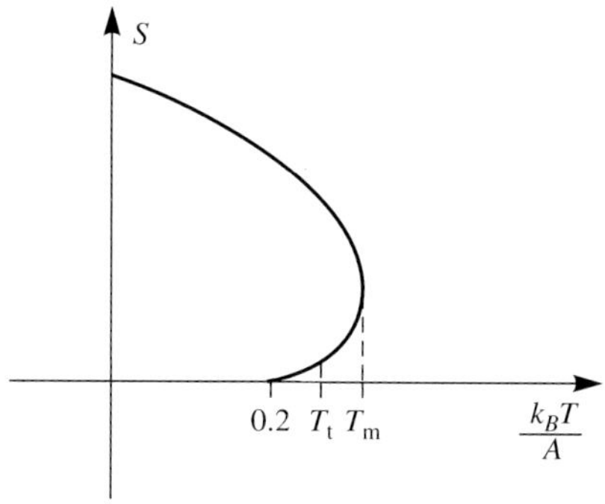

FIG. 2.8. Graph of the Maier–Saupe order parameter S versus temperature (in suitable dimensionless units). We indicate the temperature $T_{\rm m} = 0.22284\ldots$ below which there is a solution $S \neq 0$. From thermodynamic considerations, the first-order phase transition in this mean-field model occurs at $T_{\rm t} = 0.22019\ldots$, which is very close to $T_{\rm m}$.

2.5.1 *The Curie–Weiss model*

Let us first give an example of a simple statistical mechanics calculation. Consider the Curie–Weiss version of the ferromagnetic Ising model, given by the energy

$$H_{\rm CW} = -\frac{J}{2N}\sum_{i,j=1}^{N}\sigma_i\sigma_j, \tag{2.30}$$

where $J > 0$ is the exchange parameter, and $\sigma_i = \pm 1$ for all sites $i = 1, \ldots, N$. Note that there are (equal) interactions, of order $-J/N$, between all pairs of sites. This kind of expression is supposed to mimic the standard mean-field requirements of very long-range and very weak interactions (in the thermodynamic limit, $N \to \infty$). The canonical partition function associated with this Curie–Weiss model is given by

$$Z_{\rm CW} = {\rm Tr}\,\exp\left[\frac{\beta J}{2N}\left(\sum_i \sigma_i\right)^2\right], \tag{2.31}$$

where Tr is a sum over all spin configurations. Using a Gaussian identity,

$$\int_{-\infty}^{+\infty}\exp(-x^2 + 2ax)\,{\rm d}x = \sqrt{\pi}\,\exp(a^2), \tag{2.32}$$

we can also write

$$Z_{\rm CW} = {\rm Tr}\int_{-\infty}^{+\infty}\frac{{\rm d}x}{\sqrt{\pi}}\exp\left[-x^2 + 2x\left(\frac{\beta J}{2N}\right)^{1/2}\sum_{i=1}^{N}\sigma_i\right]{\rm d}x. \tag{2.33}$$

As the spin variables are linear in the argument of the exponent, it is now trivial to calculate the trace (sum over spin configurations). With a suitable change of variables, we finally have

$$Z_{\text{CW}} = \left(\frac{N\beta J}{2\pi} \right)^{1/2} \int_{-\infty}^{+\infty} \mathrm{d}m \, \exp\left[-N\beta J f(m)\right], \tag{2.34}$$

where the "free-energy functional" is given by

$$f(m) = \frac{1}{2} m^2 - \frac{1}{\beta J} \ln\left[2\cosh(\beta J m)\right]. \tag{2.35}$$

In the thermodynamic limit, $N \to \infty$, we use Laplace's method of asymptotic integration in order to write the thermodynamic free energy as the minimum of $f(m)$ with respect to m (which is just the thermal average of the spin variable). The first derivative of $f(m)$ leads to the well-known Curie–Weiss equation,

$$m = \tanh(\beta J m), \tag{2.36}$$

which has a disordered solution, $m = 0$, for all temperatures, and a pair of ordered, ferromagnetic solutions, $\pm m_0 \neq 0$, for $k_{\text{B}}T < J$ (where $T_{\text{c}} = J/k_{\text{B}}$ is the critical temperature). In the immediate neighborhood of the transition, we have the asymptotic result

$$m_0^2 \sim 3(T_{\text{c}} - T)/T_{\text{c}}, \tag{2.37}$$

where $\sim$ means $T \to T_{\text{c}} = J/k_{\text{B}}$, and which shows the continuous (second-order) character of the ferromagnetic transition.

In order to analyze the stability of these results, and to make contact with a Landau expansion, we now perform an expansion of $f(m)$ about $m = 0$,

$$f(m) = -\frac{1}{\beta J} \ln 2 + \frac{1}{2}(1 - \beta J) m^2 + \frac{(\beta J)^3}{12} m^4 + \cdots. \tag{2.38}$$

As it should be anticipated from the up-down symmetry of the model, and in full agreement with the Landau expansion, this series is restricted to even powers of m. Also, note that the coefficient of the quartic term is always positive. Therefore, the disordered solution, $m = 0$, is indeed a minimum of the free energy at high temperatures, but becomes a maximum (and unstable solution) for $T < T_{\text{c}}$. In this low-temperature range, the sketchy plot of $f(m)$, shown in Fig. 2.1(b), indicates the existence of two stable minima, corresponding to the pair of symmetric ordered solutions, $\pm m_0 \neq 0$.

As we have already pointed out, a more general expansion, in terms of the "invariants of the order parameter," is the basis of the Landau theory of (second-order) phase transitions [4]. In the case of the uniaxial ferromagnet, these invariants of the order parameter are just the even powers of m. Indeed, according to Landau, in the neighborhood of the phase transition of any uniaxial ferromagnetic system we can always write Eq. (2.1), where the coefficients $A(T)$ and $C(T)$, related to the details of the particular system under consideration, are supposed to display the same and suitable dependence on temperature. From this point of view, Landau's theory, which should work for small m, in the immediate neighborhood of the transition, is a very useful tool to deal with experimental data, and goes beyond any particular model calculation.

2.5.2 *The Maier–Saupe model*

We now go back to the Maier–Saupe model, given by the energy

$$H_{\mathrm{MS}} = -\frac{A}{2N}\sum_{i,j=1}^{N} S_i S_j, \tag{2.39}$$

which leads to the canonical partition function

$$Z_{\mathrm{MS}} = \prod_{i=1}^{N}\int \mathrm{d}\Omega_i \exp\left[\frac{\beta A}{2N}\left(\sum_{i=1}^{N} S_i\right)^2\right] \tag{2.40}$$

Using the same Gaussian identity, we linearize the exponent so that the angular integrations factorize into single-particle terms. Thus, we have

$$Z_{\mathrm{MS}} = \left(\frac{N\beta A}{2\pi}\right)^{1/2}\int_{-\infty}^{+\infty} \mathrm{d}S \exp[-N\beta A f(S)], \tag{2.41}$$

with

$$f(S) = \frac{1}{2}S^2 - \frac{1}{\beta A}\ln z, \tag{2.42}$$

where

$$z = \int \mathrm{d}\Omega_i \,\exp(\beta A S_i) = \int \mathrm{d}\Omega \exp\left[\frac{\beta A S}{2}(3\,\cos^2\,\theta - 1)\right]. \tag{2.43}$$

It is easy to see that the extrema of $f(S)$ lead to the self-consistent expression of Maier and Saupe, given by Eq. (2.29). Although it is not difficult to carry out a detailed analysis of the thermodynamic behavior of this model, it is much more interesting, and experimentally relevant, to look at the expansion of the free-energy functional in terms of the order parameter S, and to make contact with the Landau theory of phase transitions.

If we write $\beta A S = \omega$, it is easy to calculate the first few terms of an expansion of $\ln z$ in powers of ω,

$$\ln z = \ln(4\pi) + \frac{1}{10}\omega^2 + \frac{1}{105}\omega s s^3 - \frac{1}{700}\omega^4 + \cdots, \tag{2.44}$$

such that we have

$$f(S) = -\frac{1}{\beta A}\ln(4\pi) + \frac{1}{2}\left(1 - \frac{\beta A}{5}\right)S^2 - \frac{(\beta A)^2}{105}S^3 + \frac{(\beta A)^3}{700}S^4 + \cdots. \tag{2.45}$$

The presence of the cubic term, with a negative coefficient, which is not forbidden by symmetry arguments, is the striking difference with respect to the analogous expansion for the Curie–Weiss model. Note that $T_0 = A/(5k_B)$ is no longer a critical temperature. It is just a characteristic value, below which the disordered solution becomes a maximum of the free energy.

Now we show that the cubic term is indeed responsible for the discontinuous (first-order) character of the nematic–isotropic transition. In the context of Landau's theory of phase transitions, let us write a truncated expansion of the free energy,

$$f = f_0 + \frac{1}{2}aS^2 + \frac{1}{3}bS^3 + \frac{1}{4}S^4, \tag{2.46}$$

where the coefficient of the quartic term has been set to 1/4. Note that, if we adopt a general phenomenological point of view, besides depending on temperature, the coefficients a and b may depend on additional variables as well. The extrema of the free energy are given by

$$S[a + bS + S^2] = 0. \tag{2.47}$$

There is always a disordered solution, $S = 0$. For $a < b^2/4$, there are also two ordered solutions, which have the same sign, for $a > 0$, and different signs, for $a < 0$. Looking at the second derivative of $g(S)$, we see that the disordered solution is stable for $a > 0$ only. Also, we see that there is always a stable ordered solution for $a > b^2/4$ (see Fig. 2.9). In the common region of stability of both disordered and ordered solutions, $0 < a < b^2/4$, we have to calculate the respective free energies. As the physical solution is associated with the absolute minimum of the free energy, it is easy to see that there is a line of coexistence of phases (first-order transitions) at $a = 2b^2/9$.

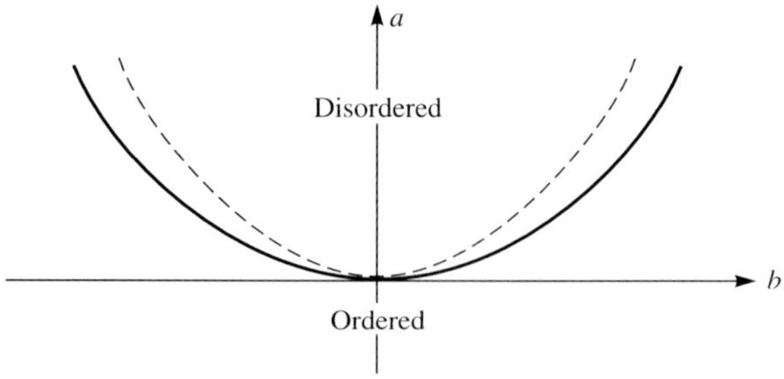

FIG. 2.9. The heavy solid line ($a = b^2/4$) is the first-order boundary between the nematic and the disordered phase. The dashed line ($a = 2b^2/9$) is the limit of stability of the ordered solution.

2.6 The uniaxial–biaxial phase transition

The description of a biaxial phase, in which we break the cylindrical symmetry of rigid rod structures, requires the choice of a more complex order parameter [4, 9, 11].

Macroscopic tensor properties, as the diamagnetic or the dielectric susceptibilities, express the typical distinctions between an isotropic liquid and the nematic phase. The magnetic moment $\overrightarrow{M}$ in terms of the applied field $\overrightarrow{H}$ may be written as

$$M_\alpha = \chi_{\alpha\beta} H_\beta, \tag{2.48}$$

where $\alpha, \beta = x, y, z$, and we are using the summation convention over repeated indices. In the uniaxial (cylindrically symmetric) nematic phase, the susceptibility tensor may be written in the diagonal, but entirely general, form

$$\boldsymbol{\chi} = \begin{pmatrix} \chi_\perp & 0 & 0 \\ 0 & \chi_\perp & 0 \\ 0 & 0 & \chi_\| \end{pmatrix}, \tag{2.49}$$

where $\chi_\perp$ and $\chi_\|$ refer to the susceptibilities perpendicular and parallel to the preferred axis, respectively. For the uniaxial nematic phase, $\chi_\perp \neq \chi_\|$. For the isotropic liquid, $\chi_\perp = \chi_\|$. However, since it does not vanish in the isotropic phase, this tensor form cannot be taken as a suitable order parameter.

Now we extract the anisotropic part of $\boldsymbol{\chi}$, and define the traceless tensor

$$\chi^{\mathrm{a}}_{\alpha\beta} = \chi_{\alpha\beta} - \frac{1}{3}\chi_{\gamma\gamma}\delta_{\alpha\beta}, \tag{2.50}$$

where $\chi_{\gamma\gamma} = (2\chi_\perp + \chi_\|)$. In a matrix form, we have

$$\boldsymbol{\chi}^a = \frac{2}{3}(\chi_\| - \chi_\perp) \begin{pmatrix} -1/2 & 0 & 0 \\ 0 & -1/2 & 0 \\ 0 & 0 & 1 \end{pmatrix}, \tag{2.51}$$

which shows that this new tensor vanishes in the (high-temperature) isotropic phase, and is thus a suitable candidate for the role of order parameter. We then define the tensor order parameter,

$$Q_{\alpha\beta} = P_2(\cos\theta) \begin{pmatrix} -1/2 & 0 & 0 \\ 0 & -1/2 & 0 \\ 0 & 0 & 1 \end{pmatrix}, \tag{2.52}$$

where the Legendre polynomial $P_2(\cos\theta) = \left[3\cos^2\theta + 1\right]/2$ had already been used as the (scalar) order parameter in the Maier–Saupe model.

A symmetric and traceless matrix $\mathbf{Q}$ turns out to be a useful order parameter for a wide class of liquid crystalline phases, including nematic and biaxial structures. If the axes are properly chosen, $\mathbf{Q}$ can be written in the (general) diagonal form

$$Q_{\alpha\beta} = \begin{pmatrix} -\frac{1}{2}(S+\eta) & 0 & 0 \\ 0 & -\frac{1}{2}(S-\eta) & 0 \\ 0 & 0 & S \end{pmatrix}, \tag{2.53}$$

where $\eta \neq 0$ is associated with a biaxial phase.

2.6.1 *An extension of the Maier–Saupe model*

First, we discuss a generalization of the Maier–Saupe model for molecules (or molecular aggregates if we refer to lyotropic mixtures) without cylindrical symmetry. Although there are a number of much more complete and detailed treatments in the literature [12, 13], we assume that the angular part of the pair potential between molecules i and j can be written as

$$u_{ij} = -B \sum_{\alpha} S_{\alpha\alpha}(i) S_{\alpha\alpha}(j), \tag{2.54}$$

where B is a positive parameter, and the elements of the tensors $S_{\alpha\beta}(i)$ and $S_{\alpha\beta}(j)$ are given by Eq. (2.53), with S and η replaced by the local values S_i and η_i, respectively. This type of interaction leads to a slight generalization of the (mean-field) Maier–Saupe model,

$$H_B = -\frac{B}{2N} \sum_{i,j=1}^{N} \sum_{\alpha} S_{\alpha\alpha}(i) S_{\alpha\alpha}(j), \tag{2.55}$$

which is associated with the canonical partition function,

$$Z_B = \prod_i \int \mathrm{d}\Omega_i \, \exp\left\{ \frac{\beta B}{2N} \sum_{\alpha} \left[\sum_i S_{\alpha\alpha}(i) \right]^2 \right\}, \tag{2.56}$$

where Ω_i stands for the Euler angles that define the direction cosines of a single molecule. If we rewrite this expression,

$$Z_B = \prod_i \int \mathrm{d}\Omega_i \, \exp\left[\frac{\beta B}{2N} \frac{3}{2} \left(\sum_i S_i \right)^2 + \frac{\beta B}{2N} \frac{1}{2} \left(\sum_i \eta_i \right)^2 \right], \tag{2.57}$$

and use twice the Gaussian identity of Eq. (2.32), it is easy to obtain the final form

$$Z_B = C \iint \mathrm{d}S \, \mathrm{d}\eta \exp\{-\beta B N f(S, \eta)\}, \tag{2.58}$$

where C is an irrelevant prefactor, and

$$f = \frac{1}{4}(3S^2 + \eta^2) - \frac{1}{\beta B} \ln z, \tag{2.59}$$

with

$$z = \int \mathrm{d}\Omega_i \, \exp\left[\beta B \frac{1}{2}(3SS_i + \eta\eta_i)\right] = \int \mathrm{d}\Omega_i \, \exp\left[\beta B \sum_{\alpha} S_{\alpha\alpha} S_{\alpha\alpha}(i)\right]. \quad (2.60)$$

Now it is convenient to define the second- and the third-order invariants of the traceless tensor order parameter,

$$\delta = \sum_{\alpha} (S_{\alpha\alpha})^2 = \frac{1}{2}(3S^2 + \eta^2), \quad (2.61)$$

and

$$\Delta = \sum_{\alpha} (S_{\alpha\alpha})^3 = \frac{3}{4} S(S^2 - \eta^2), \quad (2.62)$$

respectively. In terms of these quantities, we have

$$f = \frac{1}{2}\delta - \frac{1}{\beta B} \ln z. \quad (2.63)$$

Also, we note that $\ln z$ can be written as an expansion in powers of the parameter $\omega = \beta B$,

$$\ln z = \ln z(0) + \frac{1}{2z(0)} \left(\frac{\mathrm{d}^2 z}{\mathrm{d}\omega^2}\right)_{\omega=0} + \frac{1}{6z(0)} \left(\frac{d^3 z}{\mathrm{d}\omega^3}\right)_{\omega=0} + \cdots, \quad (2.64)$$

where the first derivative vanishes,

$$\frac{1}{z(0)} \left(\frac{\mathrm{d}z}{\mathrm{d}\omega}\right)_{\omega=0} = \frac{1}{z(0)} \int \mathrm{d}\Omega_i \frac{1}{2}(3SS_i + \eta\eta_i) = 0, \quad (2.65)$$

since we have to calculate a fully isotropic angular integral (i.e. at infinite temperature). Consider the second term,

$$\begin{aligned}\left(\frac{\mathrm{d}^2 z}{\mathrm{d}\omega^2}\right)_{\omega=0} &= \int \mathrm{d}\Omega_i \left[\sum_{\alpha} S_{\alpha\alpha} S_{\alpha\alpha}(i)\right]^2 \\ &= \sum_{\alpha,\beta} S_{\alpha\alpha} S_{\beta\beta} \int \mathrm{d}\Omega_i S_{\alpha\alpha}(i) S_{\beta\beta}(i). \end{aligned} \quad (2.66)$$

Again, we have to perform a fully isotropic integration, so that only the invariants of the tensor will survive this process. Therefore, it is not difficult to obtain the first few terms of an expansion of f in terms of the powers of the invariants of the tensor order parameter,

$$f = \frac{1}{2}\delta - \frac{1}{\beta B}\left[\ln z(0) + \frac{1}{10}\frac{3+a^2}{2}\delta + \frac{2}{35}\frac{3(1-a^2)}{4}\Delta + \cdots\right], \quad (2.67)$$

where a is assumed to gauge the degree of biaxiality in the molecular frame of reference, in which $S_{\alpha\beta}(i)$ is diagonal. In this particular model, if $a^2 = 1$, the cubic term vanishes, and the system undergoes a (second-order) transition to a biaxial phase.

It is not so relevant to carry out the calculation of the specific forms of the coefficients associated with this expansion. Even at this mean-field level, these calculations are feasible for very simple and schematic models only (and do depend on the details of the particular molecular model under consideration). A general Landau expansion, in terms of the invariants of the tensor order parameter, with coefficients to be phenomenologically determined from fittings to the available experimental data, turns out to be much more relevant for accounting to the biaxial transition. In fact, it should be remarked that there is no experimental evidence of a biaxial transition in thermotropic liquid crystals (which are usually associated with relatively simple rigid-rod molecules). On the other hand, as pointed out in Chapter 1, biaxial phases have been found in a number of lyotropic mixtures. The molecules in these lyotropic systems tend to aggregate in micelles, with shapes depending on temperature and relative concentrations. The Landau expansion, with temperature and concentration-dependent coefficients, becomes particularly suitable for describing phase transitions in these lyotropic liquid crystals.

2.6.2 *Maier–Saupe model for a mixture of prolate and oblate micelles*

Although there is no experimental evidence supporting the picture of a mixture of prolate and oblate micellar aggregates in the nematic biaxial phase, it is very instructive to consider simplified versions of the Maier–Saupe model for this kind of mixture [14]. We then present a further generalization of the mean-field Maier–Saupe model.

Given the director of a molecular aggregate, $\vec{n}_i = (n_{ix}, n_{iy}, n_{iz})$, with $|\vec{n}_i| = 1$, the local order parameter can be written as a traceless quadrupolar tensor,

$$S_{\mu\nu}(i) = \frac{1}{2}\lambda_i(3n_{i\mu}n_{i\nu} - \delta_{\mu\nu}), \tag{2.68}$$

where $\mu, \nu = x, y, z$. Note that this form is invariant under the change $\vec{n}_i \to -\vec{n}_i$. Also, the usual form of the order parameter, in the reference frame of the molecular aggregate, can be recovered if we perform an average over the standard spherical coordinate ϕ. In the uniform case, which has been discussed so far, $\lambda_i = \lambda$ for all values of i (in other words, all the molecular aggregates have the same shape). If we wish to treat a mixture of aggregates, with different shapes, it is necessary to consider a distribution of values of the parameter λ. Note that $\lambda_i > 0$ ($\lambda_i < 0$) favors a cylinder-like (disk-like) aggregate. Therefore, we keep this dependence on the aggregates, and write the Maier–Saupe interaction in the form

$$u_{ij} = -B\,\mathrm{Tr}[\mathbf{S}(i)\mathbf{S}(j)] = -B\sum_{\mu,\nu} S_{\mu\nu}(i)S_{\mu\nu}(j), \tag{2.69}$$

which leads to a slightly more general Maier–Saupe model,

$$H_{\mathrm{M}} = -\frac{B}{N} \sum_{1 \le i < j \le N} \sum_{\mu,\nu} S_{\mu\nu}(i) S_{\mu\nu}(j), \tag{2.70}$$

where the subscript M refers to mixture, and the first sum is over all pairs of distinct molecular aggregates.

We may now write

$$H_{\mathrm{M}} = -\frac{B}{N} \sum_{1 \le i < j \le N} \frac{3}{4} \lambda_i \lambda_j \left[3 \left(\sum_{\mu} n_{i\mu} n_{j\nu} \right)^2 - 1 \right], \tag{2.71}$$

which can also be written in the usual mean-field form,

$$H_{\mathrm{M}} = -\frac{3B}{8N} \left\{ 3 \sum_{\mu,\nu} \left(\sum_{i=1}^{N} \lambda_i n_{i\mu} n_{i\nu} \right)^2 - \left(\sum_{i=1}^{N} \lambda_i \right)^2 \right\}. \tag{2.72}$$

The partition function is given by

$$Z_{\mathrm{M}} = \mathrm{Tr} \exp \left\{ \frac{3\beta B}{8N} \left[3 \sum_{\mu,\nu} \left(\sum_{i=1}^{N} \lambda_i n_{i\mu} n_{i\nu} \right)^2 - \left(\sum_{i=1}^{N} \lambda_i \right)^2 \right] \right\}, \tag{2.73}$$

where the initial trace is a sum (integration) over all possible states of the director.

According to the work of Henriques and Henriques [14], we further simplify the problem by assuming that the director $\vec{n}_i$ can point along the Cartesian axes only. A similar discretization of states has been used in other contexts (e.g. by Zwanzig in order to simplify the analysis of Onsager's model of hard cylinders [15]). It is known that it drastically simplifies the calculations, without introducing significant changes in the qualitative results. Therefore, we restrict the analysis to six configurations of the director, $\vec{n}_i = (\pm 1, 0, 0)$, $\vec{n}_i = (0, \pm 1, 0)$, and $\vec{n}_i = (0, 0, \pm 1)$, which lead to an effective three-state model (and to some interesting connections with spin systems).

In this discrete version of the Maier–Saupe model, we can discard crossed terms, of the form $n_{i\mu} n_{i\nu}$, with $\mu \neq \nu$. The partition function is then written as

$$Z_{\mathrm{M}} = \sum_{\{\vec{n}_i\}} \exp \left\{ \frac{3\beta B}{8N} \left[3 \sum_{\mu,\nu} \left(\sum_{i=1}^{N} \lambda_i n_{i\mu}^2 \right)^2 - \left(\sum_{i=1}^{N} \lambda_i \right)^2 \right] \right\}. \tag{2.74}$$

This expression can be further simplified if we take into account that $|\vec{n}_i| = 1$. Instead of working with the components of the director, it is now interesting to make contact with the elements of the traceless tensor order parameter.

Considering Eq. (2.53), we introduce new variables, S_i and η_i, given by

$$-\frac{1}{2}(S_i + \eta_i) = \frac{1}{2}(3n_{ix}^2 - 1), \tag{2.75}$$

$$-\frac{1}{2}(S_i - \eta_i) = \frac{1}{2}(3n_{iy}^2 - 1), \tag{2.76}$$

and

$$S_i = \frac{1}{2}(3n_{iz}^2 - 1). \tag{2.77}$$

In terms of these new variables, the partition function is written as

$$Z_{\mathrm{M}} = \sum_{\{\vec{n}_i\}} \exp\left\{\frac{3\beta B}{8N}\left[\frac{2}{3}\left(\sum_{i=1}^{N}\lambda_i\eta_i\right)^2 + 2\left(\sum_{i=1}^{N}\lambda_i S_i\right)^2\right]\right\}. \tag{2.78}$$

We can now use the Gaussian identity, given by Eq. (2.32), and perform the sum over configurations of the director. With a convenient change of variables, we have

$$Z_{\mathrm{M}} = C\iint \mathrm{d}S\mathrm{d}\eta\exp\{-\beta BNf_{\mathrm{M}}(S,\eta)\}, \tag{2.79}$$

where C is a well-behaved factor, and

$$f_{\mathrm{M}} = (3S^2 + \eta^2) - \frac{1}{\beta BN}\prod_{i=1}^{N}\left[2\exp(3\beta BS\lambda_i)\right.$$
$$\left. + 4\exp\left(-\frac{3}{2}\beta BS\lambda_i\right)\cosh\left(\frac{3}{2}\beta B\eta\lambda_i\right)\right]. \tag{2.80}$$

Let us assume that we have a (quenched) mixture of prolate ($\lambda_i > 0$) and oblate ($\lambda_i < 0$) molecular aggregates. Thus, $\{\lambda_i\}$ may be taken as a set of independent, identical, and identically distributed, random variables, associated with a probability distribution $p(\lambda_i)$. In the thermodynamic limit ($N \to \infty$), we can invoke the law of large numbers in order to write

$$f_{\mathrm{M}} = (3S^2 + \eta^2) - t\ln 2 - tE\left\{\ln\left[\exp\left(\frac{3S\lambda_i}{t}\right)\right.\right.$$
$$\left.\left. +2\exp\left(-\frac{3S\lambda_i}{2t}\right)\cosh\left(\frac{3\eta\lambda_i}{2t}\right)\right]\right\}, \tag{2.81}$$

where $t = 1/(\beta B)$ is the temperature in convenient units, and the expectation value $E\{\cdots\}$ is taken with respect to the probability distribution $p(\lambda_i)$.

In the uniform case, $\lambda_i = \lambda$, we have

$$f = (3S^2 + \eta^2) - t\ln 2 - t\ln\left[\exp\left(\frac{3S\lambda}{t}\right) + 2\exp\left(-\frac{3S\lambda}{2t}\right)\cosh\left(\frac{3\eta\lambda}{2t}\right)\right]. \tag{2.82}$$

This expression can be easily written as an expansion in terms of the invariants of the traceless order parameter,

$$f = -t\ln 6 + 2\left(1 - \frac{3\lambda^2}{4t}\right)\delta - \frac{3\lambda^3}{2t^2}\Delta + \frac{9\lambda^4}{16t^3}\delta^2 + \frac{9\lambda^5}{16t^4}\delta\Delta - \frac{99\lambda^6}{320t^5}\delta^3 + \frac{81\lambda^6}{80t^5}\Delta^2 + \cdots, \tag{2.83}$$

where $\delta = (3S^2 + \eta^2)/2$ and $\Delta = 3(S^3 - S\eta^2)/4$ are given by Eqs. (2.61) and (2.62). Since the coefficient of the cubic invariant Δ cannot vanish, the transition is always of first order. Indeed, for $\eta = 0$, we have

$$f(\eta = 0) = -t\ln 6 + 4\left(1 - \frac{3\lambda^2}{4t}\right)S^2 - \frac{9\lambda^3}{8t^2}S^3 + \frac{27\lambda^4}{64t^3}S^4 + \cdots, \tag{2.84}$$

from which we recover the well-known results predicted by the standard Maier–Saupe model for the isotropic–uniaxial transition.

The behavior of a mixed system is much more interesting. For example, consider a bimodal distribution of prolate and oblate molecular aggregates, given by

$$p(\lambda_i) = c\delta(\lambda_i - \lambda) + (1 - c)\delta(\lambda_i + \lambda). \tag{2.85}$$

It is easy to see that the expansion of the free energy f_{M} can be written as

$$f_{\mathrm{M}} = -t\ln 6 + 2\left(1 - \frac{3\lambda^2}{4t}\right)\delta - \frac{3\lambda^3}{2t^2}(2c - 1)\Delta + \frac{9\lambda^4}{16t^3}\delta^2 + \cdots. \tag{2.86}$$

Thus, at least if $c = 1/2$, we do have a continuous, second-order, transition to a biaxial phase, with $\eta \neq 0$. A detailed analysis of this system, for bimodal and Gaussian distributions, has been performed by Henriques and Henriques [14]. A typical phase diagram, in terms of t and c, with second-order boundaries between the uniaxial nematic phases (N_1 and N_2) and a biaxial phase (BI) is sketched in Fig. 2.10.

We now turn to the (much more general) Landau phenomenological expansion in terms of the invariants δ and Δ.

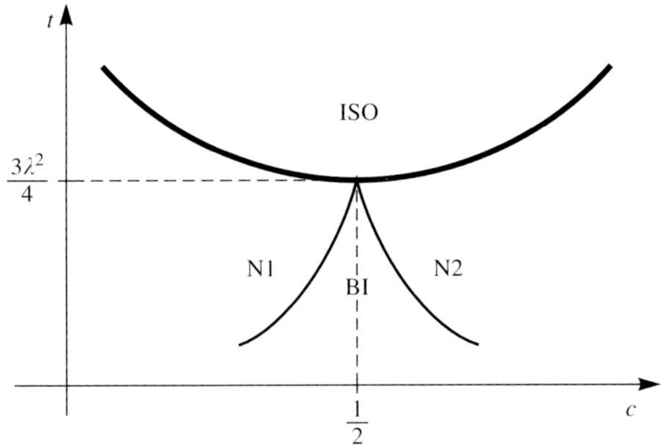

FIG. 2.10. Typical phase diagram for a mixture of prolate and oblate molecular aggregates. The heavy solid line represents a first-order transition (between an isotropic phase, ISO, and the uniaxial–nematic phases, N_1 and N_2). There are second-order transitions between the uniaxial nematic phases and the biaxial phase (BI).

2.6.3 *Landau theory of the uniaxial–biaxial transition*

The Landau expansion associated with the traceless tensor order parameter (2.53) can be written as [4]

$$F = F_0 + \frac{1}{2}A\,\mathrm{Tr}\,\mathbf{Q}^2 + \frac{1}{3}B\,\mathrm{Tr}\mathbf{Q}^3 + \frac{1}{4}C(\mathrm{Tr}\,\mathbf{Q}^2)^2 + \frac{1}{5}D(\mathrm{Tr}\,\mathbf{Q}^2)(\mathrm{Tr}\,\mathbf{Q}^3) + \frac{1}{6}E(\mathrm{Tr}\,\mathbf{Q}^2)^3 + \frac{1}{6}E'(\mathrm{Tr}\,\mathbf{Q}^3)^2 + \cdots, \tag{2.87}$$

where F_0 is a regular function, and the coefficients may depend on temperature and concentration, for example. The invariants of the tensor $\mathbf{Q}$ are given by Eqs. (2.61) and (2.62),

$$\mathrm{Tr}\,\mathbf{Q}^2 = \delta = \frac{1}{2}(3S^2 + \eta^2), \tag{2.88}$$

and

$$\mathrm{Tr}\,\mathbf{Q}^3 = \Delta = \frac{3}{4}S(S^2 - \eta^2). \tag{2.89}$$

Note that, due to symmetry requirements, we have to include two distinct sixth-order terms in this expansion for the free energy. Also, note that the definitions of δ and Δ lead to the inequality

$$(\mathrm{Tr}\,\mathbf{Q}^3)^2 \leq \frac{1}{6}(\mathrm{Tr}\,\mathbf{Q}^2)^3, \tag{2.90}$$

which is fulfilled as an equality in the uniaxial case ($\eta = 0$).

In the uniaxial case ($\eta = 0$), this Landau free energy is reduced to an expansion in terms of even and odd powers of a scalar order parameter,

$$F = F_0 + \frac{3}{4}AS^2 + \frac{1}{4}BS^3 + \frac{9}{16}CS^4 + \frac{9}{40}DS^5 + \frac{9}{16}\left(E + \frac{1}{6}E'\right)S^6 + \cdots . \qquad (2.91)$$

Therefore, with $D = E = E' = 0$, and $C > 0$, we recover Eq. (2.46), and make contact with the description of the (uniaxial) nematic–isotropic transition. If we choose $A = a(T - T_0)$, with $a > 0$, the isotropic phase is stable for $T > T_0$. According to all of the experimental data, for thermotropic and lyotropic systems as well, the nematic–isotropic transition is weakly first order (as compared to the strong first-order melting transition, for example). The transition takes place for temperatures slightly above T_0, and the inclusion of additional terms in the free energy expansion ($C < 0$ and $D > 0$, for example) may lead to interesting (pseudo-tricritical) behavior. However, if we keep $C > 0$, the inclusion of higher-order terms has no qualitative effects on this phase transition.

We now turn to the description of the biaxial phase [16]. It is known that the inclusion of the coefficient E', associated with the sixth-order invariant Δ^2, is essential for giving rise to a biaxial phase (since the expansion has to be carried out up to terms of order η^4). For simplicity, we choose $E = 0$ and $E' > 0$, which allows the truncation of the expansion at sixth order. Also, we choose $C > 0$, in order to guarantee the stability of the uniaxial structure, and $D = 0$, just for convenience. Thus, we work with the simple form

$$F_{\mathrm{BU}} = F_0 + \frac{1}{2}A\,\mathrm{Tr}\,\mathbf{Q}^2 + \frac{1}{3}B\,\mathrm{Tr}\,\mathbf{Q}^3 + \frac{1}{4}C(\mathrm{Tr}\,\mathbf{Q}^2)^2 + \frac{1}{6}E'(\mathrm{Tr}\,\mathbf{Q}^3)^2, \qquad (2.92)$$

where $C, E' > 0$. The phase diagram will be drawn in terms of the coefficients A and B, which depend on temperature and concentration, in the case of lyotropic liquid crystals. The Landau free energy F_{BU} can be minimized in terms of $\delta = \mathrm{Tr}\,\mathbf{Q}^2$ and $\Delta = \mathrm{Tr}\,\mathbf{Q}^3$, if we take into account the restrictions imposed by the inequality (2.90), $\Delta^2 \le \delta^3/6$ (and noting that the equality gives the boundaries of the biaxial phase). We can also parametrize the tensor $\mathbf{Q}$ in terms of new (polar) variables. For example, we can write

$$Q_{xx} = r\cos\left(\theta + \frac{2\pi}{3}\right); \quad Q_{yy} = r\cos\left(\theta - \frac{2\pi}{3}\right); \quad Q_{zz} = r\cos\theta, \qquad (2.93)$$

so that $S = r\cos\theta$ and $\eta = \sqrt{3}r\sin\theta$; with this parametrization, the uniaxial phases correspond to $\theta = n\pi/3$, with even n, and the inequality (2.90) holds for $\cos 3\theta < 1$. Although these parametrizations are very convenient, and do make contact with group-theoretical arguments, in this section it is easier to keep using the variables S and η.

It is straightforward to write the truncated free energy (2.92) in the form

$$F_{\mathrm{BU}} = F_{\mathrm{BU}}(S, \eta) = F_0 + \alpha(S) + \beta(S)\eta^2 + \frac{1}{2}\gamma(S)\eta^4, \tag{2.94}$$

where

$$\alpha(S) = \frac{3}{4}AS^2 + \frac{1}{4}BS^3 + \frac{9}{16}CS^4 + \frac{9}{96}E'S^6, \tag{2.95}$$

which should be compared with Eq. (2.91),

$$\beta(S) = \frac{1}{4}A - \frac{1}{4}BS + \frac{3}{8}CS^2 - \frac{9}{48}E'S^4, \tag{2.96}$$

and

$$\gamma(S) = \frac{1}{8}C + \frac{9}{48}E'S^2. \tag{2.97}$$

The extremization with respect to η leads to the equation

$$2\eta(\beta + \gamma\eta^2) = 0, \tag{2.98}$$

so that there is always a "uniaxial solution," $\eta = 0$, associated with the free energy

$$F_{\mathrm{U}} = F_0 + \alpha(S). \tag{2.99}$$

There is also the possibility of a biaxial solution,

$$\eta^2 = -\frac{\beta(S)}{\gamma(S)}, \tag{2.100}$$

which exists for $\beta(S) < 0$ (since $\gamma(S) > 0$ for our choice of coefficients), and which is associated with the biaxial free energy,

$$F_{\mathrm{B}} = F_0 + \alpha(S) - \frac{1}{2\gamma(S)}\left[\beta(S)\right]^2. \tag{2.101}$$

We now note that $\gamma(S)$, given by Eq. (2.97), is always positive. Therefore, $F_{\mathrm{B}}(S) < F_{\mathrm{U}}(S)$, which shows that the biaxial phase is a stable minimum of the free energy throughout the region of existence of a "biaxial solution." In order to finally solve the problem, we have to search for the minima of the following functions:

(i) The uniaxial free energy $F_{\mathrm{U}}(S)$, given by Eq. (2.99), if $\beta(S) > 0$;
(ii) The biaxial free energy $F_{\mathrm{B}}(S)$, given by Eq. (2.101), if $\beta(S) < 0$.

Let us consider the uniaxial–isotropic border. From the conditions

$$\frac{\partial F_{\mathrm{U}}}{\partial S} = 0 \quad \text{and} \quad F_{\mathrm{U}} = F_0, \tag{2.102}$$

we have

$$A = \frac{3}{4}CS^2 + \frac{3}{8}E'S^4, \tag{2.103}$$

and

$$B = -\frac{9}{2}CS - \frac{3}{2}E'S^3. \tag{2.104}$$

The elimination of S leads to the first-order boundary of Fig. 2.10. If $E' = 0$, we have the trivial result

$$A = \frac{1}{9C}B^2. \tag{2.105}$$

We now consider the biaxial–uniaxial transition line. From the conditions

$$\frac{\partial F_B}{\partial S} = 0 \quad \text{and} \quad \beta(S) = 0, \tag{2.106}$$

we have

$$A = -\frac{3}{2}CS^2 \quad \text{and} \quad B = -\frac{3}{4}E'S^3. \tag{2.107}$$

If $E' \neq 0$, we do have a biaxial phase in the A–B phase diagram. The second-order biaxial–nematic borders are given by

$$A = -\left[\frac{C^3}{6\,(E')^2}\right]^{1/3} |B|^{2/3}, \tag{2.108}$$

as it is sketched in Fig. 2.11. The special point at the origin of this phase diagram has been called a Landau multicritical point.

Finally, it should be mentioned that, with the exception of less symmetric transition lines, no qualitative changes in the topology of this phase diagram will

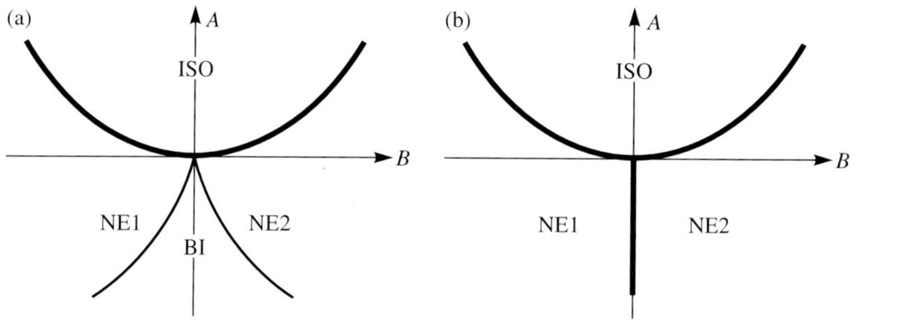

FIG. 2.11. Phase diagrams in terms of the Landau parameters A and B. The heavy solid lines are first-order boundaries. In (a), for $E' > 0$, there is a biaxial (BI) and two uniaxial nematic phases (NE1 and NE2). In (b), for $E' = 0$, there is no nematic biaxial phase.

be produced if we include the remaining coefficients $D > 0$ and $E > 0$. Of course, there will be qualitative changes, including the appearance of tricritical points, if we allow for either higher-order terms or coefficients with wrong (negative) signs.

2.7 The smectic A phase transition

The smectic phases provide some examples of phase transitions from a layered structure. In the smectic A phases the center of mass of the rigid-rod molecules sit on planes perpendicular to the preferred axis. There is nematic-type order within planes and a layered structure along the z direction. The symmetry axis of the smectic structure is parallel to the layers. These smectic phases are still optically uniaxial, but the X-ray pictures display a characteristic sharp ring corresponding to an interlayer spacing d.

The normalized distribution of the centers of mass of the rigid rods may be written as a Fourier expansion,

$$\rho(z) = \sum_{n=0}^{\infty} \alpha_n \cos\left(\frac{2\pi n z}{d}\right), \tag{2.109}$$

where

$$\alpha_n = \frac{2}{d} \int_0^d \cos\left(\frac{2\pi n z}{d}\right) \rho(z)\, \mathrm{d}z = \left\langle \frac{2}{d} \cos\left(\frac{2\pi n z}{d}\right)\right\rangle. \tag{2.110}$$

There is a disordered phase, for $\alpha_n = 0$, which corresponds to a uniform distribution, $\rho(z) = 1/d$. On the other hand, perfect order is obtained if $\alpha_n = 1$ for all n. However, as the smectic phase displays both orientational and translational order, the molecular distribution function must describe the tendency of the molecules to align along the z direction and to form layers perpendicular to z. We then introduce a more complete distribution function,

$$\rho(\cos\theta, z) = \sum_{l\,\text{even}} A_l P_l(\cos\theta) \cos\left(\frac{2\pi n z}{d}\right), \tag{2.111}$$

with the normalization

$$\int_0^{\pi} \int_0^d \rho(\cos\theta, z)\, \mathrm{d}z \sin\theta\, \mathrm{d}\theta = 1. \tag{2.112}$$

Keeping terms of lowest degree, we define the order parameters

$$\eta = \langle P_2(\cos\theta)\rangle, \quad \tau = \left\langle \cos\frac{2\pi z}{d}\right\rangle, \quad \sigma = \left\langle P_2(\cos\theta)\cos\frac{2\pi z}{d}\right\rangle. \tag{2.113}$$

Therefore, we have: (i) $\eta = \tau = \sigma = 0$, in the disordered phase; (ii) $\eta \neq 0$, $\tau = \sigma = 0$, in the nematic phase; and (iii) $\eta, \tau, \sigma \neq 0$, in the smectic phase. According to this view of the smectic phase, we can write the one-body molecular potential

$$V_1(\cos\theta, z) = u_0 + u_1\tau\cos\left(\frac{2\pi z}{d}\right) + \left[w_0\eta + w_1\cos\left(\frac{2\pi z}{d}\right)\right] P_2(\cos\theta). \tag{2.114}$$

If we further simplify this potential, by choosing $\tau = 0$, it is still possible to explain the smectic–nematic transition. We then introduce the MacMillan model [17], given by the mean-field Hamiltonian

$$\mathcal{H} = -\frac{w_0}{2N}\left(\sum_i \eta_i\right)^2 - \frac{w_1}{2N}\left(\sum_i \sigma_i\right)^2, \tag{2.115}$$

where $\eta_i = P_2(\cos\theta_i)$ and $\sigma_i = \cos(2\pi z_i/d)\, P_2(\cos\theta_i)$ are local variables. It is straightforward to write

$$Z = \mathrm{Tr}\exp\left[\frac{\beta w_0}{2N}\left(\sum_i \eta_i\right)^2 + \frac{\beta w_1}{2N}\left(\sum_i \sigma_i\right)^2\right] = \frac{\beta N}{2\pi}(w_0 w_1)^{1/2}\int\int \mathrm{d}\eta\, \mathrm{d}\sigma \exp(-\beta N f), \tag{2.116}$$

where

$$f = \frac{1}{2}w_0\eta^2 + \frac{1}{2}w_1\sigma^2 - \frac{1}{\beta}\ln Z_1, \tag{2.117}$$

with

$$Z_1 = \int_0^\pi \sin\theta\, \mathrm{d}\theta \int_0^d \mathrm{d}z \exp\left\{\beta\left[w_0\eta + w_1\sigma\cos\left(\frac{2\pi z}{d}\right)\right] P_2(\cos\theta)\right\}. \tag{2.118}$$

The equations of state are given by

$$\frac{\partial f}{\partial \eta} = w_0\eta - \frac{1}{\beta Z_1}\frac{\partial Z_1}{\partial \eta} = 0, \tag{2.119}$$

from which we have $\eta = \langle P_2(\cos\theta)\rangle$, and

$$\frac{\partial f}{\partial \sigma} = w_1\sigma - \frac{1}{\beta Z_1}\frac{\partial Z_1}{\partial \sigma} = 0, \tag{2.120}$$

which leads to $\sigma = \langle \cos(2\pi z/d) P_2(\cos\theta)\rangle$. A numerical treatment of this problem leads to a phase diagram in terms of temperature, $T = 1/(\beta w_0)$, versus the ratio $\alpha = w_1/w_2$, with isotropic ($\eta = \sigma = 0$), nematic ($\eta \neq 0$, $\sigma = 0$) and smectic ($\eta, \sigma \neq 0$) phases (see Fig. 2.12).

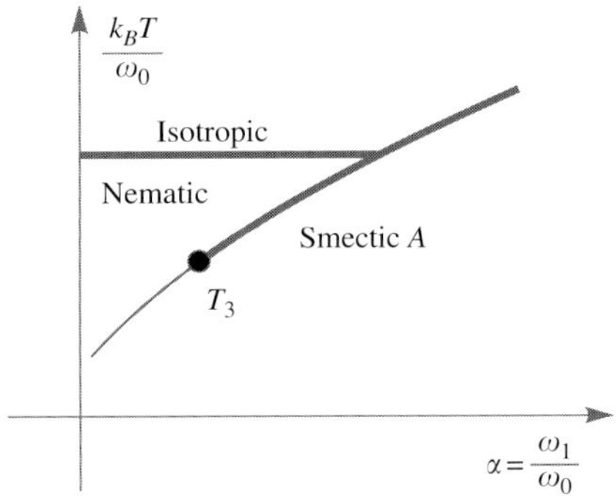

FIG. 2.12. Phase diagram of the McMillan model, with isotropic, uniaxial nematic and smectic A phases. The heavy solid lines are first-order boundaries. There is also a second-order line that ends at the tricritical point T_3.

According to an elegant approach of de Gennes [4], the smectic phase can be characterized by a complex order parameter,

$$\psi(z) = \rho(z) \exp[\mathrm{i}\phi(z)], \tag{2.121}$$

which leads to a connection with the Landau–Ginzburg description of superfluid helium. In terms of this complex order parameter, we can write the free energy expansion

$$f_s = \frac{1}{2} a |\psi|^2 + \frac{1}{4} |\psi|^4 + b \left| \frac{\partial \psi}{\partial z} \right|^2 + \cdots , \tag{2.122}$$

where $a = a_0(T - T_0)$ and $b > 0$.

The Landau theory for transitions from layered structures in lyotropic mixtures is discussed in Chapter 4.

2.8 Non-critical order parameters and the reconstruction of the phase diagrams

A few years ago, Tolédano and Figueiredo Neto [18, 19] proposed a phenomenological mechanism to explain the experimentally observed reentrant behavior of the isotropic phase in the temperature–concentration phase diagram of lyonematic liquid crystals (see the sketch in Fig. 2.13). The standard Landau free energy, given by Eq. (2.87), for example, which is used to describe a phase diagram with isotropic as well as nematic uniaxial and nematic biaxial phases, can be completed by the inclusion of an extra non-critical parameter, associated with the change in micellar form.

Before looking at the specific problem of lyonematic liquid crystals, let us give two examples of the introduction of non-critical order parameters in some magnetic model systems.

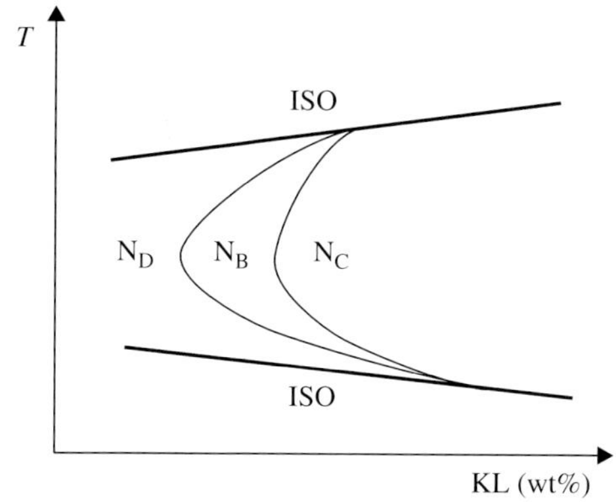

FIG. 2.13. Sketch of the typical temperature versus concentration (in this case, the potassium laurate KL amphiphile, at fixed concentration of decanol) experimental phase diagram of a ternary lyotropic mixture (KL, decanol, water). Note the reentrance of the disordered region at low temperatures.

2.8.1 *Compressible Ising model*

A simple compressible Ising model [20] is given by the Hamiltonian

$$\mathcal{H} = -J(v) \sum_{(i,j)} \sigma_i \sigma_j + \frac{1}{2} k N (v - v_0)^2, \tag{2.123}$$

where $\sigma_i = \pm 1$, for all $i = 1, \ldots, N$, the sum is over nearest-neighbor sites of a crystal lattice, and v is the specific volume. For simplicity, we assume a quadratic elastic energy term, where v_0 and $k > 0$ are constant factors. Also, we assume that the exchange interaction is given by the linear form

$$J(v) = J_0 - J_1 (v - v_0), \tag{2.124}$$

where J_0 and J_1 are arbitrary constants.

At fixed volume, the canonical partition function is given by

$$Z = \exp\left[\frac{1}{2}\beta k N (v - v_0)^2\right] Z_{\mathrm{I}}, \tag{2.125}$$

where

$$Z_{\mathrm{I}} = \mathrm{Tr} \exp\left[\beta J(v) \sum_{(i,j)} \sigma_i \sigma_j\right]. \tag{2.126}$$

The mean-field approximation can be obtained from the variational inequality

$$G(\mathcal{H}_{\mathrm{I}}) \leq G_0\left(\mathcal{H}_0\right) + \langle \mathcal{H} - \mathcal{H}_0 \rangle_0 = \Phi, \tag{2.127}$$

where $G(\mathcal{H}_\text{I}) = -\ln Z_\text{I}/\beta$, $G(\mathcal{H}_0) = -\ln Z_0/\beta$,

$$\mathcal{H}_0 = -\eta \sum_{i=1}^{N} \sigma_i \tag{2.128}$$

is a free trial Hamiltonian, the average $\langle \cdots \rangle_0$ is to be taken with respect to this trial Hamiltonian, and

$$Z_0 = \operatorname{Tr} \exp \left[\beta\eta \sum_{i=1}^{N} \sigma_i \right] = (2\cosh \beta\eta)^N. \tag{2.129}$$

Thus, we have

$$\Phi = -\frac{1}{\beta} N \ln[\cosh \beta\eta] - \frac{1}{2} J(v) q N m^2 + \eta N m, \tag{2.130}$$

where

$$\langle \sigma_i \rangle_0 = \tanh \beta\eta = m, \tag{2.131}$$

and q is the coordination of the lattice.

From these equations, with the elimination of the parameter η, we finally write

$$\begin{aligned} f(T, v, m) &= \frac{1}{2} k N (v - v_0)^2 + \frac{1}{N} \Phi \\ &= \frac{1}{2} k (v - v_0)^2 + \frac{1}{\beta} \int \mathrm{d}m \tanh^{-1} m - \frac{1}{2} J(v) q m^2, \end{aligned} \tag{2.132}$$

which is the basis for a Landau expansion in terms of the order parameter m. Note that there are just linear and quadratic volume-dependent terms. Also note that the coefficient of the quadratic elastic term is always positive, which indicates the non-critical character of the specific volume. Keeping up to fourth-order terms in the magnetization per spin, we have the Landau expansion for the free-energy density,

$$\begin{aligned} f = f(T, m, v) = f_0(T) &+ \frac{1}{2} k (v - v_0)^2 + \frac{1}{2\beta} m^2 - \frac{1}{2} J(v) q m^2 \\ &+ \frac{1}{12\beta} m^4 + O(m^6). \end{aligned} \tag{2.133}$$

Note that the coefficient of v^2 is positive, and that there are no higher-order terms in the specific volume v, which excludes the occurrence of qualitative differences as compared with the critical behavior of a simple ferromagnet. Also, note that, due to symmetry arguments, the spin–lattice coupling is given by a term of the form vm^2. If we introduce the pressure p, it may be more convenient to write a Gibbs potential,

$$g = g(T, m, p) = f + pv, \tag{2.134}$$

which should be minimized with respect to the thermodynamic densities m and v.

Keeping the most relevant terms to describe the critical behavior, we can write a Landau expansion for the Gibbs density,

$$g = am^2 + m^4 + vm^2 + bv + v^2, \tag{2.135}$$

where a and b depend on the thermodynamic fields (temperature and pressure), and some of the coefficients have been set to unity. Now it is easy to see that the usual second-order transition, at $a = 0$, with $b = 0$, turns into a line of second-order transitions, $a = b/2$, in the a–b phase diagram (see Fig. 2.14). This is an example of a reconstruction of the phase diagram in terms of an extra parameter b. In this case, the transition remains of second order. However, if we describe the system in terms of temperature and pressure (instead of a and b), and include a sixth-order term, it is easy to see the presence of a tricritical point, beyond which the continuous transition changes to first order [20].

2.8.2 *Ferromagnet in a staggered field*

Another example of the effects of a non-ordering field, associated with a non-critical order parameter, is given by a ferromagnet in a staggered field (or the analogous problem of a two-sublattice antiferromagnet in a uniform field). The spin Hamiltonian is given by

$$\mathcal{H} = -J \sum_{(i,j)} \sigma_i \sigma_j - H \sum_{i \in A} \sigma_i + H \sum_{j \in B} \sigma_j, \tag{2.136}$$

where $\sigma_i, \sigma_j = \pm 1$ are Ising spin variables, $J > 0$, and the first sum is over nearest-neighbor sites, i and j, belonging to different sublattices, A and B, respectively. Note that H is an external (staggered) magnetic field, pointing up in sublattice A and down in sublattice B. Also, note that we are not including a uniform (physical) external field, which is coupled to the magnetization order parameter.

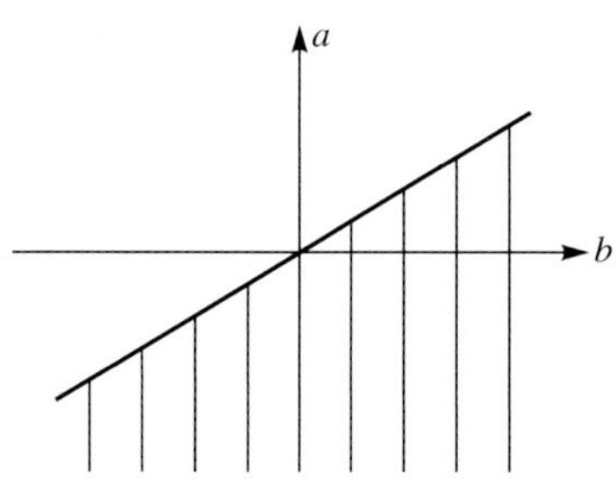

FIG. 2.14. Phase diagram, in terms of the Landau parameters a and b, of a compressible ferromagnetic model. The second-order line is given by $a = b/2$ (the hatched region corresponds to the ferromagnetically ordered phase).

We now perform a variational mean-field calculation. Using the trial Hamiltonian

$$\mathcal{H}_0 = -\eta_A \sum_{i \in A} \sigma_i - \eta_B \sum_{j \in B} \sigma_j, \tag{2.137}$$

it is straightforward to write the mean-field potential

$$\begin{aligned} f(T, H, m_A, m_B) &= -\frac{1}{2\beta} \ln[2\cosh \beta\eta_A] - \frac{1}{2\beta} \ln[2\cosh \beta\eta_B] \\ &- \frac{1}{2} H m_A - \frac{1}{2} H m_B + \frac{1}{2}\eta_A m_A + \frac{1}{2}\eta_B m_B - \frac{1}{2} J q m_A m_B, \end{aligned} \tag{2.138}$$

where

$$m_A = \tanh \beta\eta_A \quad \text{and} \quad m_B = \tanh \beta\eta_B. \tag{2.139}$$

Using the definitions of the uniform magnetization, m, and the staggered magnetization, m_s, given by

$$m = \frac{1}{2}(m_A + m_B) \quad \text{and} \quad m_s = \frac{1}{2}(m_A - m_B), \tag{2.140}$$

and making the simplifying assumption that H is a sufficiently weak field, it is not difficult to obtain the conveniently truncated (Landau) expansion

$$\begin{aligned} f &= f_0(T) + \left(\frac{1}{\beta} - Jq\right) m^2 + \frac{1}{6\beta} m^4 \\ &- 2Hm_s + \left(\frac{1}{\beta} + Jq\right) m_s^2 + \frac{1}{\beta} m_s^2 m^2 + \cdots, \end{aligned} \tag{2.141}$$

which should be compared with Eqs. (2.133) and (2.135), for the compressible Ising model.

In the free energy (2.141), with $J > 0$ for a ferromagnetic system, the coefficient of m_s^2 is always positive, which shows that the staggered magnetization m_s is a non-critical parameter (and which gives support to the truncation of the expansion at terms of quadratic order in m_s). In agreement with symmetry requirements, there is now a biquadratic coupling between m and m_s, which leads to symmetrical transition lines under the change $H \to -H$. From the minimization of this Landau expansion, we obtain the asymptotic behavior (for $H \to 0$) of the critical line in the H–T phase diagram,

$$\frac{k_B T_c}{Jq} = 1 - \frac{1}{4}\left(\frac{H}{Jq}\right)^2 + \cdots, \tag{2.142}$$

with the well-known quadratic depression of the critical temperature. Along this second-order transition line, the critical behavior remains unchanged with respect to the zero-field case.

2.8.3 *Reconstruction of the lyonematic phase diagrams*

We are now prepared to discuss the phenomenological ideas of Tolédano and Figueiredo Neto [18]. As we pointed out before, the experimental phase diagrams of lyotropic nematic liquid crystals usually display a low-temperature reentrant isotropic phase, which has not been explained by the Maier–Saupe model and its various extensions. On the basis of the preceding ideas about the effects of non-critical order parameters, let us introduce a phenomenological mechanism to account for the reentrant behavior.

In the context of a model of intrinsically biaxial micelles [21], as presented in Chapter 1 in order to account for the three nematic phases ($\mathrm{N_C}$, $\mathrm{N_B}$, and $\mathrm{N_D}$), micelles are assumed to have the same orthorhombic symmetry in all of the nematic phases. The shape anisotropy of the micelles depends on temperature and relative concentrations of the compounds of the lyotropic mixture. The macroscopic uniaxial or biaxial character of each phase is associated with orientational fluctuations of micelles as a function of temperature and relative concentrations. For example, we may sketch a micelle as a flat ellipsoid, with three different axes; the distribution of the shape anisotropies is assumed to change with temperature and relative concentrations.

In order to take into account these changes in form of the micellar aggregates, let us call n_B and n_S the number densities of "less spherical" and "more spherical" aggregates, respectively, and introduce the (non-critical) parameter

$$\tau = \frac{n_B - n_S}{n_B + n_S}. \tag{2.143}$$

This scalar non-critical parameter is supposed to represent the continuous change of shape anisotropy of the micellar population as a function of temperature. As in the examples of the compressible model and of the ferromagnet in a staggered field, this parameter preserves the existing singularities and leads to their multiplication (along a line of transitions).

Consider the Landau expansion, given by Eq. (2.92), used in our previous treatment of the biaxial–uniaxial nematic transition,

$$F_{\mathrm{BU}} = F_0 + \frac{1}{2}A\delta + \frac{1}{3}B\Delta + \frac{1}{4}C\delta^2 + \frac{1}{6}E'\Delta^2, \tag{2.144}$$

where C and E' are positive parameters, and δ and Δ have their usual meanings (traces of the square and the cube of the tensor order parameter, respectively). From a phenomenological point of view, the non-ordering parameter τ can be introduced by redefining δ ($\delta \to \delta + \tau^2$), keeping the same form of the higher-order invariant Δ, and introducing new linear and quadratic terms in τ. We then write the Landau expansion

$$F_{\mathrm{BU}}(\delta, \Delta, \tau) = F_0 + \frac{1}{2}A(\delta + \tau^2) + \frac{1}{3}B\Delta + \frac{1}{4}C(\delta + \tau^2)^2 + \frac{1}{6}E'\Delta^2 + C_1\tau + \frac{1}{2}C_2\tau^2. \tag{2.145}$$

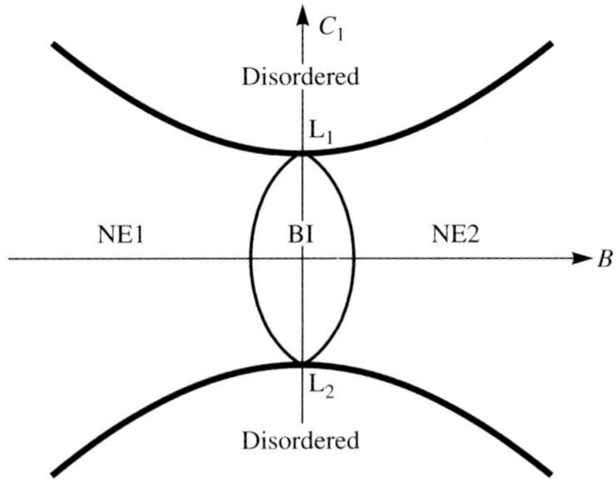

FIG. 2.15. Section of the phase diagram, in terms of the coefficients C_1 and B, with two uniaxial and a biaxial nematic phase. The heavy solid lines correspond to first-order boundaries. The biaxial–uniaxial transition remains second order. Note the re-entrance of the isotropic phase and the occurrence of two Landau points (L_1 and L_2).

The new parameter τ is non-critical if the coefficient of the quadratic term τ^2 is positive.

Assuming that $A + C_2 > 0$, we can discard quartic terms in τ, and minimize the truncated potential as usual. With the notation already used for discussing the biaxial transition, we write

$$F_{\mathrm{BU}} = F_0 + \alpha(S) + \frac{3}{4}CS^2\tau^2 + \beta(S)\eta^2 + \frac{1}{4}C\eta^2\tau^2 + \frac{1}{2}\gamma(S)\eta^4 \\ + C_1\tau + \frac{1}{2}C_2\tau^2 + \frac{1}{2}A\tau^2. \quad (2.146)$$

The minimization with respect to τ leads to the relation

$$\tau = -C_1\left[C_2 + A + \frac{1}{2}C\left(3S^2 + \eta^2\right)\right]^{-1}. \quad (2.147)$$

In the Landau potential, the biquadratic form of the coupling terms, $S^2\tau^2$ and $\eta^2\tau^2$, indicates that all transition lines are invariant under the change $C_1 \to -C_1$, which gives rise to the phenomenon of reconstruction of the phase diagram. The coefficient C_1, which is strongly dependent on temperature, gauges the changes in micellar shape. We then draw a symmetric C_1–B phase diagram, as shown schematically in Fig. 2.15.

References

[1] H. E. Stanley (1971). *Introduction to Phase Transitions and Critical Phenomena*, Oxford University Press, Oxford.

[2] J. M. Yeomans (1991). *Statistical Mechanics of Phase Transitions*, Oxford University Press, Oxford.

[3] S. R. A. Salinas (2000). *Introduction to Statistical Physics*, Springer, New York, chapters 12–14.
[4] P. G. de Gennes and J. Prost (1993). *The Physics of Liquid Crystals*, Oxford University Press, Oxford.
[5] R. B. Griffiths (1970). *Phys. Rev. Lett.* **24**, 715; M. Blume, V. J. Emery, and R. B. Griffiths (1971). *Phys. Rev. A* **4**, 1071.
[6] M. E. Fisher (1975). *AIP Conf. Proc.* **24**, 273; H. Rohrer (1975). *Phys. Rev. Lett.* **34**, 1638; N. F. Oliveira Jr., A. Paduan Filho, and S. R. Salinas (1975). *Phys. Lett.* A**55**, 293. H. Rohrer and C. Gerber (1977). *Phys. Rev. Lett.* **38**, 909.
[7] C. S. O. Yokoi, M. D. Coutinho-Filho, and S. R. Salinas (1981). *Phys. Rev. B* **24**, 4047.
[8] R. M. Hornreich, M. Luban, and S. Shtrikman (1975). *Phys. Rev. Lett.* **35**, 1678.
[9] S. Chandrasekhar (1977). *Liquid Crystals*, Cambridge University Press, Cambridge.
[10] W. Maier and A. Saupe (1958). *Z. Naturforsch. A* **13**, 564; **14**, 882; *Z. Naturfursch. A* **15**, 287 (1960).
[11] S. Singh (2000). *Phys. Rep.* **324**, 107.
[12] W. M. Gelbart (1982). *J. Phys. Chem.* **86**, 4298.
[13] J. P. Straley (1974). *Phys. Rev. A* **10**, 1881; M. J. de Oliveira and A. M. Figueiredo Neto (1986). *Phys. Rev. A* **34**, 3481.
[14] E. F. Henriques and V. B. Henriques (1997). *J. Chem. Phys.* **107**, 8036.
[15] R. Zwanzig (1963). *J. Chem. Phys.* **39**, 1714.
[16] E. F. Gramsberger, L. Longa, and W. H. de Jeu (1997). *Phys. Rep.* **135**, 195.
[17] W. L. Mcmillan (1971). *Phys. Rev. A* **4**, 1238.
[18] P. Tolédano and A. M. Figueiredo Neto (1994). *Phys. Rev. Lett.* **73**, 2216.
[19] P. Tolédano, A. M. Figueiredo Neto, V. Lorman, B. Mettout, and V. Dmitriev (1995). *Phys. Rev. B* **52**, 5040.
[20] V. B. Henriques and S. R. Salinas (1987). *J. Phys. C* **20**, 2415; S. R. Salinas (1974). *J. Phys. C* **7**, 241 (1974).
[21] Y. Galerne, A. M. Figueiredo Neto, and L. Liébert (1987). *J. Chem. Phys.* **87**, 1851.

3

PHASE DIAGRAMS OF LYOTROPIC MIXTURES

3.1 Introduction

We have already emphasized that polymorphism is one of the main features of lyotropic liquid crystals. Phase diagrams of lyotropics, in terms of temperature and relative concentrations of the components, display a wealth of stable structures and regions. Changes of pressure can also lead to stable lyotropic phases [1]. In this chapter, we initially review some general features of phase diagrams, with emphasis on phase stability and the Gibbs phase rule. We then describe the topology of the best-known and investigated phase diagrams of binary, ternary and multi-component lyotropic mixtures. Finally, we refer to phenomenological calculations for obtaining some features of these phase diagrams.

3.2 General features of phase diagrams

The thermodynamic state of equilibrium of an isotropic system with r components is characterized by the internal energy U, the volume V, and the number of moles of each component, $N_1, N_2, \ldots, N_r$ [2]. The entropy $S = S(U, V, N_1, \ldots, N_r)$ is a homogeneous function of first degree of these variables. We then write the entropy per mole,

$$s = \frac{1}{N}S = s(u, v, x_1, \ldots, x_{r-1}), \tag{3.1}$$

where $u = U/N$ and $v = V/N$, with $N = N_1 + \cdots + N_r$, are the energy per mole and the specific volume, respectively, and $x_i = N_i/N$ is the molar relative concentration of component $i = 1, \ldots, r$. Note that

$$x_1 + x_2 + \cdots + x_r = 1, \tag{3.2}$$

and that we need only $r+1$ variables to fully characterize the equilibrium thermodynamic state of a simple r-component system. Usually, it is more convenient to work in the Gibbs representation, in which temperature T and pressure p, instead of u and v, are taken as the independent variables. For example, in the Gibbs representation, with $r = 2$ components (let us call them A and B), we need three independent variables (besides T and p, we may choose the relative molar concentration $x_A = N_A/(N_A + N_B)$ of component A).

We now consider a composite system of several simple subsystems (see Fig. 3.1). At given values of temperature and pressure, if the walls of this composite system are not restrictive, the chemical potential associated with each

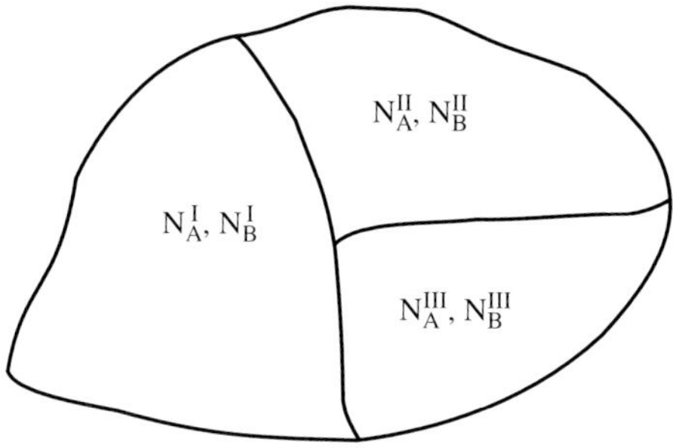

FIG. 3.1. Composite system with three subsystems (I, II, and III) of two components (A and B) each. The system is in equilibrium at fixed values of temperature and pressure. The walls between subsystems are diathermal, moveable, and permeable to all components. The outer border is impermeable.

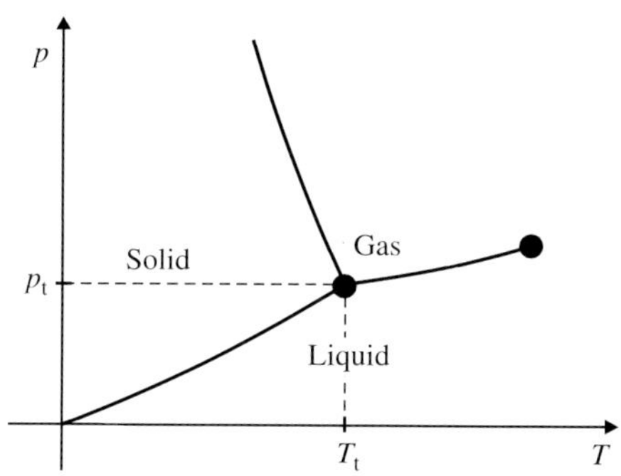

FIG. 3.2. Sketch of a T–p phase diagram of a one-component system. Solid lines indicate first-order transitions (coexistence of phases). At the triple point, T_t, p_t, there is coexistence among solid, liquid, and gas phases.

component is required to assume the same (equilibrium) value in all subsystems (phases).

For example, consider the equilibrium between a solid (S) and a liquid (L) phase of a simple system of one component only ($r = 1$). The chemical potential, which is the Gibbs free energy per mole, is a function of T and p only. It assumes the values $\mu^S(T,p)$ and $\mu^L(T,p)$ in the solid and liquid phases, respectively. In equilibrium, we have

$$\mu^S(T,p) = \mu^L(T,p), \tag{3.3}$$

which leads to a relation between temperature and pressure. Therefore, there may be coexistence between two distinct phases along a line in a T–p phase diagram (see Fig. 3.2, which illustrates the phase diagram of a pure fluid; along the solid lines there is coexistence between two distinct phases). We now look at the equilibrium among three different phases (liquid L, solid S, and gas G). In this case, there are two equations, given by

$$\mu^{S}(T,p) = \mu^{L}(T,p) = \mu^{G}(T,p), \tag{3.4}$$

which lead to the location of the triple point, at well-defined values of temperature and pressure.

As another example, consider the equilibrium between solid (S) and liquid (L) phases in a simple system of $r = 2$ components. Note that we have $r+1 = 3$ independent variables. The chemical potential of component i ($i = \mathrm{A}, \mathrm{B}$) may be written as $\mu_i^{\mathrm{L}} = \mu_i^{\mathrm{L}}(T, p, x_{\mathrm{A}}^{\mathrm{L}})$ and $\mu_i^{\mathrm{S}} = \mu_i^{\mathrm{S}}(T, p, x_{\mathrm{A}}^{\mathrm{S}})$, in the liquid and solid phases, respectively. We then write the equilibrium conditions,

$$\mu_{\mathrm{A}}^{\mathrm{L}}\left(T, p, x_{\mathrm{A}}^{\mathrm{L}}\right) = \mu_{\mathrm{A}}^{\mathrm{S}}\left(T, p, x_{\mathrm{A}}^{\mathrm{S}}\right) \tag{3.5}$$

and

$$\mu_{\mathrm{B}}^{\mathrm{L}}\left(T, p, x_{\mathrm{A}}^{\mathrm{L}}\right) = \mu_{\mathrm{B}}^{\mathrm{S}}\left(T, p, x_{\mathrm{A}}^{\mathrm{S}}\right), \tag{3.6}$$

from which we find $x_{\mathrm{A}}^{\mathrm{L}}$ and $x_{\mathrm{A}}^{\mathrm{S}}$ in terms of T and p (note that $x_{\mathrm{A}}^{\mathrm{L}} + x_{\mathrm{B}}^{\mathrm{L}} = 1$ and $x_{\mathrm{A}}^{\mathrm{S}} + x_{\mathrm{B}}^{\mathrm{S}} = 1$). In equilibrium, the coexistence between solid and liquid phases takes place for a certain range of independently assigned values of temperature and pressure. In Fig. 3.3, we draw a graph of T versus the molar concentration

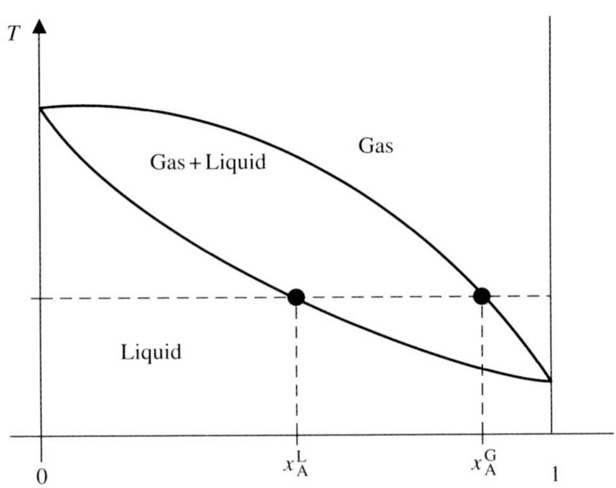

FIG. 3.3. Temperature versus concentration phase diagram, at fixed pressure, for a system with two components (A and B). The tie line (at given values of T and p) indicates the coexistence between a liquid phase, with relative molar concentration $x_{\mathrm{A}}^{\mathrm{L}}$, and a gas phase, with molar concentration $x_{\mathrm{A}}^{\mathrm{G}}$. Note that $f = 3$, in the pure liquid and pure gas phases, and $f = 2$ in the coexistence region.

$x_{\mathrm{A}} = N_{\mathrm{A}}/(N_{\mathrm{A}} + N_{\mathrm{B}})$ for a binary solution of components A and B at a given value of pressure. We now assume that there are three phases, S, L, and G, in equilibrium. We then have

$$\mu_{\mathrm{A}}^{\mathrm{L}}\left(T, p, x_{\mathrm{A}}^{\mathrm{L}}\right) = \mu_{\mathrm{A}}^{\mathrm{S}}\left(T, p, x_{\mathrm{A}}^{\mathrm{S}}\right) = \mu_{\mathrm{A}}^{\mathrm{G}}\left(T, p, x_{\mathrm{A}}^{\mathrm{G}}\right), \tag{3.7}$$

and

$$\mu_{\mathrm{B}}^{\mathrm{L}}\left(T, p, x_{\mathrm{A}}^{\mathrm{L}}\right) = \mu_{\mathrm{B}}^{\mathrm{S}}\left(T, p, x_{\mathrm{A}}^{\mathrm{S}}\right) = \mu_{\mathrm{B}}^{\mathrm{G}}\left(T, p, x_{\mathrm{A}}^{\mathrm{G}}\right), \tag{3.8}$$

from which we obtain the compositions and a relation between temperature and pressure. In this example of three-phase coexistence in a binary system, it is still feasible to freely change one of the thermodynamic variables (temperature, for example), without moving out of the region of coexistence among the three phases. In this system of two components, it is easy to see that coexistence among four different phases is restricted to well-defined values of all variables (temperature, pressure, and compositions).

3.2.1 *The Gibbs phase rule*

The Gibbs phase rule is a generalization of these arguments. Suppose that we have equilibrium among M phases in a system with r components. Each phase may be characterized by temperature, pressure, and relative molar concentrations, $x_1^{\alpha}, x_2^{\alpha}, \ldots, x_{r-1}^{\alpha}$, with $\alpha = 1, \ldots, M$. There is a total of $2 + M(r-1)$ variables, and $M - 1$ equations for the chemical potential of each component. Therefore, in this system with r components and M coexisting phases, the number f of independently assigned thermodynamic variables is given by

$$f = [2 + M(r-1)] - r(M-1) = r - M + 2. \tag{3.9}$$

This number f may be regarded as the "number of thermodynamic degrees of freedom" which can be independently assigned if we still keep the coexistence among the M phases of the r-component system. Consider again the phase diagram for a binary system ($r = 2$) as sketched in Fig. 3.3. In the gas (G) and liquid (L) phases, which are associated with $r = 2$ and $M = 1$, we have $f = 3$ (temperature, pressure, and molar fraction of component A, for instance, can be freely changed). In the coexistence region, associated with $r = 2$ and $M = 2$, we have $f = 2$. Within this liquid-gas coexistence region, given the values of T and p, we determine the molar concentration $x_{\mathrm{A}}^{\mathrm{L}}$ and $x_{\mathrm{A}}^{\mathrm{G}}$ uniquely (see the tie lines in Fig. 3.3).

In Fig. 3.4, we sketch a section of constant pressure of a typical phase diagram for a liquid–solid two-component system, with the presence of two solid phases (α and β) of distinct crystal structures. The curve BDA is called a liquidus curve, and the curves BEL and ACJ are solidus curves. Point G corresponds to a two-phase system (a liquid of relative molar concentration $x_{\mathrm{A}}^{\mathrm{L}} = x_{\mathrm{H}}$ and a solid of molar concentration $x_{\mathrm{A}}^{\alpha} = x_{\mathrm{F}}$). Point K corresponds to α-solid at J

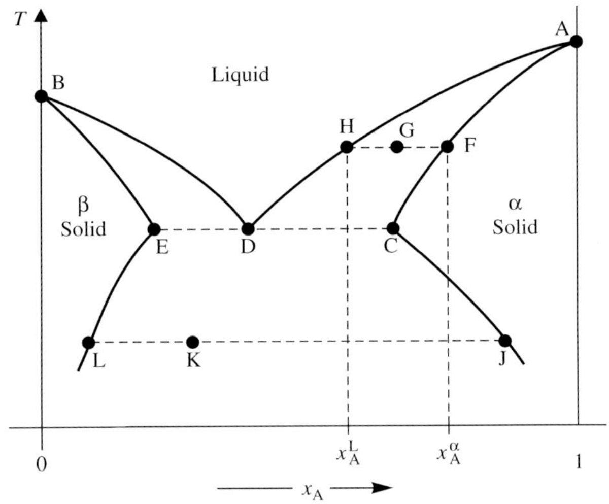

FIG. 3.4. Constant pressure section of a typical temperature versus concentration phase diagram of a binary system.

in equilibrium with β-solid at L. A liquid with composition $x_A^L = x_D$ is called eutectic.

3.2.2 *Ternary systems*

The equilibrium state of a ternary system is characterized by four thermodynamic variables. At fixed temperature and pressure, these systems can be represented by an equilateral triangle (see Fig. 3.5). From point P, we draw lines parallel to the sides of this triangle. The length of these lines represents the relative molar fractions x_A, x_B, and x_C of the three components (it is straightforward to show that $x_A + x_B + x_C = 1$). The vertices of the triangle correspond to pure substances. A line parallel to a side of the triangle corresponds to a series of ternary systems in which one of the molar fractions remains fixed. The variation of some property of a ternary system may be represented by a three-dimensional graph, in which the base is the compositional triangle, and the height is the particular thermodynamic property.

In order avoid any ambiguity, it is usual to construct an oriented triangle, which is known as a "partial isotherm" of the phase diagram (see Fig. 3.6). From each vertex of the triangle, there is an increase of the concentration of a given compound, according to the counter-clockwise direction. The compositions at a particular point are given by the (oriented) lines parallel to the sides of the triangle. Each of these lines is now oriented according to the same counter-clockwise direction of the original axes. The positions at which the arrows touch the axes give the concentrations of each component of the mixture. In Fig. 3.6, point P is associated with the concentrations (in 100 M%) $x_A = 0.7, x_B = 0.1,$

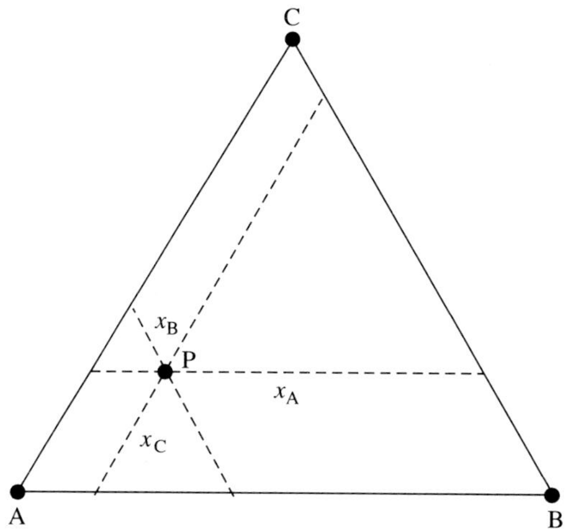

FIG. 3.5. Equilateral triangle for representing the relative molar concentrations of a ternary system (note that $x_A + x_B + x_C = 1$). The vertices correspond to pure phases.

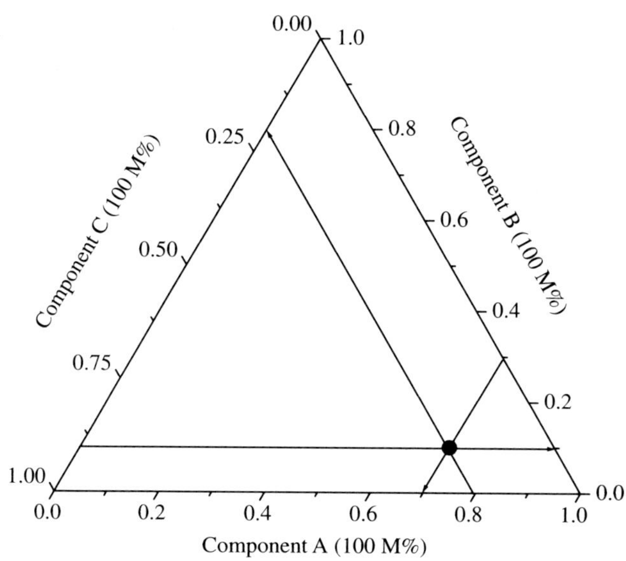

FIG. 3.6. Sketch of a triangular phase diagram (also known as partial isotherm) of a hypothetical mixture with three components. The axes are represented in molar percentages, M%.

and $x_C = 0.2$, of the three components. This representation of the axes in molar percentages, M%, is sometimes changed to a weight percentage, wt%, of each component. The topology of the phase diagrams is obviously unchanged in these two representations; only the loci of the different phases may be strongly changed from one representation to another.

In the following sections we give a number of examples of experimental realizations of these phase diagrams.

3.3 Experimental phase diagrams

Experimentalists use different, and complementary, techniques, in special X-ray and neutron scattering and diffraction, NMR spectroscopy, electrical and optical measurements, in order to obtain the phase diagrams of lyotropic liquid crystals. As we are dealing with multi-variable complex systems, the graphical representations are usually reduced to partial isotherms (for mixtures with more than two components) and surfaces of temperature versus relative concentrations.

3.3.1 *Phase diagrams of binary lyotropic mixtures*

Binary lyotropic mixtures are generally composed of one amphiphilic molecule and a (polar or non-polar) solvent. As an introductory example, consider the potassium palmitate ($C_{16}H_{31}KO_2$)/water mixture [3]. The phase diagram is usually represented in an isobaric surface of temperature versus relative concentration of the amphiphile, as shown in Fig. 3.7. Other binary mixtures composed by soaps and water display similar phase diagrams. The liquid crystalline phases

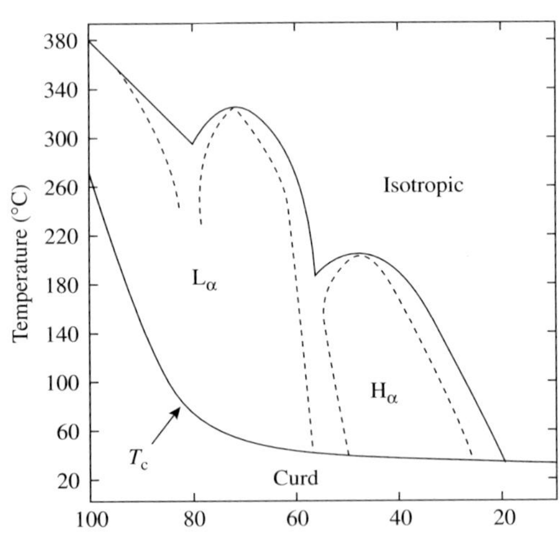

FIG. 3.7. Phase diagram of the potassium palmitate/water mixture, in terms of temperature versus concentration [3].

are in a region between the crystalline-type phases (at low temperatures) and the isotropic solution (at high temperatures). For lyotropic mixtures, we note that *low* and *high* temperatures refer to about 10 and 200°C, although these limits are highly dependent on the relative concentrations of the components; usually, larger concentrations of solvents lead to smaller values of the high-temperature limit of the liquid crystalline domain. At high temperatures, we have an isotropic micellar phase, which is formed by micelles in the solvent. However, the disordered phase can also be a mixture of isolated amphiphilic molecules and the solvent.

In the low temperature region of the phase diagram, just below the crystallization line, called T_c in Fig. 3.7, there is a region, named *curd*, with peculiar optical and structural behavior. At very high amphiphile concentrations, it gives rise to a sequence of phases named *curd fibers, waxy,* with *sub-* and *super-waxy* variants. These waxy and super-waxy phases were shown to present rectangular centered and rectangular centered/orthorhombic body-centered lattice symmetry of the amphiphiles, respectively [4,5]. This region below T_c is also characterized as the locus of the *gel* and the *coagel phases.* The gel phase is located just below T_c, where soap molecules form monolayers with an interdigitated molecular structure. Its structure is intermediate between the liquid crystalline state, where (in general) carbonic chains are in a liquid-like state, and the crystalline phase, with ordered carbonic chains. The coagel state is formed by poorly developed domains of the gel phase. Above T_c, in most of these systems, the alkyl chains of the amphiphilic molecules are in a *melted state*, without positional ordering. The orientational parameters of the different carbon atoms in the chain, measured by NMR, indicate that there is a decreasing degree of order from the atom located near the amphiphile head towards the CH_3 ending group [6].

The lamellar L_α and hexagonal H_α phases occupy a region between the T_c line and the high-temperature isotropic phase. The H_α phase is also named *middle phase* because it is located approximately in the middle (about 50 wt% of water) of the phase diagram. The lamellar phase, also called *neat phase*, is stable at a larger concentration of the amphiphile. A deformed hexagonal phase, sometimes characterized as a *ribbon phase*, and other *intermediate phases* may be present between the L_α and the H_α phases [7]. Three-dimensionally ordered phases, as micellar and bicontinuous cubic phases [8], and phases showing a *mesh structure* [9], have been identified in the so-called transition domains, between the isotropic phase and the L_α and H_α phases. A striking feature of these phase diagrams is the presence of many two-phase, or even multi-phase, domains, with the coexistence of different phases, and the associated first-order transition boundaries.

Besides these phases, which are the most common structures in binary soap/solvent mixtures, there may appear other phases in more complex amphiphile mixtures [10], as aqueous monoglycerides, mixtures of sodium bis-2-ethylhexyl sulfosuccinate (known as AOT) and water, and aqueous systems of amphiphiles of biological origin (e.g. lecithin, mitochondrial lipids, phospholipids). Some of these mixtures display phases with an inverted polymorphism. The topology of the amphiphilic molecules which favors the appearance of inverted phases is such that the polar head tends to occupy a smaller area than the alkyl chains (in some cases, two or more chains may be attached to a polar head). An inverted hexagonal phase (H^i_α) was found in the phase diagram of a mixture of sodium bis-2-ethylhexyl sulfosuccinate and water [11] (see Fig. 7.3). Uniaxial nematic phases were found in binary mixtures of Cs perfluoroctanoate and D_2O [12]. A mixture of pentaethylenglychol n-dodecyl ether (known as $C_{12}E_5$) and water [13] presents lamellar (L_α), direct (L_1) and inverted (L_2) micellar, sponge (L_3), hexagonal (H_α), and cubic (Q) phases, as a function of temperature and relative concentrations of the compounds (see Fig. 3.8). A mixture of octaethylene glycol dodecyl ether ($C_{12}EO_8$) and water [14] presents a bicontinuous $Ia\bar{3}d$ cubic phase at about 24°C.

A mixture of sodium dodecylsulfate (SDS) and water presents one of the richest phase diagrams associated with a lyotropic liquid crystal (see Fig. 3.9) [15,16]. One-dimensional (lamellar L_α), two-dimensional (hexagonal H_α and monoclinic M_α) and three-dimensional (tetragonal T_α, cubic Q, and rhombohedral Rh) phases are present in this phase diagram. Between the lamellar and the direct hexagonal phase structures, the amphiphilic aggregates change their form and spacial arrangement, giving rise to intermediate phases. If we look at decreasing SDS concentrations, from the L_α phase structure, with flat lamellae, there are periodic corrugations of the bilayers, leading to the formation of rods, of ribbon type, organized in tetragonal, cubic and rhombohedral phases. In the

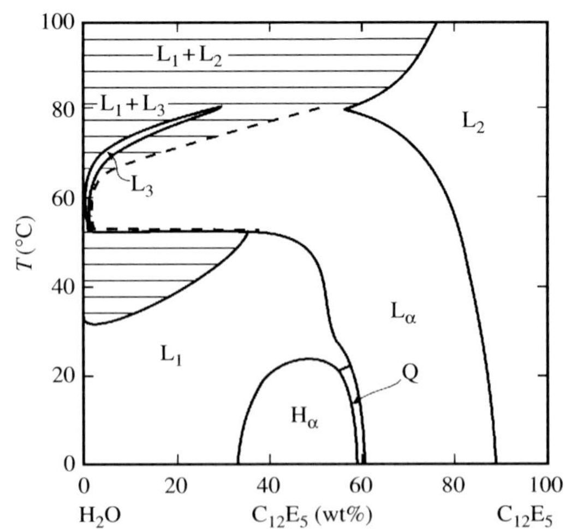

FIG. 3.8. Phase diagram of the binary $C_{12}E_5$/water mixture [13]. Labels refer to lamellar (L_α), direct and inverted micellar (L_1 and L_2), sponge (L_3), hexagonal (H_α), and cubic (Q) phases.

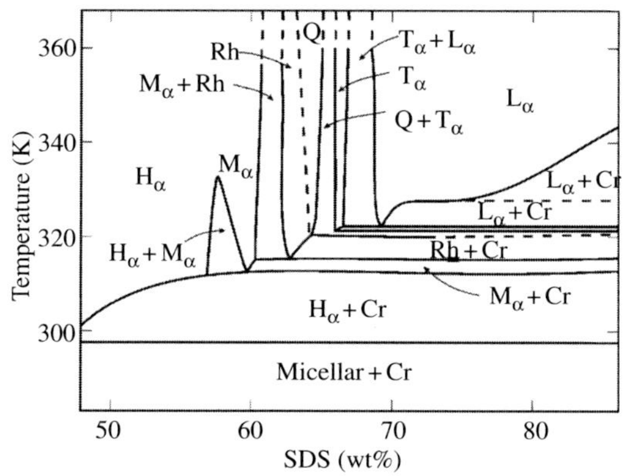

FIG. 3.9. Phase diagram of a mixture of sodium dodecylsulfate (SDS) and water [16]. Labels refer to one-dimensional lamellar (L_α), two-dimensional hexagonal (H_α), and monoclinic (M_α), three-dimensional tetragonal (T_α), cubic (Q) and rhombohedral (Rh) phases; Cr represents the SDS crystalline hydrates.

region of low SDS concentration, cylinders have a circular cross section, and the hexagonal phase is stabilized.

There are some lyotropic mixtures with more than two components, in particular with salts, which are called *pseudo-binary* mixtures, since the third component is not an amphiphile or a solvent [17–19]. A mixture of salt and water, named *brine*, is commonly used in order to prepare lyotropic liquid crystals. Mixtures of potassium laurate (known as KL, $COOK(CH_2)_{10}CH_3$), KCl and water; Cs decylsulfate, $CsNO_3$ and water; and decylammonium chloride, NH_4Cl and water, are examples of this type of system. At particular values of temperature and relative concentrations, these mixtures also present uniaxial nematic phases. However, no biaxial nematic phases were ever observed in these mixtures. A mixture of cetylpyridinium chloride (known as CPCl), hexanol ($C_6H_{14}O$) and brine (0.2 M of NaCl) [20] shows a radial-type topology of the L_1, L_α and L_3 phases, with the transition lines converging towards the vertices of the triangular isothermal section of the phase diagram, with small CPCl and hexanol concentrations. Besides L_α and L_3 phases, the AOT/brine (water with NaCl) mixture [21] also presents a cubic bicontinuous phase, Q. There is a coexistence region ($Q + L_3$) separating the sponge from the cubic phase.

3.3.2 *Phase diagrams of ternary lyotropic mixtures*

Although the polymorphism of binary lyotropic mixtures is already very rich, the addition of a third component (a solvent or another amphiphile, which is called *cosurfactant*) introduces new phases and topologies. Phase diagrams of these systems are usually represented in terms of the equilateral triangles of concentrations (partial isotherms), as described in the introductory section of this chapter. Also, we may use particular surfaces for constructing alternative representations of the four-dimensional phase diagram of ternary mixtures. For example, we may fix the molar (or weight) percentage of one of the components (usually the cosurfactant). In this *partial iso-concentration* representation, we draw a rectangular plot, in which temperature and concentration of the main amphiphile are the two coordinates.

We now consider three examples of ternary mixtures: (a) Na octylsulfate [$CH_3(CH_2)_7SO_3Na$], 1-decanol [$CH_3(CH_2)_9OH$, known as DeOH] and water [22], (b) K laurate [$COOK(CH_2)_{10}CH_3$], 1-decanol and water [23], and (c) Na decylsulfate [$CH_3(CH_2)_9OSO_2$ ONa, known as SdS], 1-decanol and water [24].

In Fig. 3.10, we sketch a partial isotherm, at 20°C, of a mixture of Na octylsulfate, 1-decanol and water. The role of the alcohol in the "skeleton" formed by the main amphiphile consists in reducing the interfacial curvature of the surfaces. In the middle of the triangle (for DeOH and water concentrations of about 40 and 50 wt%, respectively), there is a large region of a lamellar L_α phase, extending towards larger concentrations of water. The H_α phase region is located at concentrations of about 60 and 10 wt% of Na octylsulfate and DeOH, respectively. Direct and inverse micellar isotropic phases are present at relative concentrations of about 8 wt% DeOH and 80 wt% water, in the case of the L_1 phase, and 80 wt% DeOH and 10 wt% water, for the L_2 phase. Other lyotropic phases may appear if we change temperature and relative concentrations of the different components of the mixture. In general, the same features are present in most of the partial isotherms of ternary mixtures composed by a soap, or detergent,

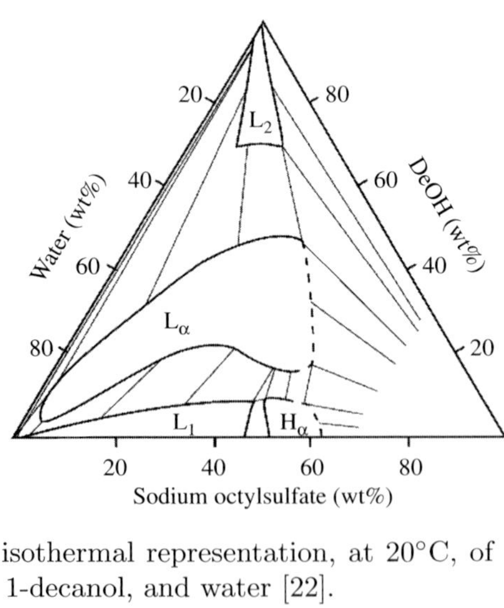

FIG. 3.10. Partial isothermal representation, at 20°C, of a ternary mixture of Na octylsulfate, 1-decanol, and water [22].

an alcohol and water. If the main amphiphile is a molecule whose topology favors the formation of inverted structures, as aerosol OT (di-2-ethylhexyl Na sulfosuccinate), there may appear lower-symmetry inverted phases, as the reversed hexagonal structure.

A K laurate/1-decanol/water mixture is one of the most interesting and studied lyotropic system. This system was used by Yu and Saupe [23] to show the existence of a biaxial nematic phase (N_B) located between the known uniaxial calamitic (N_C) and discotic (N_D) phases. The phase diagram, at partial iso-concentration of 6.24 wt% DeOH has already been sketched and discussed in Chapter 1. A similar topology has been observed in other lyotropic mixtures with a surfactant, a cosurfactant and water, as the SdS/DeOH/water mixture, for example [25]). The range of existence of the N_B phase is strongly dependent on the ratio between the molar concentrations of surfactant and cosurfactant. It is important to remark that the only observations of a biaxial nematic phase have been reported in mixtures of this nature, with at least two amphiphiles. Micellar (direct) isotropic phases can be present at higher or even lower temperatures, with respect to the nematic domains, in partial iso-concentration diagrams of ternary lyotropic mixtures. The low-temperature isotropic phase is explained in terms of modifications of the shape anisotropy of micelles, which tend to increase the thickness of the double layer as the temperature decreases. This is a consequence of the reduction of the probability of *gauche* configurations of the alkyl chains for decreasing temperatures. Usually, the nematic domain in the partial isotherm of the phase diagram of ternary mixtures is very small. In other partial iso-concentration diagrams, we can also find different phases with lower symmetries.

The partial isotherm of the SdS/DeOH/water mixture, at 23°C, with the localization of a nematic domain, is sketched in Fig. 3.11. This nematic region is located at relative concentrations of SdS/DeOH/water of about 40/10/50 wt% (see Fig. 3.11(a); see the zoom view sketched in Fig. 3.11(b)). This region is limited by the following phases: L_α, at the upper right side; R_α and H_α, at the lower right side; L_1 at the lower left side. There are large coexistence domains between the nematic phases and their neighboring phases. In many cases, there is a coexisting isotropic micellar phase. As we pointed out before, the increase of DeOH concentrations favors flatter surfaces of the amphiphilic structure. This explains the $H_\alpha \rightarrow R_\alpha \rightarrow L_\alpha$ phase sequence along the line at a fixed concentration of SdS, of about 0.45 wt%, as we decrease the concentration of water.

In analogy with the case of pseudo-binary mixtures, there are also *pseudo-ternary* mixtures, in which a salt is added to the original ternary mixture. A mixture of Na dodecylsulfate (known as SDS), dodecane ($C_{12}H_{26}$), pentanol ($C_5H_{12}O$) and a brine (water with NaCl) [26] displays a L_3 phase in the oil and water rich domains of the phase diagram. A mixture of SDS, pentanol and brine (water NaCl) [27,28] displays an interesting topology on a surface of the phase diagram at constant temperature and NaCl concentration. Besides regions of L_α phase, sponge phase regions have also been observed, with a second-order

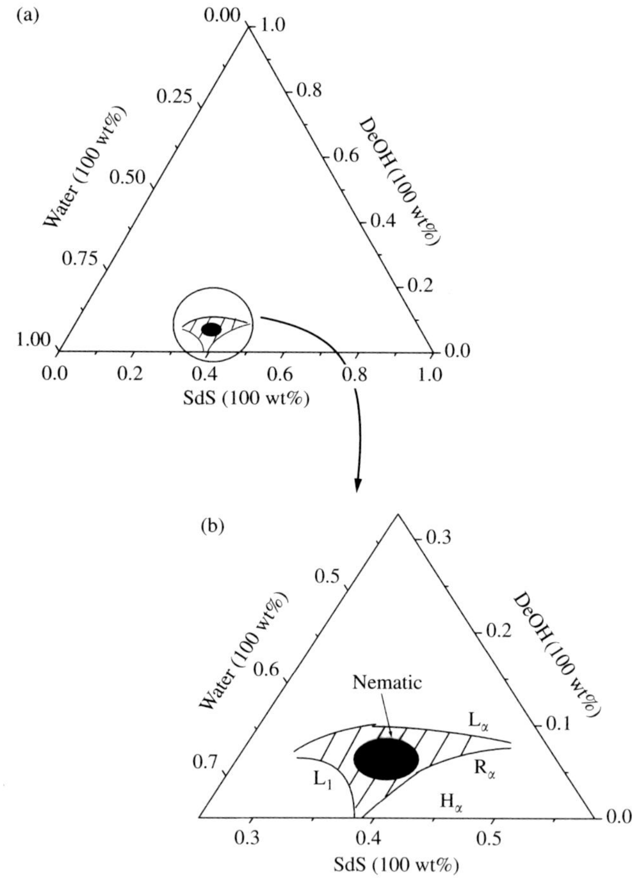

FIG. 3.11. Sketch of the partial isothermal representation, at 23°C, of the SdS/DeOH/water mixture [24]: (a) localization of the nematic domain; (b) zoom of the nematic domain and its neighboring phases (micellar isotropic, L_1, hexagonal, H_α, rectangular, R_α, and lamellar, L_α).

symmetric–asymmetric phase transition. This concept of symmetric state is related to the sponge phase, in which case the inside and the outside volumes occupied by the solvent, and separated by the amphiphilic bicontinuous structure, are identical; the asymmetric state is the situation in which this symmetry is broken. Regions of two and three coexisting phases are also present in this phase diagram.

Pseudo-ternary mixtures, with at least one chiral component, give rise to lyotropic cholesteric phases (Ch_C or Ch_D). As examples, we mention mixtures of KL, KCl, water, and cholesterol ($C_{27}H_{45}OH$) [29], and di-sodium N-lauroyl aspartate, NH_4Cl, water, and DeOH [30].

3.3.3 *Phase diagrams of quaternary lyotropic mixtures*

One of the striking features of lyotropics is the possibility of mixing a large number of components. In most of these cases, the topology of the phase diagrams changes with each new addition. Let us look at the phase diagrams of some quaternary mixtures. A complete description of these systems with $r = 4$ components requires the use of five independent thermodynamic variables. Therefore, we are forced to introduce several types of constraints in order to draw two-dimensional representations of the phase diagrams.

In Fig. 3.12(a), we sketch a surface of the phase diagram of a mixture of SDS, 1-pentanol (known as POH), cyclohexane (C_6H_{12}, known as C_6) and water [31]. There are large L_α and isotropic domains. A triangular representation is sketched in Fig. 3.12(b) [32]. Besides L_α and isotropic phases, of micellar or microemulsion character, there is a hexagonal phase in the left corner of the triangle (at small pentanol concentrations, not shown in Fig. 3.12(b)). The phase sequence hexagonal $\rightarrow$ lamellar, for increasing concentrations of pentanol, has already been observed in other lyotropic mixtures, which reinforces the proposal that the presence of alcohol reduces the interfacial curvature of the surfaces. Another interesting surface of this phase diagram, at 21.5°C and at a molar ratio 0.17 between the partial molar concentrations of SDS and pentanol, is sketched in Fig. 3.12(c) [33]. In this representation, at a molar concentration [SDS+POH] $\sim$ 7.5 M%, there is an almost linear domain of the L_3 phase. At [SDS + POH] $\sim$ 15 M%, there is an extensive inverted micellar isotropic phase, L_2, parallel to the sponge domain. Near the left vertex of the triangular representation, there are several two-phase coexistence regions, always involving an isotropic micellar phase and either L_3, a microemulsion, or a (thermodynamically unstable) emulsion phase. In the left corner of the triangle, in the neighboring region of the L_3 phase, there is coexistence between a microemulsion and a micellar isotropic phase.

Another interesting quaternary system is a mixture of SDS, dodecane, pentanol, and water [34,35]. A particular partial isotherm, at room temperature and at [water wt%] /[SDS wt%] = 1.2, is sketched in Fig. 3.13. Again, the sides of the triangle represent POH, with increasing weight percentage concentrations from the bottom to the top of the figure. In the poor-dodecane region, there is a phase sequence hexagonal $\rightarrow$ lamellar $\rightarrow$ L_1 for increasing amounts of POH. A small domain of the L_3 phase is found in the dodecane-rich vertex of the triangle.

If one of the components of the quaternary mixture is an optically active molecule, there may exist cholesteric phases in the phase diagram. Examples of particular surfaces of these phase diagrams are given in Fig. 3.14(a), for KL/DeOH/water/brucine sulfate heptahydrate ($C_{46}H_{68}N_4O_{19}S$, known as BS) [36], and in Fig. 3.14(b), for KL/DeOH/water/L-*N*-lauroyl potassium alaninate (known as l-LAK) [37]. There are cholesteric phases, Ch_C, Ch_D, and Ch_B, in these surface phase diagrams. Cholesteric phases

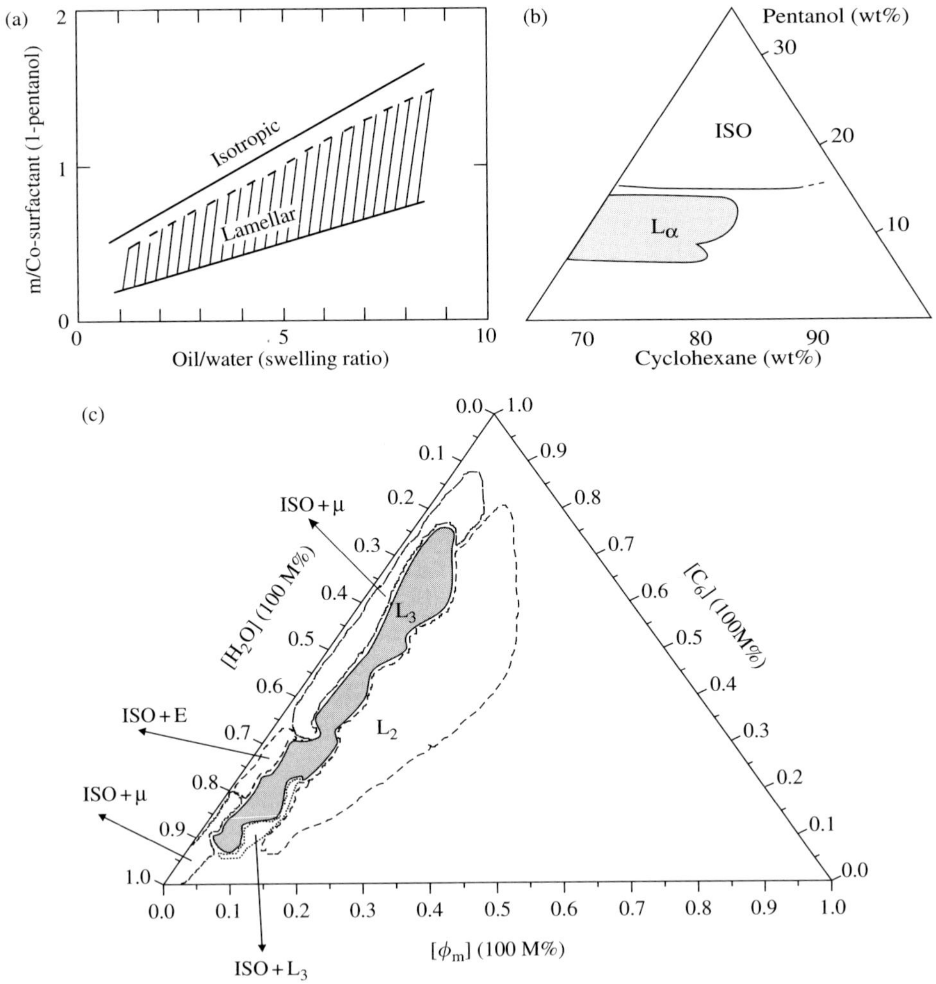

FIG. 3.12. Sketch of a particular surface of the phase diagram of a quaternary mixture of SDS, 1-pentanol (POH), cyclohexane and water: (a) at room temperature, the water volume is fixed, and the volume ratio of water/SDS is 2.5 [31]; (b) the volume ratio of water/SDS is fixed at 2.5 [32]. The sides of the triangle represent the pentanol concentration, which increases from the bottom towards the upper vertex; L_α and ISO stand for the lamellar and isotropic phases, respectively; (c) representation at 21.5°C and molar ratio [SDS]/[POH] = 0.17. The labels Iso, E, μ, L_2 and L_3 refer to isotropic, emulsion, micro-emulsion, inverted micellar, and sponge phases, respectively; ϕ_m = [SDS + POH] is the molar concentration of SDS and POH.

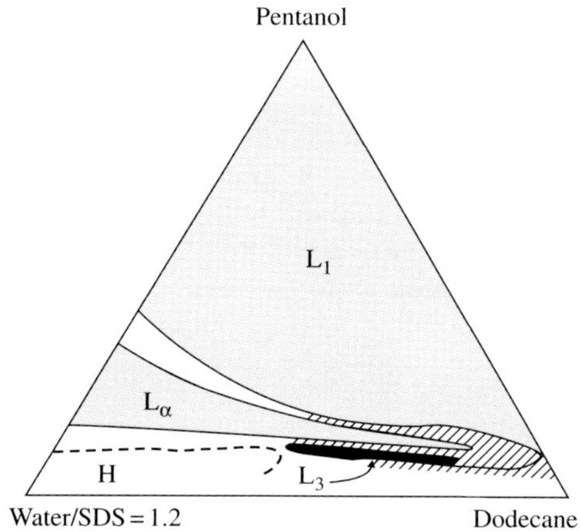

FIG. 3.13. Sketch of a particular partial isotherm, at room temperature and [water wt%] /[SDS wt%] = 1.2, of a mixture of SDS, dodecane, pentanol, and water [35].

were also found in *pseudo-quaternary* phase diagrams of mixtures of SdS, Na_2SO_4, water, cholesterol, and glycine (NH_2CH_2COOH), and cesium decylsulfate, Na_2SO_4, water, tartaric acid [HOOCCH (OH)CH(OH)COOH], and glycine [29].

3.3.4 *Specific features of the topology of phase diagrams of lyotropic mixtures*

The analysis of the different phase diagrams described in the previous sections, together with the examples of lyotropic mixtures displaying specific phase sequences, lead to some interesting conclusions about the polymorphism in these systems.

In Chapter 1 we described almost 30 different phases that may be displayed by lyotropic mixtures. Almost all of these structures are already observed in *binary mixtures.* However, there are some particular phases that have been observed in mixtures with a surfactant and a cosurfactant only. One of these exceptions is the *biaxial nematic* phase, which has been found in phase diagrams of some ternary mixtures only. The two-dimensional tetragonal phases have also been found in mixtures with at least three components. The cholesteric phases require the addition of a chiral compound to a previously nematic phase. They exist in pseudo-ternary and quaternary mixtures.

Lyotropic phase diagrams are also characterized by the remarkable possibility of reconstructive phase transitions between ordered structures [38,39]. In this kind of transition, molecular aggregates can drastically modify their geometry and organization in space, as a function of relative concentrations of the different

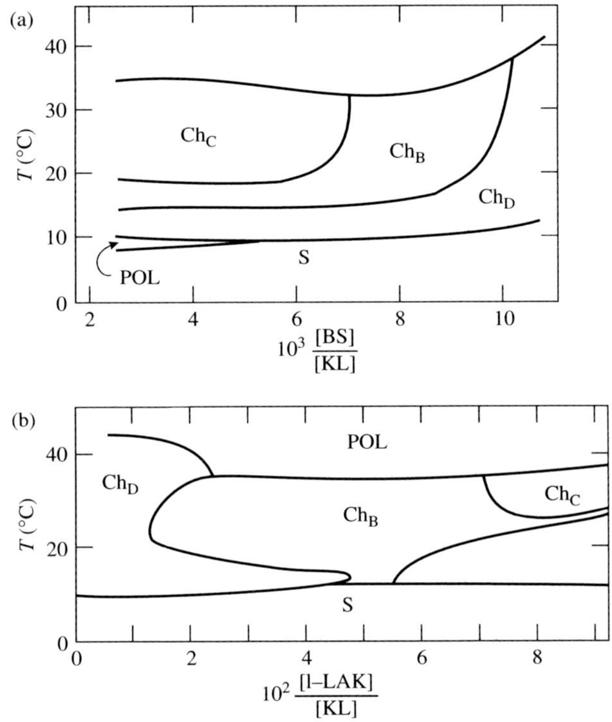

FIG. 3.14. Sketch of particular surfaces of phase diagrams of quaternary mixtures of cholesteric systems: (a) KL/DeOH/water/BS; the number of moles of KL, water and DeOH are fixed; POL is a polyphase and S is a gel region. The horizontal axis represents the relative molar ratio of BS with respect to KL [36]; (b) KL/DeOH/water/L-LAK; the number of moles of KL, water and DeOH is fixed [37]. The horizontal axis represents the relative molar ratio of L-*N*-lauroyl potassium alaninate with respect to KL.

compounds of the mixture, temperature and pressure. These modifications of shape of the aggregates may lead to strong first-order phase transitions, with large phase coexistence domains between isolated phases. Another important feature of the phase diagrams of lyotropic liquid crystals is the stabilization of inverted (or reverse) phases [10]. In the direct–inverted transition, there may occur an interchange of roles of the polar and nonpolar parts of the structure. For example, in the case of direct micellar phases, we can define the inner and outer parts of the micelles, and the nonpolar medium is located inside the micelles. In the case of inverted structures, the nonpolar medium is outside of the micelles. We may apply to lyotropics a symmetry operation, named *fluid-reversal symmetry* [39], which transforms a direct to an inverted phase.

3.4 Calculations for the phase diagrams of binary lyotropic mixtures

Consider a binary lyotropic mixture formed by a solvent (W) and a surfactant (S). Let us call $x_W = N_W/N$, and $x_S = N_S/N$, with $N = N_W + N_S$, the relative molar concentrations of solvent and surfactant, respectively, and define the dimensionless relative density

$$\rho = x_S - x_W. \tag{3.10}$$

Assuming that the volume and the total number of moles are fixed, the internal energy per mole can be written as

$$u = u(s, \rho), \tag{3.11}$$

where s is an entropy per mole. We then write

$$\mathrm{d}u = T\mathrm{d}s + \mu \mathrm{d}\rho, \tag{3.12}$$

where μ is a (relative) chemical potential associated with the density ρ.

It is now convenient to use a Legendre transformation in order to define a Helmholtz free energy per mole,

$$f = f(T, \rho) = u - Ts, \tag{3.13}$$

where

$$\mathrm{d}f = -s\mathrm{d}T + \mu \mathrm{d}\rho, \tag{3.14}$$

so that

$$\mu = \left(\frac{\partial f}{\partial \rho}\right)_T. \tag{3.15}$$

A second Legendre transformation leads to the grand thermodynamical potential per mole,

$$\varphi = \varphi(T, \mu) = f - \mu\rho. \tag{3.16}$$

At a given temperature, two distinct phases are in equilibrium if they have the same chemical potential. We then write the equilibrium condition

$$\left[\left(\frac{\partial f}{\partial \rho}\right)_T\right]_1 = \left[\left(\frac{\partial f}{\partial \rho}\right)_T\right]_2, \tag{3.17}$$

where the subscripts 1 and 2 refer to the distinct phases. In order to guarantee a continuous thermodynamic potential, we also require that

$$\varphi_1(T, \mu) = \varphi_2(T, \mu), \tag{3.18}$$

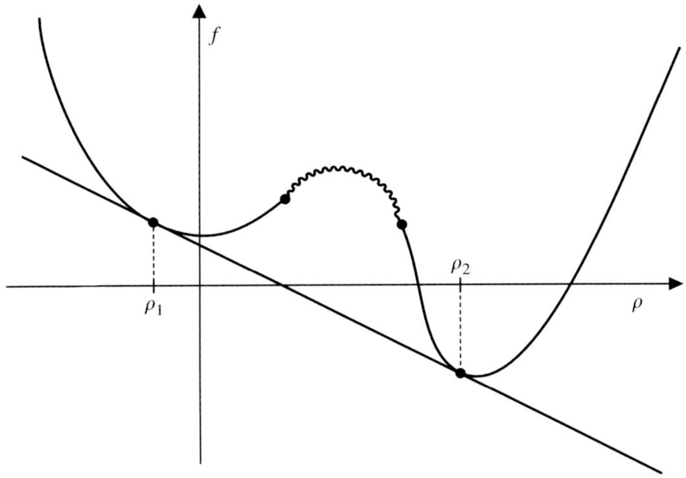

FIG. 3.15. Graph of the Helmholtz free energy per mole f versus the density ρ for sufficiently low temperatures. Note that convexity is recovered by the double-tangent construction. The curly line indicates the unstable solutions.

which is equivalent to

$$\left[f(T,\rho) - \left(\frac{\partial f}{\partial \rho}\right)_T \rho\right]_1 = \left[f(T,\rho) - \left(\frac{\partial f}{\partial \rho}\right)_T \rho\right]_2 . \tag{3.19}$$

If we draw a graph of f in terms of ρ, it is an easy exercise to see that these equations correspond to the well-known double-tangent construction (see the illustration in Fig. 3.15).

3.4.1 *Simple example of a binary phase diagram*

According to Tolédano, *et al.* [41], we now assume that $f(T,\rho)$ can be written as an expansion in powers of the dimensionless relative density ρ,

$$f(T,\rho) = \sum_{n=2}^{p} \frac{1}{n} a_n \rho^n, \tag{3.20}$$

which is very much reminiscent of the Landau expansions of Chapter 2 (since ρ can be associated with the order parameter). Also, for a simple binary system, and still being quite general, we assume that $a_2 = a = T - T_0$, $a_3 = -1$, $a_4 = 1$, and $a_n = 0$ for $n \geq 5$. Thus, we have

$$f(T,\rho) = \frac{1}{2}(T - T_0)\rho^2 - \frac{1}{3}\rho^3 + \frac{1}{4}\rho^4. \tag{3.21}$$

At sufficiently large temperatures, $f(T,\rho)$ is a convex function of ρ, with a minimum at $\rho = 0$, which prevents the possibility of coexistence between two phases (with different values of ρ). Looking at the second derivative of f,

$$\frac{\partial^2 f}{\partial \rho^2} = T - T_0 - 2\rho + 3\rho^2, \tag{3.22}$$

we see that there is a violation of convexity, $\partial^2 f/\partial\rho^2 < 0$, if $T - T_0 < 1/3$, and

$$\frac{1}{3} - \frac{1}{3}\sqrt{1 - 3(T - T_0)} \leq \rho \leq \frac{1}{3} + \frac{1}{3}\sqrt{1 - 3(T - T_0)}, \tag{3.23}$$

which leads to the existence of a critical point at $T - T_o = 1/3$, with $\rho_1 = \rho_2 = 1/3$, and which corresponds to $\partial^2 f/\partial\rho^2 = 0$.

The ρ–T phase diagram of Fig. 3.16 comes from the solutions of Eqs (3.17) and (3.19), which can be written as

$$a\rho_1 - \rho_1^2 + \rho_1^3 = a\rho_2 - \rho_2^2 + \rho_2^3, \tag{3.24}$$

and

$$-\frac{1}{2}a\rho_1^2 + \frac{2}{3}\rho_1^3 - \frac{3}{4}\rho_1^4 = -\frac{1}{2}a\rho_2^2 + \frac{2}{3}\rho_2^3 - \frac{3}{4}\rho_2^4, \tag{3.25}$$

where $a = T - T_0$. Discarding the trivial homogeneous (unstable) solutions,

$$\rho_1 = \rho_2 = \frac{1}{3}(1 - \sqrt{1 - 3a}), \tag{3.26}$$

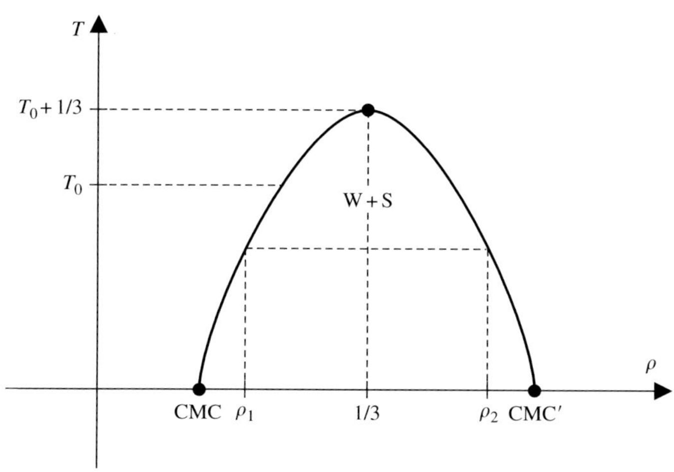

FIG. 3.16. Phase diagram, in terms of density ρ and temperature T, for the simple solvent–surfactant model; CMC and CMC′ indicate critical micellar concentrations for direct and inverted micelles.

it is easy to find a pair of physical non-trivial solutions,

$$\rho_1 = \frac{1}{3}\left[1 - \sqrt{3(1-3a)}\right] \quad \text{and} \quad \rho_2 = \frac{1}{3}\left[1 + \sqrt{3(1-3a)}\right]. \tag{3.27}$$

In Fig. 3.15, we give an illustration of the double-tangent construction. Besides indicating the physical solutions ρ_1 and ρ_2, we also indicate the region of instability. The simple phase diagram in Fig. 3.16 can be trivially obtained from these solutions. Also, it is not difficult to consider a negative term $-a_3 \neq -1$. We then have the exact solutions

$$\rho_{1,2} = \frac{1}{3}\left[a_3 \pm \sqrt{3(a_3^2 - 3a)}\right]. \tag{3.28}$$

3.4.2 *Additional examples*

A phase diagram with a eutectic point (Fig. 3.17) can be obtained from the expansion

$$f(T,\rho) = \frac{1}{2}(T - T_{\mathrm{E}})(\rho - \rho_{\mathrm{E}})^2 + \frac{1}{3}a_3(\rho - \rho_{\mathrm{E}})^3 + \frac{1}{4}a_4(\rho - \rho_{\mathrm{E}})^4, \tag{3.29}$$

where T_E is the eutectic temperature, and the coefficients a_3 and $a_4 > 0$ should be chosen to fit the slopes of $T(\rho)$ at the eutectic and the limit compositions. Using the same approach, with the inclusion of higher-order powers of the density ρ, we can draw much more complex phase diagrams.

Anisotropic liquid crystalline phases can be described by the inclusion of additional symmetry-breaking order parameters. For example, we can write a free energy of the form

$$\varphi(T,\mu) = -\mu\rho + f(T,\rho) + f_{\mathrm{L}}(T,\eta) + f_{\mathrm{GL}}(\rho,\eta), \tag{3.30}$$

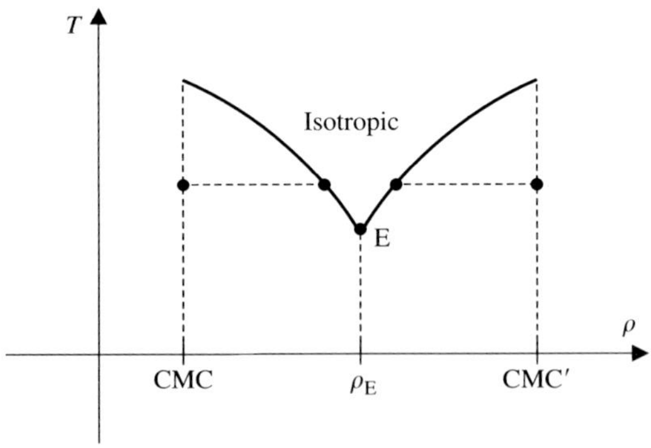

FIG. 3.17. Section of the phase diagram, in terms of temperature versus density, for a binary system with a eutectic point (E).

where

$$f_{\mathrm{L}}(T,\eta) = \frac{1}{2}(T - T_{\mathrm{c}})\eta^2 + \sum_{n=3}^{p'} \frac{1}{n}\alpha_n \eta^n, \tag{3.31}$$

and

$$f_{\mathrm{GL}}(\rho,\eta) = \delta_{12}\rho\eta^2 + \delta_{22}\rho^2\eta^2 + \cdots + \delta_{mn}\rho^m\eta^n. \tag{3.32}$$

In order to illustrate this procedure, consider the example

$$\varphi(T,\mu) = -\mu\rho + \frac{1}{2}(T - T_0)\rho^2 + \frac{1}{4}\rho^4 + \\ + \frac{1}{2}(T - T_{\mathrm{c}})\eta^2 + \frac{1}{4}\eta^4 + \delta_{12}\rho\eta^2, \tag{3.33}$$

in which the cubic term has been eliminated by a shift of the density ρ.

3.4.3 *An illustrative example: Phase diagram of a mixture of sodium laurate and water*

The phase diagram of a mixture of sodium laurate ($C_{12}H_{23}NaO_2$) and water, which is topologically similar to the previously described diagram of a mixture of potassium palmitate and water, is sketched in Fig. 3.18 [40]. The liquid crystalline phases are in a region between the crystalline-type phases (at low-temperatures) and the isotropic solution (at high temperatures). For smaller water concentrations, there are larger temperature domains of the liquid crystalline phase. The isotropic micellar phase is formed by the micelles in the solvent. At high temperatures, however, the isotropic phase is a mixture of isolated amphiphilic molecules and the solvent. In the low-temperature region, below the crystallization line (T_{c} in Fig. 3.18), there is a *curd* domain. At high concentrations of amphiphiles, of about 90 wt%, there appear *supercurd*, *waxy*, and *superwaxy* phases. The lamellar L_α and hexagonal H_α phases are located between the T_{c} line and the high-temperature isotropic phase. The H_α phase is located at about 50 wt% of water. The domain of the lamellar phase, which extends to higher temperatures, is stable at larger concentrations of amphiphile. The transitions between the different liquid crystalline phases and between them and the isotropic phase are characterized by large phase coexistence regions. In the H_α and L_α phases, the carbon chains are in a liquid-like state.

A phenomenological description of some of the main features of this phase diagram can be obtained as follows [41]. First, we note that the equilibrium transition line $\mathrm{T}(\rho)$ separating the isotropic micellar region from the ordered lyotropic phases displays two maxima and an intermediate eutectic-type minimum. Therefore, a power series expansion of ρ should include terms up to third order in $(\rho - \rho_{\mathrm{E}})$, where ρ_{E} is the eutectic density, and the expansion for the

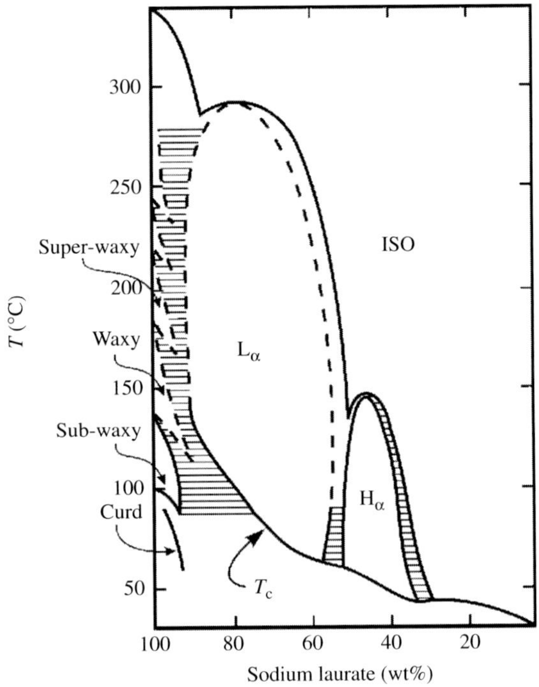

FIG. 3.18. Temperature versus concentration phase diagram of a mixture of sodium laurate and water [40].

free-energy density should be carried out up to the fifth-order term,

$$f_G(T,\rho) = \frac{1}{2}(T - T_E)(\rho - \rho_E)^2 + \frac{1}{3}a_3(\rho - \rho_E)^3 + \frac{1}{4}a_4(\rho - \rho_E)^4 + \frac{1}{5}a_5(\rho - \rho_E)^5. \tag{3.34}$$

From the double-tangent construction, we have $\partial T/\partial\rho = -a_3$, at $\rho = \rho_E$, which shows that the coefficient a_3 can be determined from the slope of $T(\rho)$. The effective values of the remaining coefficients come from the slopes of $T(\rho)$ at the minimum and maximum values of the density ρ.

A second step consists in choosing the symmetry-breaking parameters which are suitable for describing the transitions from the isotropic micellar to the L_α, H_α, and, eventually, to the R_α phases. We then make the following considerations:

(i) The isotropic–lamellar transition may be characterized by a pair of wave vectors, $\vec{k}_\pm$, with $|\vec{k}_\pm| = \pi/d$, where d is the period along the lamellae. In equilibrium, we should have $k_+ = k_- = \eta$, which is associated with the effective free energy

$$f_{\mathrm{L}_\alpha} = \frac{1}{2}(T - T_\mathrm{c})\eta^2 + \frac{1}{3}\alpha_3\eta^3 + \frac{1}{4}\alpha_4\eta^4. \qquad (3.35)$$

(ii) The two-dimensional cylindrical mesophases H_α and R_α require two pairs of wave vectors, $\pm\vec{k}_1$ and $\pm\vec{k}_2$, with $|\vec{k}_1| = |\vec{k}_2|$. Thus, if we call γ the angle between vectors $\vec{k}_1$ and $\vec{k}_2$, we have hexagonal ($\gamma = 120°$) or C-centered rectangular ($\gamma = 90°$) possibilities (space groups). In equilibrium, all the four components of the order parameter assume the same value (ζ for the H_α phase, and ζ' for the R_α phase). The corresponding free energy densities, f_{H_α} and f_{R_α}, are given by Eq. (3.35), with η and T_c replaced by ζ and $T_{\mathrm{c}1}$, and ζ' and $\mathrm{T}_{\mathrm{c}2}$, respectively. We then describe the phase diagram with the free energy

$$\varphi(T,\mu) = -\mu\rho + f_\mathrm{G}(T,\rho) + f_{\mathrm{L}_\alpha}(T,\eta) + f_{\mathrm{H}_\alpha}(T,\zeta) + f_{\mathrm{R}_\alpha}(T,\zeta')$$
$$+ \rho[\delta_1\eta^2 + \delta_2\zeta^2 + \delta_3\zeta'^2] + \mu_1\eta^2\zeta^2 + \mu_2\eta^2\zeta'^2 + \mu_3\zeta^2\zeta'^2. \qquad (3.36)$$

In Fig. 3.19, we sketch some diagrams which come from this approach if we assume different values for the free parameters. Of course, we are still far from a closer contact with the enormous variety of experimental possibilities.

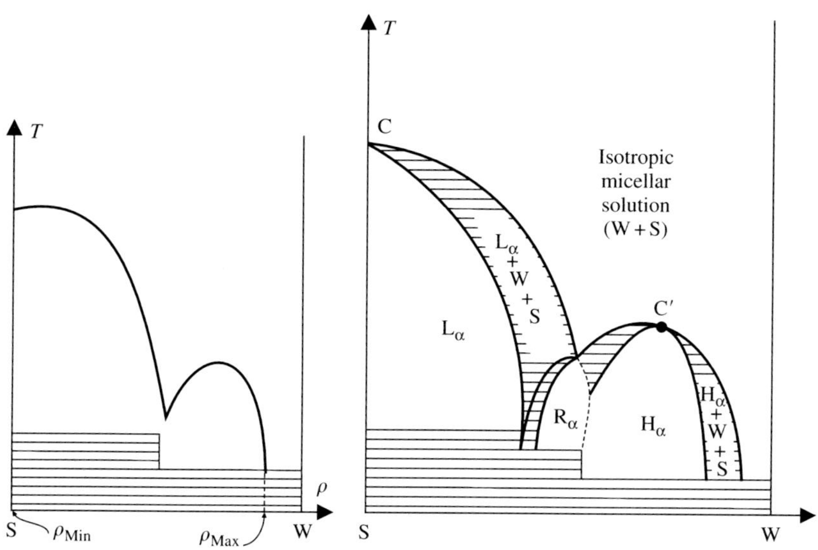

FIG. 3.19. Some theoretical predictions from a phenomenological approach to the description of the mixture of sodium laurate and water [41].

References

[1] W. N. Knisely and P. H. Keyes (1986). *Phys. Rev. A* **34**, 717.
[2] H. B. Callen (1985). *Thermodynamics and an Introduction to Thermostatistics*, 2nd edn, John Wiley, New York.
[3] J. W. McBain and W. C. Sierichs (1948). *J. Am. Oil Chem. Soc.* **25**, 221.
[4] A. E. Skoulios and V. Luzzati (1961). *Acta Crystallogr.* **14**, 278.
[5] V. Luzzati (1968). In *Biological Membranes*, Ed. D. Chapman, Academic Press, New York.
[6] B. J. Forrest and L. W. Reeves (1981). *Chem. Rev.* **1**, 1.
[7] J. Charvolin (1983). *J. Chim. Phys.* **80**, 15.
[8] V. Luzzati, A. Tardieu, and T. Gulik-Krzywicki (1968). *Nature* **217**, 1028.
[9] J. Burgoyne, M. C. Holmes, and G. J. T. Tiddy (1995). *J. Phys. Chem.* **99**, 6054.
[10] P. Ekwall (1975). In *Advances in Liquid Crystals*, Ed. G. H. Brown, Academic Press, New York, vol. 1.
[11] J. Rogers and P. A. Winsor (1969). *J. Colloid Interface Sci.* **30**, 247.
[12] N. Boden, P. H. Jackson, K. McMullen, and M. C. Holmes (1979). *Chem. Phys. Lett.* **65**, 476.
[13] R. Strey, R. Schomäcker, D. Roux, F. Nallet, and U. Olsson (1990). *J. Chem. Soc. Faraday Trans.* **86**, 2253.
[14] M. I. Clerc and A. M. Levelut (2001). *Eur. Phys. J. E* **4**, 209.
[15] P. Kékicheff and B. Cabane (1987). *J. Phys. (France)* **48**, 1571.
[16] P. Kékicheff and B. Cabane (1988). *Acta Crystallogr. B* **44**, 395.
[17] K. Radley, L. W. Reeves, and A. S. Tracey (1976). *J. Phys. Chem.* **80**, 174.
[18] F. Fujiwara, L. W. Reeves, M. Suzuki, and J. A. Vanin (1979). In *Solution Chemistry of Surfactants*, Ed. by K. L. Mittal, Plenum Press, New York, vol. 1.
[19] A. Saupe (1984). *Il Nuovo Cimento* **3D**, 16.
[20] G. Porte, R. Gomati, O. El Haitami, J. Appell, and J. Marignan (1986). *J. Phys. Chem.* **90**, 5746.
[21] B. Balinov, U. Olsson, and O. Söderman (1991). *J. Phys. Chem.* **95**, 5931.
[22] P. Ekwall, L. Mandell, and K. Fontell (1969). *Mol. Cryst. Liq. Cryst.* **8**, 157.
[23] L. J. Yu and A. Saupe (1980). *Phys. Rev. Lett.* **45**, 1000.
[24] Y. Hendrikx, J. Charvolin, M. Rawiso, L. Liébert, and M. C. Holmes (1983). *J. Phys. Chem.* **87**, 3991.
[25] R. Bartolino, T. Chiaranza, M. Meuti, and R. Compagnoni (1982). *Phys. Rev. A* **26**, 1116.
[26] D. Gazeau, A. M. Bellocq, D. Roux, and T. Zemb (1989). *Europhys. Lett.* **9**, 447.
[27] G. Guerin and A. M. Bellocq (1988). J. Phys. Chem. **92**, 2550.
[28] C. Coulon, D. Roux, and A. M. Bellocq (1991). *Phys. Rev. Lett.* **66**, 1709.
[29] M. R. Alcantara and J. A. Vanin (1984). *Mol. Cryst. Liq. Cryst.* **107**, 333.

[30] M. R. Alcantara, M. V. M. C. Melo, V. R. Paoli, and J. A. Vanin (1983). *Mol. Cryst. Liq. Cryst.* **90**, 335.

[31] J. M. di Meglio, M. Dvolaitzky, and C. Taupin (1987). In *Physics of Complex and Supermolecular Fluids*, Eds S. A. Safran and N. A. Clark, *Exxon Monograph Series*, John Wiley & Sons, New York, p. 361.

[32] C. Quilliet, V. Ponsinet, and V. Cabuil (1994). *J. Phys. Chem.* **98**, 3566.

[33] M. Magalhães, A. M. Figueiredo Neto, and P. Tolédano (2000). *Phys. Rev. E* **62**, 5847; M. Magalhães and A. M. Figueiredo Neto, and A. C. Tromba, *J. Phys. Chem. B*, in press.

[34] O. Diat and D. Roux (1995). *Langmuir* **11**, 1392

[35] C. A. Miller, M. Gradzielski, H. Hoffmann, U. Kramer, and C. Thanic (1991). *Prog. Colloid Polym. Sci.* **84**, 243.

[36] A. M. Figueiredo Neto and M. E. Marcondes Helene (1987). *J. Phys. Chem.* **91**, 1466.

[37] M. E. Marcondes Helene and A. M. Figueiredo Neto (1988). *Mol. Cryst. Liq. Cryst.* **162B**, 127.

[38] J. C. Tolédano and P. Tolédano (1987). *The Landau Theory of Phase Transitions*, World Sci. Lect. Notes in Physics, vol. 3, World Scientific, Singapore.

[39] P. Tolédano and B. Mettout (1998). In *Phase Transitions in Complex Fluids*, Eds P. Tolédano and A. M. Figueiredo Neto, World Scientific, Singapore.

[40] J. W. McBain and W. W. Lee (1943). *Oil Soap (Chicago)* **20**, 17.

[41] P. Tolédano, C. E. I. Carneiro, and A. M. Figueiredo Neto (2001) *Liq. Cryst.* **28**, 1547.

4

PHASE TRANSITIONS BETWEEN PERIODICALLY ORGANIZED LYOTROPIC PHASES

4.1 Introduction

We now discuss phase diagrams and phase transitions in lyotropic liquid crystals from the point of view of the symmetry transformations between periodically ordered mesophases. The presentation in this chapter is based on two articles by Mettout et al., "Symmetry breaking interface undulation for reconstructive phase transitions in complex fluids" [1], and "Theory of reconstructive phase transition in lyotropic complex fluids" [2], and an article by Tolédano and Mettout, "Symmetry aspects of reconstructive phase transitions in lyotropic complex fluids" [3].

One of the key features of lyotropics is the possibility of changing the structure and shape anisotropy of self-assembled amphiphilic aggregates, including bicontinuous structures, depending on external variables as temperature and relative concentrations of the components.

In some particular cases, amphiphilic aggregates exhibit discontinuous modifications of geometry, with a drastic reorganization of the aggregates, which leads to strong first-order phase transitions. These *reconstructive phase transitions*, commonly observed in atomic or molecular crystals, are characterized by the absence of a group–subgroup relationship between the newly formed and the initial structures. In lyotropic mixtures, there are examples of reconstructive phase transitions between lamellar, hexagonal, cubic, and sponge phases [4,5]. In other cases, amphiphilic aggregates (specially micelles) may undergo continuous modifications of their shape anisotropy. The equilibrium structures are related by a group–subgroup relationship, and the transitions may be of (weak) first or second order, as in the respective cases of nematic–isotropic and nematic–nematic transitions [6].

Second-order and weak first-order transitions are quite well described within the classical framework of the Landau theory of phase transformations. According to this approach, the phase of lowest symmetry is characterized by the onset of a few measurable quantities, which cancel in the phase of highest symmetry. These quantities are the order-parameter components and have well-defined transformation properties with respect to the parent-phase symmetry group. The order parameter may have either macroscopic or microscopic interpretations. For instance, at the transition between a square phase (formed by micellar-type cylindrical aggregates located at the nodes of a two-dimensional lattice) and a rectangular phase, in which the quadratic unit cell becomes slightly orthorhombic, the order parameter is a second-rank tensor. It can be interpreted as the (macroscopic) strain tensor of the square lattice. On the other hand, it can also be regarded as a geometrical parameter describing the (microscopic) orthorhombic deformation of a single cylindrical aggregate.

Due to the lack of a high-symmetry phase, the standard Landau approach is not enough for describing reconstructive phase transitions. A common ingredient of a number of treatments which have been proposed to deal with this problem is the key role attributed to the microscopic form of the amphiphile–solvent interface. For instance, in the random surface model [7], the statistical average of the position and form of the interface is driven by its surface tension and curvature energy. In the frustrated surface model [8], the geometry and the topology of the interface result from the compromise between the surface curvature and the inter-surface distance. In these treatments, there are predictions of several stable high-symmetry phases, although symmetry considerations do not play an important role.

In contrast to these treatments, there is the possibility of an extension of the Landau approach, as presented in this chapter, based on a symmetry analysis of the interfacial deformations, and overlooking the microscopic origin of the free energy terms. The order parameter consists of a small number of variables which account for the main features of the interface. In the simplest (scalar) cases, these variables describe the shape of the surface; in the general (vector or tensor) case, they can also take into account the internal structure. The changes of the order-parameter components with temperature and concentration describe the modifications of geometry, symmetry, and topology in lyotropic periodic systems. In addition, the characteristic self-assembling property of amphiphiles in solvents gives rise to remarkable structures. Depending on the external variables, the drastic change of curvature of an amphiphile–solvent interface inverts the structure completely. These structures are then called *reversed* or *inverted*. Transitions between direct and reversed phases are typically reconstructive phase transitions which are naturally explained in terms of this extension of the Landau theory for reconstructive phase transitions.

4.2 The lamellar–tetragonal transition

From a theoretical point of view, the lyotropic systems can be regarded as being composed of two regions, A and S, referring to the amphiphilic aggregates and the solvent, respectively. These regions are separated by characteristic interfaces. In the lamellar phase the interfaces are parallel planes; in the tetragonal phase they form a two-dimensional periodic lattice of almost cylindrical closed surfaces. Changing the external variables, as temperature and relative concentrations, there appear modifications in these interfaces, which trigger the transitions from the lamellar to the tetragonal structures. In the phase diagram, along the path from the lamellar to the tetragonal phases, the change in the form of the interface (see Fig. 4.1) is characterized by two main features :

1. A two-step modification of the symmetry groups. In the first step, the deformation of the lamellae breaks the continuous translational symmetry of the planar interface. In the second step, the formation of the cylinders increases the rotational symmetry of the system. Therefore, no group–subgroup relationship can be found between the two phases. In addition, there appears a third (intermediate) phase when the lamellar symmetry is broken and the tetragonal symmetry has not yet been established in the system;
2. A topological modification of the system which transforms from a bicontinuous lamellar structure (called *connected state*) to a structure of cylindrical aggregates (*disconnected state*).

In order to choose an order parameter for this transition, let us give a pictorial description of the interface deformation. Starting from the lamellar phase, the symmetry-breaking mechanisms, responsible for the onset of the transition, are represented by periodic undulations of the interfaces. In Fig. 4.1 we sketch a

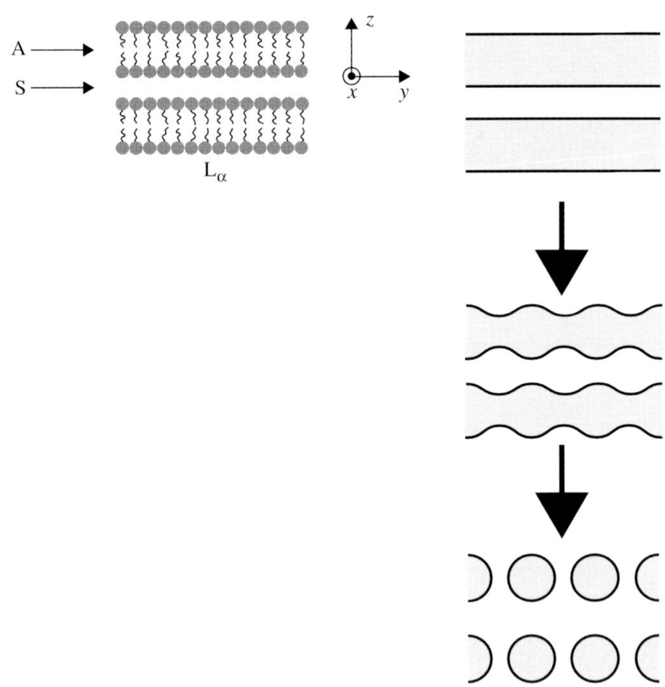

FIG. 4.1. Sketch of the lamellar–tetragonal phase transition generated by undulations of the lamellae in the x–y plane.

lamellar structure on the x–y plane and the undulations which generate the tetragonal phase [9–11]. For weak undulation amplitude, the continuous translations along the lamellae are broken, but the topology of the interface remains bicontinuous. At larger values of the amplitude, there is a merging of two adjacent lamellae, and the topology of the system becomes disconnected. At a certain large value of the amplitude, the symmetry of the disconnected aggregates changes to tetragonal.

This simple process can be easily formalized if we consider the equation of the interface on the x–y plane, $\Psi(x, y) = 0$, where Ψ is assumed to be positive in the A region and negative in the S region. The function Ψ can be expanded as a Fourier series,

$$\Psi(x, y) = \sum_{n,p=-\infty}^{+\infty} \Psi_{np} \exp\left[\frac{2\mathrm{i}\pi}{a}(nx + py)\right], \tag{4.1}$$

where n and p are integers, and a is the lattice parameter of the tetragonal phase structure. The four components of the order parameter associated with the transitions between the lamellar and tetragonal phases are the coefficients of the first harmonics of this Fourier series. In terms of Goldstone variables, θ_1 and θ_2, we write

$$\Psi_{10} = \rho_1 \exp(\mathrm{i}\theta_1), \quad \Psi_{01} = \rho_2 \exp(\mathrm{i}\theta_2), \quad \Psi_{\bar{1}0} = \Psi_{10}^*, \quad \Psi_{0\bar{1}} = \Psi_{01}^*, \tag{4.2}$$

where ρ_1 and ρ_2 represent the two effective components of the order parameter. The free energy of this system can be written as

$$F(\rho_1, \rho_2) = a_1 I_1 + a_2 I_1^2 + \cdots + b_1 I_2 + b_2 I_2^2 + \cdots + c_{12} I_1 I_2 + \cdots, \tag{4.3}$$

where I_1 and I_2 are the two invariants of the order parameter, given by

$$I_1 = \rho_1^2 + \rho_1^2 \quad \text{and} \quad I_2 = \rho_1^2 \rho_2^2, \tag{4.4}$$

and the phenomenological coefficients, a_1, a_2, b_1, b_2, and c_{12}, depend on temperature and concentration. As usual, the equations of state are obtained from

$$\frac{\partial F}{\partial \rho_1} = 0 \quad \text{and} \quad \frac{\partial F}{\partial \rho_2} = 0. \tag{4.5}$$

The stability regions of the phases associated with the solutions of these equations are given by the conditions

$$\frac{\partial^2 F}{\partial \rho_1^2} \geq 0 \quad \text{and} \quad \frac{\partial^2 F}{\partial \rho_1^2}\frac{\partial^2 F}{\partial \rho_2^2} - \left(\frac{\partial^2 F}{\partial \rho_1 \partial \rho_2}\right)^2 \geq 0. \tag{4.6}$$

A typical phase diagram [12], obtained from the minimization of the free energy (4.3) with respect to ρ_1 and ρ_2, for $c_{12} < 0$, is shown in Fig. 4.2. The lamellar phase, L, is stable for $\rho_1 \neq 0, \rho_2 = 0$ (or $\rho_1 = 0, \rho_2 \neq 0$); the tetragonal phase, C, is stable for $\rho_1 = \rho_2 \neq 0$; the intermediate phases, which we call IP,

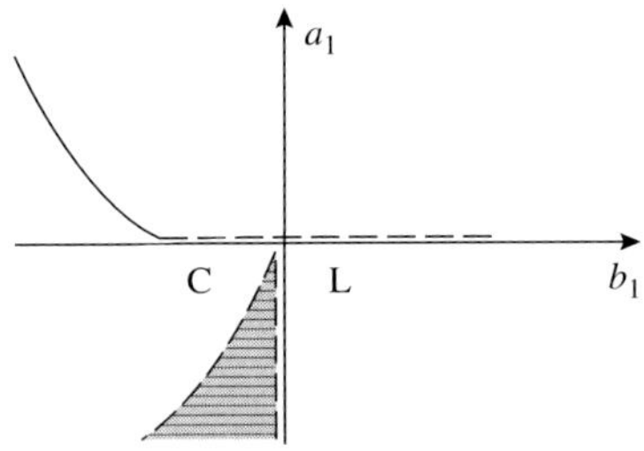

FIG. 4.2. Typical phase diagram [12] obtained from the minimization of Eq. (4.3) with respect to ρ_1 and ρ_2; L and C are lamellar and tetragonal phases. The locus of the intermediate phases corresponds to the region between dashed lines. Dashed and full lines represent second- and first-order transitions, respectively.

corresponds to $\rho_1 \neq 0$, and $\rho_2 \neq 0$ (region between dashed lines in Fig. 4.2). Dashed and full lines in this figure correspond to second- and first-order phase transitions, respectively. The lamellar–tetragonal phase transition takes place either across the intermediate phase, through two second-order transitions, or across a strong first-order boundary, depending on the sign of c_{12}.

The main symmetry and topological aspects of the interface are well described by Eq. (4.1), even if we take into account first harmonic components only. In this approximation, the equation of the interface is given by

$$\Psi(x, y) = \Psi_{00} + \rho_1 \cos\left(\frac{2\pi x}{a}\right) + \rho_2 \cos\left(\frac{2\pi y}{a}\right) = 0, \tag{4.7}$$

which describes the interface in the stable domain defined by the conditions $\theta_1 = (\theta_2, \pi)$. For small values of ρ_1 and ρ_2, the interface disappears and the system contains a single region, either A or S, according to the sign of Ψ_{00}. This simple approach, however, cannot account for the stability of this region and the corresponding phase transitions. Indeed, for $a_1 > 0$, the phase diagram of Fig. 4.2 is non-physical. In general, the interpretation of this abstract phase diagram in terms of physical quantities requires the establishment of relationships between the order parameter and the concentration. The following sections are devoted to this question.

4.2.1 *The effective thermodynamic potential*

We now assume that the temperature of the system is fixed. Changes of relative concentrations of the amphiphilic molecules and the solvent are then responsible for the phase transitions. In the previous approach, the concentration c of the amphiphile depends on the order parameter components ρ_1 and ρ_2. Thus, a term of the form $\mu c(\rho_1, \rho_2)$, where μ is related to the chemical potential of oil in water, and $c = c(\rho_1, \rho_2)$, cannot be neglected in the expression of the free energy to be minimized, even at fixed concentration, where μ plays the role of a Lagrange multiplier.

An expression for the concentration c can be obtained from a calculation, on the x–y plane, of the ratio between the area of the amphiphilic region, S_{A}, and the total area of the domains, S_{t}. Thus, we have

$$c = \frac{S_{\mathrm{A}}}{S_{\mathrm{t}}} = \left(\frac{2}{a}\right)^2 \int_0^{a/2} x(y)\mathrm{d}y, \tag{4.8}$$

where $x(y)$ comes from Eq. (4.7),

$$x(y) = \frac{a}{2\pi} \arccos\left\{-\frac{1}{\rho_1}\left[1 + \rho_2 \cos\left(\frac{2\pi y}{a}\right)\right]\right\}. \tag{4.9}$$

We then have

$$c(\rho_1, \rho_2) = 1 - \frac{1}{\pi^2} \int_{-1}^{z_0} \arccos\left\{\frac{(1 + \rho_2 z}{\rho_1}\right\} (1 - z^2)^{-1/2}\,\mathrm{d}z, \tag{4.10}$$

with

$$z = \cos\left(\frac{2\pi y}{a}\right) \quad \text{and} \quad z_0 = \frac{\rho_1 - 1}{\rho_2}, \tag{4.11}$$

if $\rho_1 - \rho_2 > 1$, whereas $z_0 = 1$, if $\rho_1 - \rho_2 < 1$.

The thermodynamic potential of this system can be finally written as

$$\Phi(\rho_1, \rho_2) = F(\rho_1, \rho_2) - \tilde{c}(\rho_1, \rho_2)(\mu - \mu_0), \tag{4.12}$$

where $\tilde{c} = c - 1/2$ is the reduced concentration, μ is the exchange chemical potential of regions A and S in the solution [13], and μ_0 is the exchange chemical potential of the independent fluids at the same temperature and densities.

4.3 Phase transitions between direct and reversed mesophases

The formalism introduced in the last section, on the basis of an order parameter derived from the equation describing the form of the interface, provides a natural description of the transitions between direct and reversed mesophases. If we change the sign of Ψ_{00} for identical values of ρ_1 and ρ_2 in Eq. 4.7, there is an interchange of regions A and S; in other words, there is a transformation from a direct to an inverted phase. Indeed, we can define a symmetry operation, F_{r}, called *fluid-reversal symmetry*, which transforms a direct mesophase into its inverted analog,

$$F_{\mathrm{r}}(\Psi_{00}, \rho_1, \rho_2) = (-\Psi_{00}, \rho_1, \rho_2). \tag{4.13}$$

We now discuss the effects of F_{r} on the thermodynamical potential Φ, given by Eq. (4.12). It is more convenient to express the order-parameter components in terms of spherical coordinates,

$$\rho_1/\Psi_{00} = \tan(\theta)\cos(\varphi) \quad \text{and} \quad \rho_2/\Psi_{00} = \tan(\theta)\sin(\varphi). \tag{4.14}$$

The action of F_r on the angles θ and φ, which fully determine the shape of the interface, is given by

$$F_r(\theta, \varphi) \to (\pi - \theta, \pi + \varphi). \tag{4.15}$$

Assuming that the free energy is invariant under F_r, we can construct two independent invariants,

$$I_1'(\theta, \varphi) = \cos^2\theta \quad \text{and} \quad I_2'(\theta, \varphi) = \sin^2 2\varphi, \tag{4.16}$$

which lead to the interfacial free energy

$$\begin{aligned} F(\theta, \varphi) = {} & a_1 \cos^2 \theta + a_2 \cos^4 \theta + \cdots + b_1 \sin^2 2\varphi + \cdots \\ & + c_{12} \cos^2 \theta \sin^2 2\varphi + \cdots \end{aligned} \tag{4.17}$$

If the free energy is not invariant under F_r, I_1' has to be replaced by $\cos\theta$, and F contains odd powers of $\cos\theta$. In real space, since F_r exchanges the volume of the amphiphilic part of the system into the volume of the solvent, $F_r(V_A) = V_S$, the concentration is not invariant under the fluid reversal symmetry,

$$F_r(c) = 1 - c, \tag{4.18}$$

and the reduced concentration $\tilde{c}$ is antisymmetric with respect to F_r,

$$F_r(\tilde{c}) = F_r(c) - \frac{1}{2} - c = -\tilde{c}. \tag{4.19}$$

Note that $\tilde{c}$ can actually be written in the general form

$$\tilde{c} = G(I_1', I_2') \cos\theta, \tag{4.20}$$

where G is an integral function depending on I_1' and I_2' only. Thus, in spherical coordinates, the thermodynamic potential can be expressed as

$$\Phi(\theta, \varphi) = F(I_1', I_2') - (\mu - \mu_0) G(I_1', I_2') \cos\theta. \tag{4.21}$$

4.3.1 *F_r-non-invariant systems*

In general, a lyotropic system is not invariant with respect to the fluid reversal symmetry, because the two sides of the interface are not equivalent (in contrast, for instance, with a bilayer membrane within a single fluid). This property yields the spontaneous curvature of an isolated membrane towards only one of the A or S regions. The minimization of the thermodynamic potential with respect to θ and φ gives the same stable states obtained from the minimization of Eq. (4.3). Each phase is determined by specific values of φ, whereas the values of θ distinguish between two configurations with different interfacial shapes within a single phase. Thus, we have L, with $\varphi = (0, \pi/2)$, C, with $\varphi = \pi/4$, and intermediate phases, with $\varphi \neq 0, \pi/4, \pi/2$.

It is possible to obtain stable phases with concentrations ranging from $c = 0$, for pure solvent, to $c = 1$, for the pure amphiphile ($\theta < \pi/2$ and $\theta > \pi/2$ yield $c < 1/2$ and $c > 1/2$, respectively). Each solvent-rich configuration, characterized

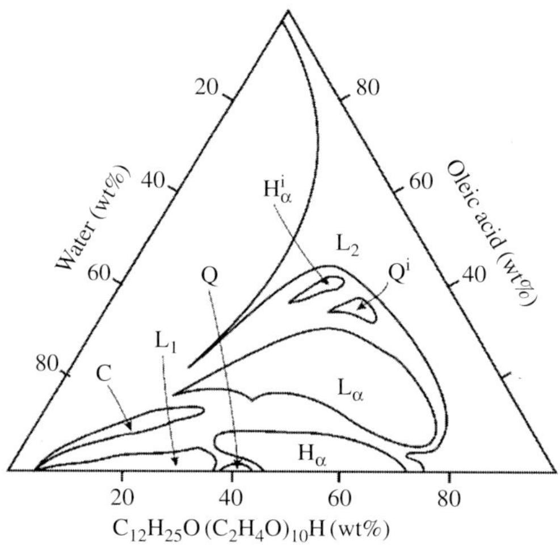

FIG. 4.3. Partial isothermal section, at 20°C, of the phase diagram of a mixture of decaoxyethylene glycol monolauryl ether, oleic acid, and water [14].

by $\theta < \pi/2$, is associated with its reversed amphiphile-rich image, $\theta > \pi/2$. However, these two configurations are energetically non-equivalent. In general, they will not be simultaneously present in the temperature–concentration phase diagram of this system. The absence of direct and reversed mesophases in the same phase diagram is the signature of a non-invariant fluid. However, there may exist exceptions in which the two phases are present. An illustrative example of a F_r-non-invariant system is the mixture of decaoxyethylene glycol monolauryl ether [$C_{12}H_{25}O(C_2H_4O)_{10}H$], oleic acid and water [14], in which there appear direct and reversed cubic phases, Q and Q^i, as shown in Fig. 4.3. The phase domains are located almost symmetrically with respect to the water concentration (of about 40 wt%). The non-invariant character is clear because the stability domains of the reversed and direct phases have different form (and area); also, at concentrations c and $1 - c$, direct and inverse interfaces have different forms.

4.3.2 *F_r-invariant systems*

Both sides of the interface of an invariant system are equivalent. In an infinite system with concentration $c = 1/2$, if an isolated interface is spontaneously curved, it can be indifferently curved towards the A or the S regions. In other words, if a flat interface undergoes a process of spontaneous curvature, the two possible curvatures (towards A or S) have the same energy and can happen with the same probability. Within this scheme, the direct and reversed mesophases correspond to two domains of the same equilibrium phase, transforming into each other by the application of F_r. In this case, the minimization of Φ, given by Eq. (4.21), with respect to θ and φ shows that, besides L, C, and IP, there are other phases, L^*, corresponding to $\theta = \pi/2$, $\phi = (0, \pi/2)$, and C^*, corresponding to $\theta = \pi/2$, $\phi = \pi/4$. In these "starred" phases, half of the symmetry operations come from F_r, so that their concentration is necessarily $c = 1/2$. In the μ–T phase diagrams, all the phases have extended domains of stability separated by second or first-order transition lines. In the corresponding c–T phase diagrams, the starred phases appear either on the $c = 1/2$ line or within two-phase regions at concentrations $c \neq 1/2$. These latter regions are associated with first-order transition borders in the μ–T diagram. In the phase diagram of F_r-invariant systems, the direct and reversed mesophases appear in distinct regions, symmetrically located with respect to $c = 1/2$. Since they correspond to the same equilibrium state, their phase boundaries are symmetric with respect to this value of the concentration.

We now give two illustrative examples of this type of system in ternary mixtures. The presence of a third component brings new ingredients to the problem. Note that we have been describing systems with only two types of molecules, so that there may be deviations from the predicted theoretical behavior if we add a third component. Consider the mixtures of sodium caprylate, decanol and water, and sodium caprylate, nonanol and water [4], shown in Fig. 4.4, with the presence of hexagonal direct (H) and reversed (H^i) mesophases around an alcohol concentration of about 40 wt%. Also, note that the regions limiting the H and H^i phases are almost symmetric with respect to the this concentration.

F_r can never be an exact symmetry. An F_r-invariant system is in fact defined by the weakness of the non-invariant terms (proportional to $\cos\theta$ in the interfacial

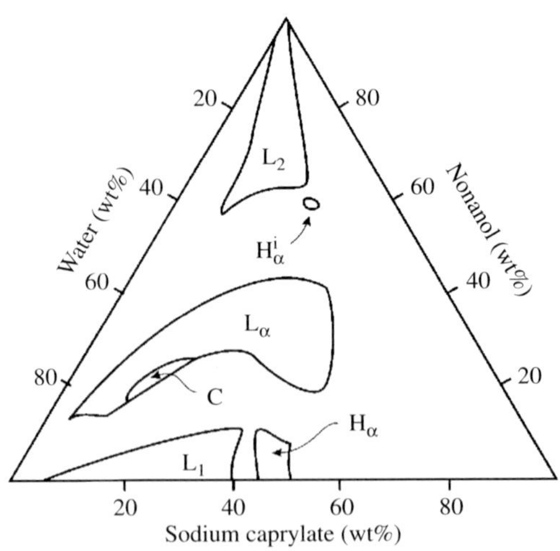

FIG. 4.4. Partial isothermal section, at 20°C, of the phase diagram of a mixture of sodium caprylate, nonanol, and water [14].

energy). Taking into account these small terms, we have to change the preceding considerations: (i) phase transitions between starred and non-starred phases must be replaced by fast but smooth variations of the interface configuration; (ii) at concentrations c and $1-c$, the direct and reversed mesophases have slightly different shapes; (iii) the phase diagram is not exactly symmetric with respect to the $c = 1/2$ axis.

In order to summarize this discussion, we point out that the existence of reversed mesophases in lyotropic fluids can be understood by considering the additional fluid-reversal symmetry operation F_r. We have predicted the existence of new types of mesophases, in which half of the symmetry elements are combined with F_r.

4.3.3 *Influence of the F_r-symmetry on some experimental phase diagrams*

We have been considering a lamellar–tetragonal transformation that corresponds to a realistic situation for a number of ternary systems [4], such as mixtures of sodium caprylate, water and decanol, sodium caprylate, water, caprylic acid and water, and decaoxyethylene glycol monolauryl ether, oleic acid and water. The L and C phases are separated by an undulated lamellar phase corresponding to the IP mesophase in our model. Analogously, the IP phase of the disconnected type, shown in Fig. 3.8, is similar to the rectangular phase reported in potassium oleate/water/p-xylene mixtures [4].

In Chapter 3, we have already given a general view of binary and ternary phase diagrams in lyotropic mixtures. We now check the effects of the F_r-symmetry on the topology of the phase diagrams reported in Ekwall's review [4]. There is no example of binary, amphiphilic/water, phase diagram with the coexistence of a direct ordered mesophase and its reverse analog. The experimental situation supports the proposal that a crossover from direct to reversed micellar ordered phases is considerably improved if a third suitable component is added to the amphiphilic/water system. However, note that direct, L_1, and reversed, L_2, micellar phases are found to coexist in a number of binary compounds such as Aerosol OT/water at 100°C, Emu 09/water at 20°C (Emu 09 is a commercial product, probably decaoxyethylene glycol monylphenol ether [4]), and decaoxyethilene glycol monolauryl ether/water at 20°C. We cannot exclude the coexistence of direct and reversed ordered mesophases in these compounds.

The analysis of the ternary phase diagrams reported by Ekwall leads to the characterization of three main classes of phase diagrams. In the first class, L_1 and L_2 micellar phases occupy equivalent areas in the phase diagram. In this case, there may appear both direct and reverse mesophases, located almost symmetrically with respect to an axis from the middle of the amphiphilic-water side of the triangular phase diagram to the solute apex. The loci of the direct and reversed phases can be symmetrically or asymmetrically shaped with respect to this axis; from the point of view of our theoretical approach, this property reflects the influence or absence of the fluid-reversal symmetry F_r. A second class of phase diagrams corresponds to the existence of a single micellar phase with a continuous crossover between direct and reversed micelles. In the context of our treatment, it corresponds to F_r-invariant systems, with direct and reversed mesophases within the same sequence. In the third class of phase diagrams, one of the L_1 or L_2 micellar phase regions is predominant, and there exist either direct or reversed micelles, but not both of them simultaneously.

4.4 Lyotropic phases with oriented interfaces

We have implicitly assumed that, as an average, the amphiphilic stick-like molecules are perpendicular to an interface. Strictly speaking, this happens only in the lamellar phase, in which the alkyl chain configurations correspond to a two-dimensional isotropic liquid-like state. In these configurations, the interface presents no spontaneous internal structure, except the amphiphile/solvent structure as described above for non-F_r-invariant systems. If this structureless planar surface curves during the phase transition; molecules can tilt with respect to the surface layer without breaking the symmetry of the curved interface. For instance, in the orthorhombic phase described above, molecules remain normal to the interface, at positions with local symmetry C_{2v} (at the intersection of the interface with the orthorhombic mirror planes). Indeed, they are slightly tilted at non-symmetric positions. This tilting process is a secondary effect, induced by the primary deformation of the interface. Taking into account these secondary processes, our theoretical treatment leads to a more detailed description of the interface, without any changes of the stable phases, of their symmetry groups or of their thermodynamic properties and phase diagrams. Actually, as shown in Chapter 1, this corresponds to the most usual case of ordered mesophases in lyotropic fluids. However, there are some examples, as the rippled phases found in lipid membranes [15], where molecules are found to be tilted with respect to the surface in a smectic-C type configuration [16]. In these cases, the interface presents a spontaneous internal structure, even in the lamellar phase, with a homogeneous interface, but loses its isotropic symmetry. If this surface is curved, the average molecular orientation is affected by the interface deformation and by its underlying spontaneous structure. This process can no longer be accounted for by a simple scalar interface function, $\Psi(x, y, z)$, but needs vector or tensor functions describing the coupled deformations of both shape and structure of the interface. Figure 4.5 shows examples of tilting which may occur in the amphiphilic molecules within a lamella, that is, with the same orientation for the two molecular layers (Fig. 4.5(a)) or with a chevron-type ordering (Fig. 4.5(b)).

The rank of the tensor function taking into account the internal structure of the interface depends on the molecular symmetry. The simplest case occurs for polar molecules (with symmetry group Cn). In an orthogonal configuration, the dipoles are normal to the planar interface for non-F_r-invariant systems, and cancel for invariant systems. In a structured interface, there may appear a dipole

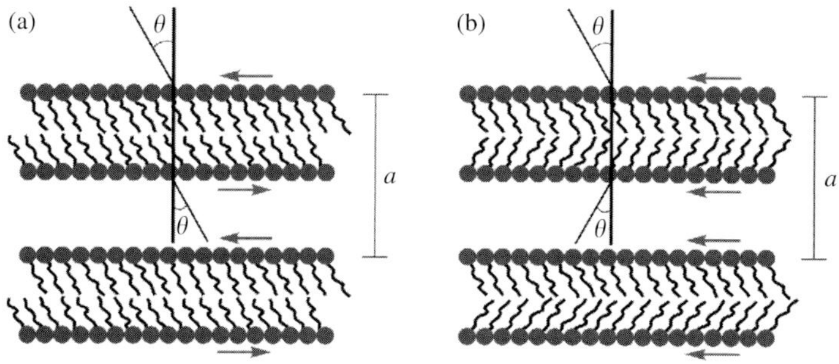

FIG. 4.5. Sketch of two tilted lamellar phases, with oriented interfaces: (a) antiparallel surfaces in a lamellae; (b) parallel surfaces or chevron-type.

component parallel to the interface if the molecules are tilted. A polar vector function is then necessary for the description of the curvature. For molecules with symmetry *Cnh*, that is, forbidding polarization, the tilting process may give rise to the onset of an axial vector parallel to the molecular axis (as in the smectic-C phase). In this case, the interface curvature requires the definition of an axial-vector function. For each molecular symmetry, various types of tensors are possible, leading to a rich variety of structured interfaces.

4.4.1 *Symmetry-breaking undulation mechanism*

The undulation of the interfaces sketched in Fig. 4.1 presents the following properties: (i) two neighboring surfaces limiting the same, A or S, region are dephased by π; (ii) two consecutive A (or S) regions behave identically, that is, they are in phase, with a minimal period a along the direction perpendicular to the interfaces.

These two features are not arbitrary and represent necessary conditions for the occurrence of the sequence of phases shown in Fig. 4.1. The absence of dephasing between two consecutive regions would not allow a symmetry-breaking mechanism, giving rise to the tetragonal configuration. The requirement that A and S regions play a symmetric role is also a necessary condition for the existence of the fluid-reversal symmetry.

There is a different situation if the interfaces are oriented. An identical orientation for two consecutive A and S regions excludes a phase opposition in these neighboring regions. In order to obtain this dephasing, we have to double the periodicity along a direction perpendicular to the interfaces. Note that, in this case, A and S regions do not behave symmetrically. It then follows the absence of F_r-symmetry for systems with oriented interfaces.

4.4.2 *Field lines and oriented domains*

Let us use an academic example, as sketched in Fig. 4.5(a), in order to give an illustration of field lines and oriented domains. The interface is oriented by the projection of the molecular vector (which is oriented from the alkyl chain towards the head of the amphiphile). We then consider a vector field $\vec{\Psi}(x, y)$, which corresponds to the projection of the molecular vector onto a given interface. A unit molecular vector measures the tilt angle with respect to the normal to the interfaces. The infinite set of field lines of $\vec{\Psi}$ represent the possible positions of the interface. Since these lines form a one-parameter family, we need an additional scalar degree of freedom for determining which line corresponds to the actual interface. The symmetry of the vector function $\vec{\Psi}$ can be analyzed by expanding its components $\Psi_x(x, y)$ and $\Psi_y(x, y)$ in a Fourier series,

$$\Psi_x(x, y) = \sum_{n,p=-\infty}^{+\infty} \Psi^x_{np} \exp\left[\frac{2i\pi}{a}(nx + py)\right], \tag{4.22}$$

and

$$\Psi_y(x, y) = \sum_{n,p=-\infty}^{+\infty} \Psi^y_{np} \exp\left[\frac{2i\pi}{a}(nx + py)\right]. \tag{4.23}$$

In analogy with our previous treatment of the scalar quantity $\Psi(x, y)$, we consider the first harmonic components only, Ψ^*_{01}, Ψ^y_{10}, $\Psi^x_{0\bar{1}} = (\Psi^x_{01})^*$, and $\Psi^y_{\bar{1}0} = (\Psi^y_{10})^*$. Also, we write $\Psi^x_{01} = \rho_1 \exp(i\theta_1)$ and $\Psi^y_{10} = \rho_2 \exp(i\theta_2)$. In a stable region, these equations are reduced to

$$\Psi^x_{01} = \rho_1 \cos\frac{2\pi y}{a} \quad \text{and} \quad \Psi^y_{01} = \rho_2 \cos\frac{2\pi x}{a}. \tag{4.24}$$

The equations of the field lines are given by

$$\rho_2 \sin\left(\frac{2\pi x}{a}\right) - \rho_1 \sin\left(\frac{2\pi y}{a}\right) = \pm k, \tag{4.25}$$

where the value of the constant k fulfills the condition

$$-(\rho_1 + \rho_2) \leq k \leq \rho_1 + \rho_2. \tag{4.26}$$

Note that the field lines corresponding to the same absolute values of k, but with opposite signs, are oriented in opposite directions, and determine the same A or S regions. Therefore, the full set of stable structures and their orientations are determined by the values of ρ_1 and ρ_2 and by the absolute value of the constant k.

4.4.3 *Symmetry of the mesophases with oriented interfaces*

The interfacial energy F depends on three invariants, $I_1 = \rho_1^2 + \rho_2^2$, $I_2 = \rho_1^2\rho_2^2$, and $I_3 = k^2$. The minimization of F with respect to ρ_1, ρ_2, and k, gives the same equilibrium conditions as obtained for non-oriented interfaces, L ($\rho_1 = 0, \rho_2 \neq 0$, or $\rho_1 \neq 0, \rho_2 = 0$), C ($\rho_1 = \rho_2 \neq 0$), and IP ($\rho_1 \neq \rho_2 \neq 0$). However, due to the orientation of the surfaces, these phases have different symmetries. The two-dimensional space groups of the oriented interfaces are non-symmorphic: *P2gm* for phase L, $P2gg$ for phase IP, and $P4gm$ for phase C, where g represents glide reflection planes.

4.5 The lamellar–hexagonal phase transition

Hexagonal phases are often present in phase diagrams of lyotropic mixtures. In the preceding chapters we have shown a number of examples with this type of stable mesophase. The most general phase sequence in which the H phase appears involves the L phase. These sequences are usually separated by phase coexistence domains, that is, by a first-order reconstructive transition, or by intermediate mesophases (of rectangular, monoclinic, or cubic symmetries). Three types of hexagonal phases were identified in phase diagrams of lyotropic mixtures: direct, H_α, inverted, H^i_α, and complex, H_c, hexagonal phases. In this section we show that the same phenomenological model leads to a description of the lamellar–hexagonal transition and to the prediction of the possible types of related intermediate mesophases. Also, we show that the assumption of oriented interfaces provides a basis for the description of tilted hexagonal mesophases.

4.5.1 *Phenomenological description of the lamellar–hexagonal transition*

In Fig. 4.6 we sketch a two-dimensional view of a transition from the lamellar L to the hexagonal H phase structures. As in the lamellar–tetragonal case, the symmetry-breaking mechanism of this transition consists of a periodic undulation of the interfaces. The L–H transition takes place through two types of intermediate phases, which are separated by a topological transition. According to the phenomenological treatment of this chapter, the equation of the interface can be written as

$$\Psi(x, y) = \sum_{n,p=-\infty}^{+\infty} \Psi_{np} \exp\left[\frac{2\mathrm{i}\pi}{a}(nx + py)\right], \tag{4.27}$$

where n and p are integers, and x and y represent hexagonal coordinates. The complex components of the order parameter associated with the undulation mechanism are the first harmonics of this Fourier expansion, $\Psi_{10} = \rho_1 \exp(\mathrm{i}\theta_1)$, $\Psi_{01} = \rho_2 \exp(\mathrm{i}\theta_2)$, $\Psi_{1,\bar{1}} = \rho_3 \exp(\mathrm{i}\theta_3)$, $\Psi^*_{01} = \Psi_{0\bar{1}}$, $\Psi^*_{10} = \Psi_{\bar{1}0}$, and $\Psi^*_{1\bar{1}} = \Psi_{\bar{1}1}$. They can be split into two Goldstone angles, θ_1 and θ_2, and four energetic components, ρ_1, ρ_2, ρ_3, and $\theta = \theta_1 - \theta_2 + \theta_3$.

We then construct four independent polynomial invariants,

$$I_1 = \rho_1^2 + \rho_2^2 + \rho_3^2, \tag{4.28}$$

$$I_2 = \rho_1^4 + \rho_2^4 + \rho_3^4, \tag{4.29}$$

$$I_3 = \rho_1^2\rho_2^2\rho_3^2, \tag{4.30}$$

and

$$I_4 = \rho_1\rho_2\rho_3 \cos\theta. \tag{4.31}$$

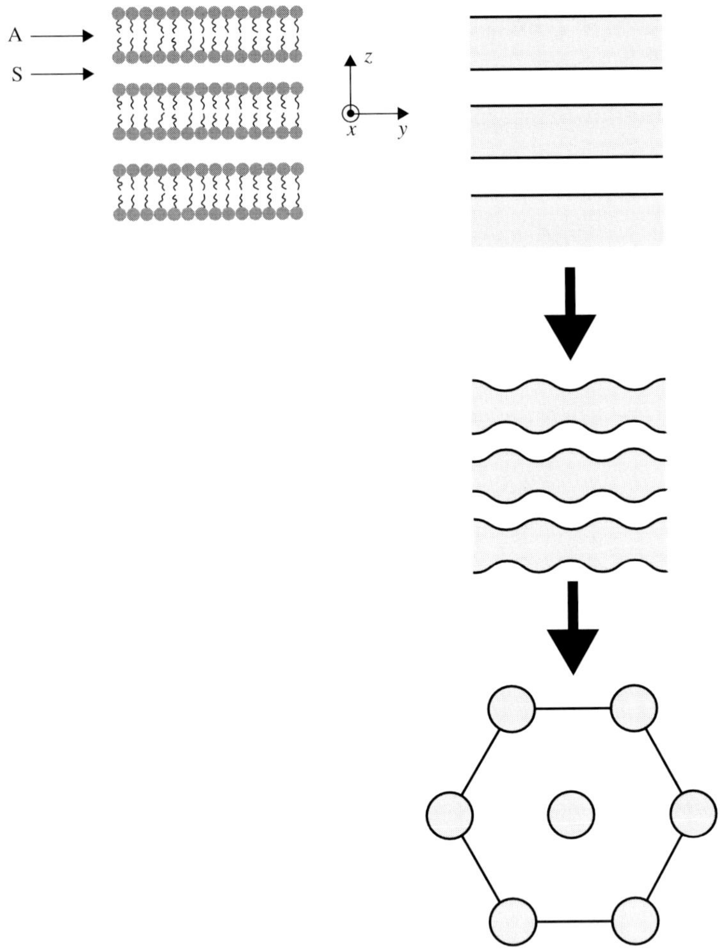

FIG. 4.6. Two dimensional sketch of the transition from the lamellar, L, to the hexagonal, H, phase, generated by undulations of the lamellae on the x–y plane.

The corresponding free-energy expansion is written as

$$F(\rho_{\rm i},\theta) = a_1 I_1 + a_2 I_1^2 + \cdots + b_1 I_2 + b_2 I_2^2 + \cdots + c_1 I_3 + \cdots$$
$$+ d_1 I_4 + d_2 I_4^2 + \cdots + e_{12} I_1 I_2 + e_{13} I_1 I_3 + e_{14} I_1 I_4 + \cdots . \quad (4.32)$$

The minimization of F with respect to $\rho_{\rm i}$ and θ yields at most seven possible stable states [12]. In addition to the L, H, $\tilde{\rm L}$, and R phases, we may have four additional intermediate stable mesophases (of rhombohedral and monoclinic nature).

This approach accounts for the topological modifications when the interface undergoes a lamellar to hexagonal transition. If the hexagonal phase results from a direct first-order transformation, there is a sudden topological change, which is triggered by the phase transition. On the other hand, if the hexagonal phase is stabilized after a series of second-order transformations through a set of intermediate low-symmetry phases, the topological change is smooth and happens at a given temperature, within an intermediate structure, as two adjacent interfaces get into contact. The same changes may take place between direct and reverse hexagonal phases when the micelles form a compact hexagonal structure. In this case, the symmetry breakdown of the undulation is not involved in the topological process.

Modifications in the topology of the interface can be described by keeping only the first harmonic components in Eq. (4.27). Fixing the global position of the interface lattice by assuming $\theta_1 = \theta_2 = 0$, we then write

$$\Psi(x, y) = \Psi_{00} \left\{ \eta_1 \cos\left(\frac{2\pi x}{a}\right) + \eta_2 \cos\left(\frac{2\pi y}{a}\right) + \eta_3 \cos\left[\frac{2\pi}{a}(x-y) + \theta\right] + 1 \right\}$$
$$= 0, \qquad (4.33)$$

where $\eta_{\rm i} = \rho_{\rm i}/\Psi_{00}$, for i $= 1, 2, 3$. In the hexagonal phase, the equilibrium conditions are given by $\rho_1 = \rho_2 = \rho_3$ and $\theta = 0$. Since ρ_1 depends on concentration, the topology of the interface changes at a concentration corresponding to $\rho_1 = 1$.

The undulation can produce topological modifications without changing the symmetry of the phases, which is an indication of the connections between symmetry and the geometric aspects of this mechanism. However, although the symmetry breakdown is fairly well described at the thermodynamic level, with this approach we do not explain the singularities of the free energy associated with topological modifications. Indeed, the free energy given by Eq. (4.27) displays a regular behavior at the topological boundaries in the order-parameter space (e.g. as $\rho_1 = 1$, in the hexagonal phase for instance), although the curvature of the interfaces diverges at the points of contact. This is the main limitation of this global Landau-type approach with respect to local theories.

4.5.2 *Tilted hexagonal phases*

We now turn to the description of hexagonal and lamellar phases with oriented interfaces. A possible scenario in this type of topological evolution of the amphiphile aggregate is a transition from an oriented lamellar phase structure to vesicles forming a hexagonal array.

The orientation of the interfaces can be expressed by the vector field

$$\vec{\Psi}(\vec{r}) = \Psi_a(x, y)\vec{a} + \Psi_b(x, y)\vec{b}, \qquad (4.34)$$

where $\vec{r} = x\vec{a} + y\vec{b}$ is written in terms of hexagonal coordinates. The vector $\vec{\Psi}$ is either polar or axial, depending on the (polar or non-polar) symmetry of the molecules. The components Ψ_a and Ψ_b can be expressed as a Fourier series,

$$\Psi_{a,b}(x, y) = \sum_{n,p=-\infty}^{+\infty} \Psi_{n,p}^{a,b} \exp\left[\frac{2\mathrm{i}\pi}{a}(nx + py)\right]. \tag{4.35}$$

As in the case of untilted systems, the symmetry of the ordered phases is completely determined by the first harmonics of Ψ_a and Ψ_b, and their complex conjugates, $\Psi_{01}^a = \rho_1 \exp(\mathrm{i}\theta_1)$, $\Psi_{10}^b = \rho_2 \exp(\mathrm{i}\theta_2)$ and $\Psi_{\bar{1}1}^a + \Psi_{\bar{1}1}^b = \rho_3 \exp(\mathrm{i}\theta_3)$. The angles θ_1 and θ_2 indicate the position of the two dimensional lattice. If they are set to zero by a suitable choice of the origin of coordinates, we have the truncated expressions for polar and axial vector fields,

$$\begin{aligned} \mathrm{i}\vec{\Psi}(\vec{r}) = &\left\{\rho_2 \sin\left(\frac{2\pi y}{a}\right) + \rho_3 \sin\left(\frac{2\pi}{a}(y-x) - \theta\right)\right\}\vec{a} \\ &+ \left\{-\rho_1 \sin\left(\frac{2\pi x}{a}\right) + \rho_3 \sin\left(\frac{2\pi}{a}(y-x) - \theta\right)\right\}\vec{b}, \end{aligned} \tag{4.36}$$

$$\begin{aligned} \mathrm{i}\vec{\Psi}(\vec{r}) = &\left\{\rho_1 \sin\left(\frac{2\pi x}{a}\right) - \rho_3 \sin\left(\frac{2\pi}{a}(y-x) - \theta\right)\right\}\vec{a} \\ &+ \left\{\rho_2 \sin\left(\frac{2\pi y}{a}\right) + \rho_3 \sin\left(\frac{2\pi}{a}(y-x) - \theta\right)\right\}\vec{b}. \end{aligned} \tag{4.37}$$

Using group-theoretical arguments, it is possible to show [2] that the transverse polar vector field, given in Eq. (4.36), and the longitudinal axial vector field, given by Eq. (4.37), are the only tensor waves inducing a reconstruction of the lamellar structure into a hexagonal array. In the polar case, the field parallel to the interface represents the projection of the molecular polarization. The axial field is normal to the interface. It may represent the projection (onto the surface normal) of a vector lying along the two-fold axis of a molecule with, for instance, symmetry C2h. In both cases, the equation of the interfaces is given by

$$\left\{\rho_1 \cos\left(\frac{2\pi x}{a}\right) + \rho_2 \cos\left(\frac{2\pi y}{a}\right) + \rho_3 \sin\frac{2\pi}{a}(y-x) - \theta\right\}^2 = K, \tag{4.38}$$

where K is a constant representing an additional degree of freedom of the system. This parameter K is a secondary, non-symmetry-breaking, order parameter which describes the form of the interfaces. In this case, it is possible to construct four independent invariants, I_1, I_2, I_3, and $I_4^{'} = \rho_1^2\rho_2^2\rho_3^2 \cos(\theta)$, which give rise to the free energy

$$F(\rho_i, \theta) = a_1 I_1 + \cdots + b_1 I_2 + \cdots + c_1 I_3 + \cdots + d_1 I_4^{'} + \cdots + e_{12} I_1 I_2 + \cdots. \tag{4.39}$$

The minimization of this expression with respect to the order parameters, supplemented by the stability conditions, leads to eight stable states with different forms and orientations of the interfaces [12]: (i) tilted lamellar, L, and undulated tilted lamellar, $\tilde{L}$, configurations; (ii) an orthorhombic cylindrical phase; (iii) cylindrical phases; and (iv) either a cylindrical or a lamellar configuration with a monoclinic symmetry.

4.6 The lamellar–cubic phase transition

Cubic phases, called Q in Chapter 1, are present in many phase diagrams of lyotropic mixtures, in the vicinity of micellar isotropic, hexagonal, and lamellar phases (see, e.g. Fig. 4.3). The lamellar–cubic phase transition is a remarkable example of a crossover between one-, two-, and three-dimensional spatially organized structures. There are several theoretical proposals to describe the cubic to isotropic micellar transition on the basis of scalar order parameters, as the shape of interfaces and density fluctuations [7,17–19]. These proposals, however, do not take into account the reconstructive aspects of this phase transition.

The same scheme of the preceding sections can also be used to describe the transitions between the cubic phases and their neighboring structures. Most of the cubic phases [20] can be obtained from a lamellar phase using a single symmetry-breaking order parameter which consists in a vector or a tensor field. Also, a full description of the cubic configurations requires the consideration of the interface undulations as a *secondary* (non-symmetry breaking) scalar order parameter.

4.6.1 *Symmetry basis of the model*

Consider the transition between a lamellar and a simple cubic mesophase. The symmetry-breaking undulation of the interfaces produces both the high-symmetry cubic phase and a set of intermediate configurations. We then have the undulated lamellar phase $\tilde{L}$, a micellar tetragonal phase, a cylindrical phase with a two-dimensional tetragonal lattice, a micellar orthorhombic phase, cylindrical orthorhombic and cylindrical undulated phases.

In analogy with the case of two-dimensional mechanisms, we introduce an interface equation, $\Psi(x, y, z) = 0$, where the real scalar field Ψ is assumed to be positive in the A regions and negative in the S regions. We then have the Fourier series,

$$\Psi(x, y, z) = \sum_{n,p,q=-\infty}^{+\infty} \Psi_{npq} \exp\left[\frac{2i\pi}{a}(nx + py + qz)\right], \tag{4.40}$$

where n, p, and q are integers, and the coefficients of the first harmonics are written as $\Psi_{001} = \rho_1 \exp(i\theta_1)$, $\Psi_{00\bar{1}} = \Psi^*_{001}$, $\Psi_{010} = \rho_2 \exp(i\theta_2)$, $\Psi_{0\bar{1}0} = \Psi^*_{010}$, and $\Psi_{100} = \Psi^*_{\bar{1}00} = \rho_3 \exp(i\theta_3)$. As usual, θ_1, θ_2, and θ_3, are Goldstone variables, and ρ_1, ρ_2, and ρ_3, represent the three effective components of the order parameter. We now write three independent invariants,

$$I_1 = \rho_1^2 + \rho_2^2 + \rho_3^2, \tag{4.41}$$

$$I_2 = \rho_1^2\rho_2^2 + \rho_1^2\rho_3^2 + \rho_2^2\rho_3^2, \tag{4.42}$$

and

$$I_3 = \rho_1^2\rho_2^2\rho_3^2, \tag{4.43}$$

and the corresponding expansion for the free energy,

$$F(\rho_1, \rho_2, \rho_3) = a_1 I_1 + a_2 I_1^2 + a_3 I_1^3 + \cdots$$
$$+ b_1 I_2 + \cdots + c_1 I_3 + \cdots + d_{12} I_1 I_2 + \cdots . \tag{4.44}$$

The minimization of this free energy with respect to the components of the order parameter, supplemented by the requirements of stability, leads to seven different structures, with cubic, tetragonal, and orthorhombic space groups. The structures of the corresponding interfaces are obtained by assuming $\theta_1 = \theta_2 = \theta_3 = 0$ and replacing the equilibrium values of ρ_1, ρ_2, and ρ_3, in the interface equation,

$$\Psi(\rho_1, \rho_2, \rho_3) = \Psi_{000} \left[1 + \eta_1 \cos\left(\frac{2\pi z}{a}\right) + \eta_2 \cos\left(\frac{2\pi y}{a}\right) + \eta_3 \cos\left(\frac{2\pi x}{a}\right) \right] = 0, \tag{4.45}$$

where $\eta_{\mathrm{i}} = \rho_{\mathrm{i}}/\Psi_{000}$, for i $= 1, 2, 3$. In the cubic phase, given by $\rho_1 = \rho_2 = \rho_3$ and $\theta = 0$, the shape and the topology of the interface is determined by ρ_1, which depends on concentration but is almost independent of temperature. At low concentrations, direct micelles filled with oil are located at the nodes of a simple cubic lattice. At intermediate concentrations, micelles merge into a single labyrinth, usually called *plumber's nightmare*; at higher concentrations, water remains located inside the micelles, forming the inverted image of the low-concentration structure.

4.6.2 *Group-theoretical considerations*

The parent symmetry G_{o} of the model discussed in the preceding section is $O_h \times R_3$. If we recall that the parent symmetry is the minimum space group which contains all the symmetry operations leaving invariant the stable phases and their domains, G_{o} contains all the three-dimensional continuous translations and the point symmetries of the cube. The order parameter spans a representation of G_{o} associated with the wave vector $\vec{k}_1 = (2\pi/a, 0, 0)$. The corresponding star $\vec{k}_1^*$ has six branches, which are formed by the three pairs $\pm\vec{k}_1$, $\pm\vec{k}_2 = (0, \pm 2\pi/a, 0)$, and $\pm\vec{k}_3 = (0, 0, \pm 2\pi/a)$. The ordered structures are obtained if we consider a periodic behavior along one direction (lamellar phase), two directions (cylindrical phase with a two-dimensional tetragonal lattice) or three directions (tetragonal and cubic phases); in other words, there is a freezing of the fluctuations corresponding to one, two, or three pairs of branches of $\vec{k}_1^*$. The irreducible representation $A_1(\vec{k}_1^*)$ which induces the $Pm3m$ cubic symmetry is constructed from the identity representation of the tetragonal invariance group of $\vec{k}_1$, $G_{k_1} = C4v$. This representation corresponds to the scalar field presented in Eq. (4.40). Also, there are four additional representations of G_o, called A_2, B_1, B_2, and E, which can be deduced from the non-identity representations of G_{k_1}. They lead to different primitive cubic space groups, which are listed in column (5), in the upper part of Table 4.1. We have $Pn3n$, $Pm3n$, and $Pn3m$ symmetries for the six-dimensional representations A_2, B_1, and B_2, respectively, whereas the 12-dimensional representation E is associated with the enantiomorphous-symmetries $P4_132$ and $P4_332$.

Face-centered (fcc) and body-centered (bcc) cubic symmetries can also be obtained from the consideration of the wave vectors $\vec{k}_2 = (2\pi/a, 2\pi/a, 2\pi/a)$ and $\vec{k}_3 = (2\pi/a, 2\pi/a, 0)$. The fcc and bcc space groups, corresponding to the irreducible representations (IR) of G_o for the 8- and 12-branch stars $\vec{k}_2^*$ and $\vec{k}_3^*$, are listed in the middle and lower sections of Table 4.1. Note that only the cubic space groups are indicated in column (5). According to the IR of G_o, there may appear other stable structures, associated with lower symmetries, corresponding to various intermediate phases between the lamellar an cubic structures. In column (6) of Table 4.1, we list the tensorial character of the order parameter associated with the transformation to the cubic phases. Hence, we have scalar (0^+), axial (1^+) or polar (1^-) vectors, and second-rank symmetric (2^+) or anti-symmetric (2^-) tensors. The simplest cases of scalar waves (e.g. the case treated in the last section) are associated with structureless interfacial configurations. The resulting phases form vector or tensor fields, with internal structures arising from the ordering of the molecular units within the interface.

Seven different cubic symmetries have been observed in lyotropic fluids:

1. The $Pn3m$ space group corresponds to the bicontinuous cubic phase in lipid/water monoglycerides [21] and systems of tetraether lipids [22].
2. The $Pm3n$ micellar cubic phase is observed in several lipid/water and lipid/water/oil systems [23,24], in a concentration range between hexagonal and micellar solutions.
3. The $P4_332$ micellar cubic phase has been found in monoglyceride/water systems [25].
4. The $Fd3m$ micellar cubic structure is also observed in monoglyceride (fatty acid)/water systems, and in a glycolipid/water mixture [20].
5. The $Fm3m$ symmetry was recently found in a physical gel [25], resulting from the aggregation process of the association of triblock copolymers into spherical micelles.
6. The $Im3m$ space group is associated with the structure of several monoglyceride (protein salt)/water mixtures [20]. The $I\bar{4}3m$ and $Im3$ space groups, also proposed for these systems, are listed in column (5) of Table 4.1.

Table 4.1 Order-parameter symmetries corresponding to the different cubic structures. Columns give the following indications: (1) the Bravais lattices and the wave vectors k_i; (2) the labels of the IRs; (3) the dimensions of the IRs; (4) the effective order-parameter components in the cubic phases; (5) the cubic space groups; the underlined groups correspond to experimentally reported symmetries; (6) the rank and parity of the tensor fields spanning each IR; ϕ_i, ϕ'_1 and $\rho_\mathrm{i}, \rho'_\mathrm{i}$ are the phases and moduli of the order-parameter components

(1)	(2)	(3)	(4)	(5)	(6)
	A_1	6	ρ_1, ρ_2, ρ_3	$Pm3m$	0+
P	A_2	6	ρ_1, ρ_2, ρ_3	$Pn3n$	1+
k_1	B_1	6	ρ_1, ρ_2, ρ_3	$\underline{Pm3n}$	2±
	B_2	6	ρ_1, ρ_2, ρ_3	$\underline{Pn3m}$	2±
	E	12	$\rho_i, \rho'_i, i = 1-3$ $\omega_i = \phi_i - \phi'_i$	$P4_132(\omega_i = -\pi/2)$ $P4_332(\omega_i = \pi/2)$	1± 1±
F	A_1	8	$\rho_1, \rho_2, \rho_3, \rho_4$ $\phi = \rho_1 - \rho_2 + \rho_3 - \rho_4$	$\underline{Fm3m}(\phi = 0)$ $\underline{Fd3m}(\phi = \pi)$	0+
k_2	A_2	8	$\rho_1, \rho_2, \rho_3, \rho_4$ $\phi = \rho_1 - \rho_2 + \rho_3 - \rho_4$	$Fd3c(\phi = 0)$ $Fm3c(\phi = \pi)$	1+
	E	16		No cubic phases	1±
I	A_1	12	$\rho_i,\ i = 1-6$	$\underline{Im3m}(\omega_i = 0)$ $I4_132(\omega_i = \pi/2)$	0+
k_3	A_2	12	$\omega_1 = \phi_1 - \phi_4 - \phi_5$	$\underline{Ia\bar{3}d}(\omega_i = \pi/2)$ $I432(\omega_i = 0)$	1+
	B_1	12	$\omega_2 = \phi_2 + \phi_3 + \phi_5$	$\underline{Im3}(\omega_i = 0)$ $I\bar{4}3d(\omega_i = \pi/2)$	1±
	B_2	12	$\omega_3 = -\phi_1 - \phi_2 + \phi_6$	$Ia3(\omega_i = \pi/2)$ $I\bar{4}3m(\omega_i = 0)$	1±

However, unlike $Im3m$ (A_1), they are identical to other representations of G_o (A_2 and B_1). Therefore, it is important to take into the physical mechanisms for establishing distinctions among the three cubic symmetries.

7. The $Ia3d$ space group has been the first observed example of a bicontinuous lipid/water structure [26]. It is found in a large variety of systems: hydrous soaps of divalent cations [27], soap and detergent/water in the intermediate region between the lamellar and hexagonal phase, diacil lipid/water mixtures at low hydration and high-temperature [28],

monoglycerides/water [19] and tetraether diacilated lipids/water mixtures, at large water content [22].

4.6.3 *The bicontinuous cubic phase*

We now use a bicontinuous (gyroid) cubic mesophase, of symmetry $Ia\overline{3}d$, in order to illustrate the connection between a given order-parameter symmetry and the corresponding cubic configurations. The microstructure of this phase can be modeled in terms of three-dimensional periodic *minimal surfaces*, free from self-intersections, as the Schoen gyroid surface [29,30]. In Table 4.1, we see that the $Ia\overline{3}d$ space group is associated with a 12-dimensional order parameter corresponding to an axial vector field. The real scalar function defined by Eq. (4.37) is then replaced by an imaginary axial vector field which can be expressed as the sum of 12 longitudinal waves,

$$\vec{\Psi}(\vec{r}) = \sum_{j=1}^{6} \left[\eta_j \exp\left(-\mathrm{i}\vec{k}^j \cdot \vec{r}\right) - \eta_j^* \exp\left(\mathrm{i}\vec{k}^j \cdot \vec{r}\right)\right] \vec{k}'^j, \tag{4.46}$$

where $\pm k^j$, $j = 1, \ldots, 6$, are the 12 branches of the star $\vec{k}_3^*$, and $\vec{k}'^j$ is an axial vector proportional to $\vec{k}^j$. The equilibrium relationships between the order-parameter components η_i in the Ia$\overline{3}$d phase, given in column (4) of Table 4.1, lead to the effective form of $\vec{\Psi}$ in the natural orthonormal frame of axial vectors,

$$\vec{\Psi} = \mathrm{i}\eta \begin{pmatrix} -\cos x \sin z + \sin x \sin y \\ \sin y \cos z - \cos x \cos y \\ \cos y \sin z - \sin x \cos z \end{pmatrix}, \tag{4.47}$$

with space coordinates in units of $a/2\pi$. This function describes the orientational order of the molecules within the curved interfaces. For example, in the case of molecules of individual symmetry C_2, the *oriented* molecular axis, as an average, is parallel to the imaginary part of $\vec{\Psi}$. For a molecule of symmetry C_S, this molecular axis is replaced by an oriented axis perpendicular to the molecular mirror symmetry. By contrast, a non-axial molecular symmetry (as D_2 or Cnv) is incompatible with the order parameter and cannot induce the $Ia\overline{3}d$ structure.

The primary, symmetry-breaking, vector-field order parameter $\vec{\Psi}$ expresses the molecular ordering within the interfaces and allows the determination of the symmetries of the induced phases. However, it is not enough to describe the concrete structures. In order to determine the form of the interfaces, we have to introduce a scalar field $\bar{\Psi}(x, y, z)$, which can be regarded as a secondary, non-symmetry-breaking, order parameter. In equilibrium, the values of $\bar{\Psi}$ are induced by the coupling with the vector field $\vec{\Psi}$. For the $Ia\overline{3}d$ symmetry, and restricting to the first three harmonic components, we can write

$$\bar{\Psi} = 1 + \xi_1\{\sin 2z \sin x \cos y + \sin 2x \sin y \sin z + \sin 2y \cos x \cos z\}$$
$$+ \xi_2\{\cos 2x \cos 2y - \cos 2x \cos 2z - \cos 2y \cos 2z\}, \quad (4.48)$$

where the effective components of the secondary order parameters, ξ_1 and ξ_2, transform as scalar waves, respectively associated with the wave vectors $(2\pi/a, 2\pi/a, 2\pi/a)$ and $(4\pi/a, 4\pi/a, 0)$. The physical structures of the stable phases depend on the values of both primary and secondary order parameters. However, the shape of the interface depends only on ξ_1 and ξ_2. Thus, the symmetry of the shape is higher than the symmetry of the phase. For instance, in the cubic stable state, the symmetry of the phase is $Ia\bar{3}d$ while the symmetry of the interface shape is $Im3m$, with an interface distance twice smaller than the lattice parameter of the full structure. There is a wide variety of structures associated with the $Ia\bar{3}d$ symmetry, which depend on the values of ξ_1 and ξ_2 in Eq. (4.48). For example, if we assume that $|\xi_1| \gg |\xi_2|$, there are five different topologies which are determined by the parameter $\zeta = -1/\xi_1$, with $-1 < \zeta < 3/2$. For $1 < \zeta < 3/2$, the structure is micellar, with 16 micelles in the conventional cubic cell. For $0 < \zeta < 1$, the micelles connect and form two separated crossed labyrinths, in agreement with a model proposed by Luzzati and collaborators [20]. For $-0.64 < \zeta < 0$, these labyrinths merge into a single labyrinth, which coexists with a single water labyrinth. For $-0.81 < \zeta < -0.64$, the preceding water labyrinth splits into two disconnected parts. Finally, for $-1 < \zeta < -0.81$, there is an inverse micellar structure with 24 water micelles in the cubic unit cell.

The lamellar phase can be described within the same framework, in the limit where $\xi_1 = 0$ in Eq. (4.48), and with $\bar{\Psi}(x, y, z) = 1 + \xi_2 \cos[2(x + y)]$. This limit corresponds to a non-symmorphic orthorhombic symmetry, D_{2h}, with molecules of consecutive lamellae along opposite directions, which doubles the periodicity normal to the lamellae.

For $|\xi_1| \ll |\xi_2|$, each cubic unit cell contains 16 micelles, half of which are non-equivalent. Changes of the parameter $\zeta = -1/\xi_2$, from -1 to 3, lead to cubic micellar structures with water labyrinths, for $-1 < \zeta < 0$, and to bicontinuous structures, for $0 < \zeta < 3$. The molecular aggregates form three-dimensional labyrinths surrounding 16 inverted micelles.

In conclusion, depending on the cubic symmetry, the order parameter describing the transformation of a lamellar into a cubic structure can be a scalar, a vector or a second-rank tensor field. In the vector or tensor cases, we have to introduce, as a secondary order parameter, a scalar quantity reflecting the form of the cubic interfaces. The connections between the symmetry of the order parameters and of the actual cubic configurations, which have been illustrated in the case of the gyroid structure, show that bicontinuous and micellar cubic structures do not correspond to *qualitatively* different systems. There exists a crossover between the two types of topologies for specific values of the secondary order parameters.

References

[1] B. Mettout, P. Tolédano, H. Vasseur, and A. M. Figueiredo Neto (1997). *Phys. Rev. Lett.* **78**, 3483.
[2] B. Mettout, P. Tolédano, H. Vasseur, and A. M. Figueiredo Neto (1997). *Phys. Rev. E* **56**, 6889.
[3] P. Tolédano and B. Mettout (1998). In *Phase Transitions in Complex Fluids*, Eds P. Tolédano and A. M. Figueiredo Neto, World Scientific, Singapore.
[4] P. Ekwall (1975). In *Advances in Liquid Crystals*, Ed. G. H. Brown, Academic Press, New York, vol. 1.
[5] J. M. di Meglio, M. Dvolaitzky, and C. Taupin (1987). In *Physics of Complex and Supermolecular Fluids*, Eds S. A. Safran and N. A. Clark, *Exxon Monograph Series*, John Wiley & Sons, New York, p. 361.
[6] L. J. Yu and A. Saupe (1980). *Phys. Rev. Lett.* **45**, 1000.
[7] D. A. Huse and S. Leibler (1988). *J. Phys. (Paris)* **49**, 605.
[8] J. Charvolin and J. F. Sadoc (1987). *J. Phys. (Paris)* **48**, 1559.
[9] W. Helfrich (1994). *J. Phys. Condens. Matter* **6**, A79.
[10] J. Meunier (1987). In *Physics of Amphiphilic Layers*, Eds J. Meunier, D. Langevin, and N. Boccara, Springer Verlag, Berlin.
[11] D. Roux and C. R. Safinya (1987). In *Physics of Amphiphilic Layers*, Eds J. Meunier, D. Langevin, and N. Boccara, Springer Verlag, Berlin.
[12] J. C. Tolédano and P. Tolédano (1987). *The Landau Theory of Phase Transitions*, World Scientific, Singapore, Chap. 4.
[13] P. G. de Gennes (1979). *Scaling Concepts in Polymer Physics*, Cornell University Press, Ithaca, NY.
[14] P. Ekwall, L. Mandell, and K. Fontell (1969). *Mol. Cryst. Liq. Cryst.* **8**, 157.
[15] G. Chidichimo and F. P. Nicoletta (1998). In *Phase Transitions in Complex Fluids*, Eds P. Tolédano and A. M. Figueiredo Neto, World Scientific, Singapore.
[16] P. G. de Gennes and J. Prost (1993). *The Physics of Liquid Crystals*, Clarendon Press, Oxford.
[17] J. Charvolin and J. F. Sadoc (1988). *J. Phys. (Paris)* **49**, 52.
[18] R. Bruinsma (1992). *J. Phys. II* **2**, 425.
[19] M. Laradji, A. C. Shi, J. Noolandi, and R. C. Desai (1997). *Macromolecules* **30**, 3242.
[20] V. Luzzati, R. Mariani, and T. Grilik-Krzywicki (1987). In *Physics of Amphiphilic Layers*, Eds J. Meunier, D. Langevin, and N. Boccara, Springer Verlag, Berlin.
[21] S. T. Hyde, S. Anderson, B. Ericsson, and K. Larsson (1984). *Z. Kristallogr.* **168**, 213.

[22] A. Gulik, V. Luzzati, M. De Rosa, and A. Gambacorta (1985). *J. Mol. Biol.* **182**, 13.
[23] R. R. Balmbra, J. S. Chenie, and J. F. Goodman (1969). *Nature* **222**, 1159.
[24] P. O. Eriksson, G. Lidblom, and G. Arvidson (1988). *J. Phys. Chem.* **89**, 1050.
[25] O. Diat, G. Porte, and J. F. Berret (1996). *Phys. Rev.* **54** , 14869.
[26] V. Luzzati and P.A. Spegt (1967). *Nature* **215**, 701.
[27] V. Luzzati, A. Tardieu, and T. Gulik-Krzywicki (1968). *Nature*, **217**, 1028.
[28] V. Luzzati, A. Tardieu, T. Gulik-Krzywicki, E. Rivas, and F. Reiss-Husson (1968). *Nature* **220**, 485.
[29] B. Halle, S. Ljunggren, and S. Lidin (1992). *J. Chem. Phys.* **97**, 1401.
[30] V. E. Podneks and I. W. Hamley (1996). *JETP Lett.*, **64**, 617.

5

THE ISOTROPIC MICELLAR AND BICONTINUOUS PHASES

5.1 Introduction

From the point of view of optical measurements, isotropic phases of lyotropic mixtures may be grouped according to the packing of amphiphilic molecules. If the amphiphilic molecules form micelles, we have the micellar direct and inverted isotropic phases, L_1 and L_2, and the micellar cubic phases. If they form bicontinuous structures, we have the L_3 sponge phase and the cubic bicontinuous phase. In this chapter both categories will be discussed, particularly L_1, L_2, and L_3 phases.

5.2 The micellar L_1 and L_2 isotropic phases: Experimental facts

L_1 and L_2 isotropic micellar phases are found in phase diagrams of binary, ternary, and quaternary lyotropic mixtures. Consider the example of a mixture of an amphiphile, water, and an organic solvent. In the L_1 phase, there are direct micelles, and the polar solvent (water) is in the intermicellar medium. The polar heads of the amphiphilic molecules are located on the surface of the micelles, in contact with the polar solvent, and the aliphatic chains are kept inside the micelles. In the inverted (or reverse) L_2 phase, the intermicellar medium is the organic solvent, and the polar heads of the amphiphilic molecules enclose polar solvent droplets. In the phase diagrams, at about 25°C, L_1 and L_2 phases are usually located at specific regions of a triangular representation; the L_1 phase is located at the organic solvent-poor ($\lesssim$10 wt%) and water-rich ($\gtrsim$60 wt%) region; L_2 is located at a water-poor ($\lesssim$50 wt%) and an organic solvent-rich ($\gtrsim$30 wt%) region [1]. This topology, including re-entrant domains, depends on the particular mixture and the temperature.

A schematic representation of a partial isotherm of an amphiphile/water/oil mixture is drawn in Fig. 5.1, where LC represents liquid crystalline phases [2]. The L_2 phase is located near the lower-right vertex of the triangular representation (oil-rich and amphiphile-poor region), but it can also be found near the upper vertex (amphiphile-rich region) of the representation.

The shape of the micelles is strongly dependent on the type and concentration of the components. Spherical micelles are found in the dilute regime of binary mixtures (for concentrations of amphiphile $c \gtrsim$ CMC, where CMC is the critical micellar concentration [3,4]). Ternary mixtures with amphiphile concentrations near the nematic domain ($c \gg$ CMC) may have micelles with orthorhombic

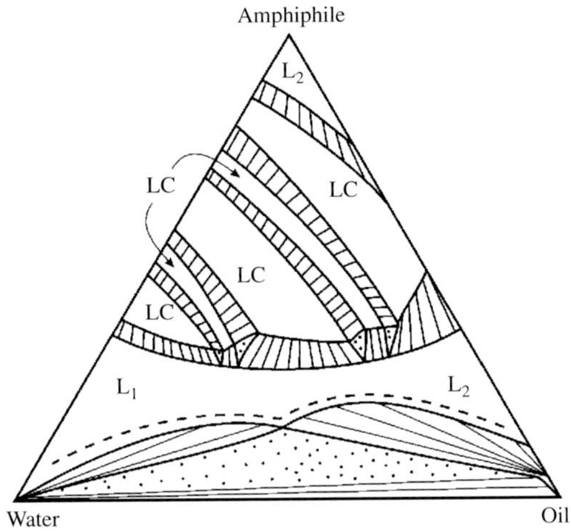

FIG. 5.1. Sketch of the partial isothermal representation of a mixture of an amphiphile, water, and oil [2]. The symbol LC represents liquid crystalline phases; the lateral sides of the triangle represent the amphiphile concentration.

symmetry, locally piled up in small, randomly oriented [5], correlation volumes with a pseudo-lamellar structure, leading to the isotropic phase.

5.2.1 *Self-assembling of amphiphiles in dilute solutions*

As we have already mentioned, amphiphiles display a double character from the electrostatic point of view. Amphiphilic molecules have a polar water-soluble group attached to a water-insoluble hydrocarbon chain. For amphiphilic concentrations $c <$ CMC, the amphiphilic molecules remain isolated in the solution of amphiphiles and a solvent. Above CMC, amphiphilic molecules begin to self-organize as micelles; the fraction of isolated amphiphilic molecules remains almost constant, and the concentration of micelles increases with c. The *hydrophobic–hydrophilic* effect is the most relevant mechanism associated with micellar formation [4]. For $c \gtrsim$ CMC, we have a *dilute regime*, in which the micelles can still be treated as isolated objects. Several parameters, as aggregation number (mean number of amphiphilic molecules in a micelle), micellar size, and micellar shape, are strongly dependent on temperature, pH, ionic strength, and amphiphile concentration.

Micelles are dynamic objects, forming and dissociating at time scales ranging from microseconds to milliseconds [6,7]. There are different types of equilibrium processes in micellar solutions: monomer exchange (if a molecule moves from the micelle to the bulk or vice versa); formation–dissociation; partial breakdown; size redistribution, keeping the number of micelles changed or unchanged [8]. The height of the free-energy barrier for the extraction of an amphiphile from a micelle is of the order of $5k_BT$, where k_B is the Boltzmann constant and T is the temperature [7]. The energy cost for the insertion of an amphiphile in a micelle has been estimated at about k_BT, so the amphiphile association rate is controlled by diffusion.

5.2.1.1 *Transformation from a molecular to a micellar solution: The critical micellar concentration* From the experimental point of view, some physico-chemical properties (as detergency, equivalent conductivity, HF conductivity, surface tension, osmotic pressure, and interfacial tension) of amphiphile–solvent solutions are characterized by a peculiar behavior as the concentration c approaches CMC [3] (see Fig. 5.2). In actual mixtures, there is no unique concentration of amphiphiles at which all of these properties display drastic changes. Usually, CMC is defined in an interval of concentrations of about 0.02 M. For $c \gtrsim$ CMC, the concentration of isolated amphiphilic molecules remains almost constant, but the concentration of micelles increases with c. For example, in the case of the potassium laurate/water

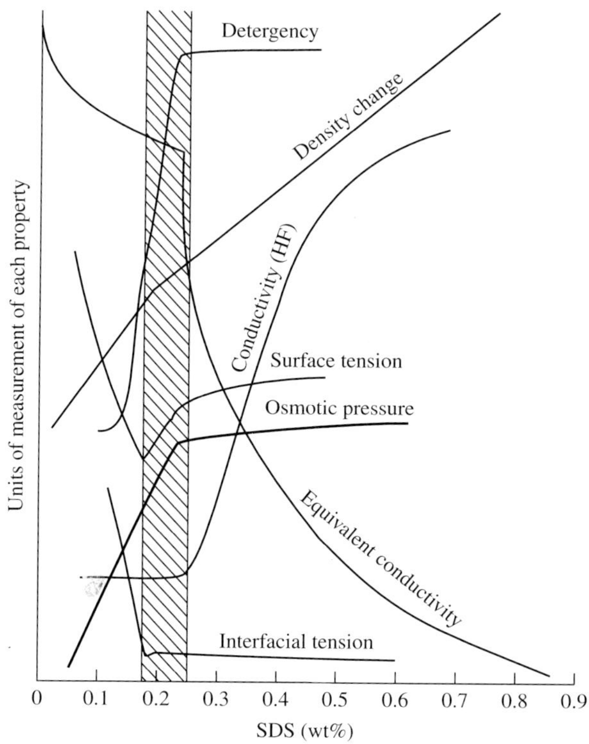

FIG. 5.2. Sketch of the behavior of some physico-chemical properties of the SDS/water mixture around CMC [3]. The hatched region corresponds to the range of CMC.

mixture, we have CMC = 0.008 M [9]; for a mixture of *N*-dodecyl-*N*,*N* dimethyl-*N*-[3-(β-methyl-β-nitrovinyl)-6-methoxybenzyl] ammonium chloride and water, CMC = 0.2 mM, with an enthalpy of micellization $\Delta H^0 = -11.3$ kJ/mol [10]. In the case of surfactants and lipids, CMC decreases with the increase of the number of carbon atoms of the amphiphile [11]. For example, at 25°C, CMC is 33.2 mM, for SdS, 8.1 mM, for SDS, and 2.0 mM, for tetradecyl sodium sulfate, $CH_3(CH_2)_{13}$ OSO_2ONa.

Although some physico-chemical properties of the solution behave differently above and below CMC, this behavior cannot be characterized as a phase transition. Monte Carlo simulations indicate that the micellar phase can be regarded as an intermediate region between a microscopically homogeneous molecular solution and a solid phase [12].

Even in this regime of low amphiphile concentration, if the temperature is raised above a *critical mixing temperature* $T_{\rm cm}$, which depends on amphiphile concentration, the single micellar system undergoes a phase separation between two isotropic phases. This behavior was observed, for example, in a mixture of pentaethylene glycol-mono-*n*-dodecyl ether ($C_{12}E_5$) and water [13]. The static and dynamic critical behavior of this system, in the neighborhood of $T_{\rm cm}$, were investigated by measurements of viscosity, intensity of scattered light, and relaxation rate of the order-parameter fluctuations [14]. These experimental results show differences with respect to the behavior of both simple or molecular binary fluids.

5.2.1.2 *Molecular organization of the aggregates; equilibrium micellar configurations and distributions* Consider a mixture of a soap and water. Just after mixing, the soap molecules are pushed towards the free surface, with the polar heads in contact with the solvent (water), and with the aliphatic, non-polar, chains sticking upwards into the atmosphere, according to a topology which reduces the surface tension. Increasing the amount of amphiphilic molecules, the need to keep a reduced surface tension leads to a transition from a system of isolated amphiphilic molecules to a self-organized micellar system. Usually, this micellar system presents aggregates of different sizes. This polydispersity is reflected by the aggregation number n, defined as the number of amphiphilic molecules per micelle. In Fig. 5.3, at given values of temperature and amphiphile concentration, for $c \gtrsim$ CMC, we illustrate the typical behavior of the size distribution of micelles [15–17]. This distribution has a maximum for small values of n; as we increase the aggregation number, it has a minimum at n_1 followed by a local maximum at $n_2 > n_1$. The region $n < n_1$ is called *submicellar*. Micellar polydispersity can be obtained from measurements of the diffusion coefficient (from quasielastic light scattering experiments, for example). For a SDS/brine mixture, with concentrations of NaCl ranging from 0.15 to 0.6 M, the width of the distribution of aggregation numbers is ±70% of the mean value [18], which indicates the high degree of polydispersity of this micellar phase.

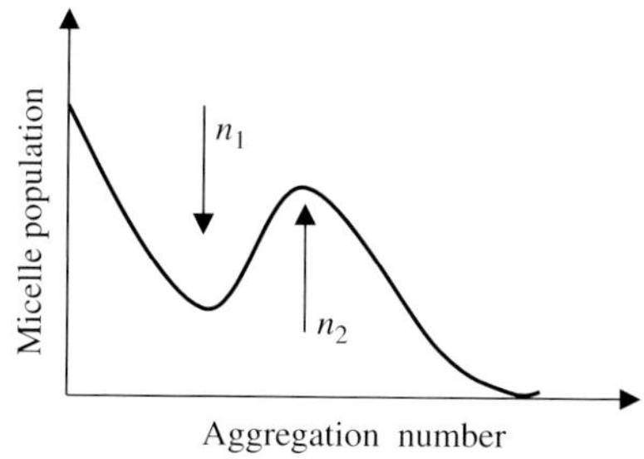

FIG. 5.3. Schematic representation of the micellar population as a function of the aggregation number, at fixed temperature and for $c \gtrsim$ CMC.

At lower amphiphile concentrations, we anticipate a spherical, or almost spherical, shape for the micelles. The diameter of the aggregates is about twice the amphiphile length l in the *trans* (extended) configuration. This length l and the volume v occupied by the chain can be obtained from the expressions

$$l(\text{nm}) \sim (0.15 + 0.13n_\text{c}),$$

and

$$v(\text{nm}^3) \sim (27.4 + 26.9n_\text{c}) \times 10^{-3},$$

where n_c is the number of carbons in the saturated hydrocarbon chain, with the exclusion of the CH_3 ending group [4].

The most common experimental techniques to investigate sizes and shape of micelles are light scattering, tracer diffusion [19], ultracentrifugation [20], and X-ray scattering and diffraction. Most of the amphiphiles with charged head groups form spherical micelles. This tendency is attributed to the electrostatic repulsion between heads, which leads to large values of the surface per head. The addition of ions (from a salt, for example) can reduce the surface per amphiphile head, inducing the stabilization of non-spherical micellar shapes, as cylindrical, spherocylindrical, and prolate and oblate ellipsoidal forms [11,17].

5.2.1.3 *Thermodynamics of micellization* We now look at the basic thermodynamic ideas underlying the process of micellization [11,21]. Consider a solution of a fixed number of N_a amphiphilic and N_w water (solvent) molecules in a volume V. There are many micellar aggregates, with different sizes (number of amphiphilic molecules), associated with the size distribution $\{N_n; n = 1, 2, \ldots\}$, and subjected to the constraint

$$\sum_{n=1,2,\ldots} nN_n = N_a. \tag{5.1}$$

The free energy of this system may be written in terms of three terms,

$$F = F_{\mathrm{mix}} + F_{\mathrm{mic}} + F_{\mathrm{w}}. \tag{5.2}$$

The free energy of mixing is given by the (ideal gas) expression,

$$F_{\mathrm{mix}} = \frac{1}{\beta}\left[N_{\mathrm{w}} \ln X_{\mathrm{w}} + \sum_{n=1,2,\ldots} N_n \ln X_n + N_{\mathrm{w}} + \sum_{n=1,2,\ldots} N_n\right], \tag{5.3}$$

with $k_{\mathrm{B}}T = 1/\beta$, $X_{\mathrm{w}} = N_{\mathrm{w}}/V$, and $X_n = N_n/V$. The free energy of the micelles is written as

$$F_{\mathrm{mic}} = \sum_{n=1,2,\ldots} N_n F_n, \tag{5.4}$$

where F_n is the free energy of a single micellar aggregate of size n. The contribution of water, given by F_{w}, is not so relevant at this point.

The equilibrium concentrations come from the minimization of the total free energy, $F = F(\{N_n\})$, with the constraint, given by Eq. (5.1), of fixed number of amphiphiles. We then introduce a Lagrange multiplier, and minimize the function

$$\widetilde{F}(\{N_n\}) = F(\{N_n\}) + \alpha \sum_{n=1,2,\ldots} nF_n, \tag{5.5}$$

from which we have

$$\mu_n = -\alpha n = \mu_1 n, \tag{5.6}$$

where $\alpha = -\mu_1$, and μ_n is the chemical potential of a micellar aggregate of size n,

$$\mu_n = \frac{\partial F}{\partial N_n} = \frac{1}{\beta} \ln X_n + F_n. \tag{5.7}$$

Using these equations, we have

$$X_n = \exp[-\beta(F_n - n\mu_1)], \tag{5.8}$$

which can also be written as a "law of mass action,"

$$X_n = X_1^n \exp[-\beta \Delta G(n)], \tag{5.9}$$

where $\Delta G(n)$, given by

$$\Delta G(n) = F_n - nF_1, \tag{5.10}$$

can be regarded as the difference between the free energies of amphiphiles belonging to a micellar aggregate and free amphiphiles in water.

The quantity $\Delta G(n)$ depends on several ingredients [11,21]: the energy of binding within the micelles; the surface tension; the arrangements of the hydrophobic tails. Different forms of $\Delta G(n)$ have been proposed in the literature, even in very recent publications [22]. In order to keep calculations at a very simple level, let us assume that

$$-\beta\Delta G(n) = (n - n_0)\epsilon, \tag{5.11}$$

where n_0 and ϵ are fixed parameters, and which is supposed to account for the amount of energy necessary for filling a hydrophobic core [11]. We then have

$$X_n = X_1^n \exp[\epsilon(n - n_0)], \tag{5.12}$$

and

$$X_{\rm a} = \frac{N_{\rm a}}{V} = \frac{X_1 \exp[\epsilon(1 - n_0)]}{[1 - X_1 \exp(\epsilon)]^2}, \tag{5.13}$$

which can also be written as

$$X_1 = \frac{2X_{\rm a} \exp(\epsilon n_0) + 1 - [4X_{\rm a} \exp(\epsilon n_0) + 1]^{1/2}}{2X_{\rm a} \exp(\epsilon n_0 + \epsilon)}. \tag{5.14}$$

Eq. (5.14) shows that $X_{\rm a} \sim X_1$ in the regime of low dilution ($X_{\rm a} \ll 1$). At large amphiphile concentrations, $X_{\rm a} \exp(\epsilon n_0) \gg 1$, we reach an asymptotic regime, $X_1 \sim \exp(-\epsilon)$, which suggests that the critical micellar concentration should be given by

$$X_{\rm a}(\mathrm{CMC}) \sim X_1 \sim \exp(-\epsilon). \tag{5.15}$$

In a recent publication, Chandler and coworkers [22] obtained excellent fittings of the temperature dependence of the critical micellar concentration with an expression of the form

$$-\beta\Delta G(n) = n\beta\Delta\mu - \beta g n^{2/3} + h n^{5/3}, \tag{5.16}$$

which accounts for several stages of the formation of micelles. We should also mention the existence of several numerical simulations of the processes of micellization. For example, a Monte Carlo simulation of a schematic model of amphiphiles and water molecules [21], represented by three- and one-site molecules on a square lattice, already leads to all of the characteristic properties of micellization (including polydipersity and a critical micellar concentration).

5.2.2 *Self-organization of amphiphiles in semi-dilute and concentrated regimes*

We now consider semi-dilute and concentrated micellar regimes, with amphiphile concentrations larger than CMC but far from the liquid crystalline region of the phase diagram.

5.2.2.1 *Changes in size and shapes of micelles* Size and shape of micelles depend on concentration of amphiphiles, temperature and ionic strength of the mixture. The mean aggregation number of micelles has been shown [11] to behave as $n \propto \sqrt{c}$, where c is the concentration of amphiphiles. Parameters as concentration, temperature, and ionic strength, affect the competition between attractive and repulsive forces, which characterize the intermolecular interactions. The competition between different interactions defines the *surface per amphiphile head*, which gives some information about the permeability of the micellar surface with respect to the solvent. For larger values of the surface per amphiphile head, the molecules of the solvent have a larger probability of penetrating into the micelles. Different types of micellar aggregates have typical values of the surface per amphiphile head. As a rule, the sequence spheres $\rightarrow$ cylinders $\rightarrow$ disks presents decreasing values of the surface per amphiphile head. The particular shape of the micelles also depends on other features of the amphiphile, as the length of the alkyl chain and the volume occupied by the chain.

As an introductory example, we consider the case of a mixture of SDS and brine (with NaCl), studied by quasielastic light scattering spectroscopy [23], with amphiphile and NaCl concentrations ranging from 1.7×10^{-2} to 6.9×10^{-2} M and from 0.15 to 0.6 M, respectively, and temperatures between 10 and 85°C [16]. Spherical micelles, with a typical radius of about 2.5 nm, were shown to be stabilized at high temperatures and for all values of NaCl concentration, almost independently on the SDS concentration. As the temperature increases, the mean hydrodynamic radius of the micelles tends to the asymptotic value 2.5 nm. At fixed NaCl concentration, larger than 0.3 M, the sizes of the micelles grow as the temperature decreases. These micelles approach a prolate ellipsoidal shape with a minor axis of about 5.0 nm and a major axis increasing with salt concentration but decreasing with temperature. A typical value of this major axis is about 130 nm. The ellipsoidal shape, however, is not the only possibility for representing a non-spherical micelle; another proposal is a spherocylindrical shape [24]. The mean diffusion coefficient, $\overline{D}$, has been shown to depend on both temperature and NaCl concentration. Typical values of $\overline{D}$ range from about 10^{-7} to 10^{-6} cm^2/s, decreasing with concentration of NaCl (at constant temperature) and increasing with temperature (at constant NaCl concentration). At fixed values of salt concentration and temperature, $\overline{D}$ decreases about 10% as the concentration of SDS increases by a factor of 2. The mean aggregation number of the micelles ranges from about 60, for spherical micelles, to 1500, for highly anisotropic ellipsoidal micelles.

The size growth of micelles in the presence of electrolytes is heavily dependent on the type of counterion of the system. For example, small-angle neutron scattering investigations of the cetyltrimethyl ammonium chloride/water mixture with the addition of KCl and KBr have shown that the diameter of the micelles increases with the concentration of KBr (from 0.03 to 1 M), but remains at about the same typical value with increasing amounts of KCl (from 0.05 to 1 M) [25]. According to this result, the presence of the *counterion* is relevant, but the effects of the *co-ion* are negligible. Micellar growth is enhanced in the presence of larger hydrated counterions.

Steady-state fluorescence quenching is another experimental technique for investigating the size of micelles [26]. Considering again the SDS/water mixture as a working example, the presence of additives (ethanol, *n*-propanol, *n*-butanol,

Table 5.1 Micellar radius (R) and surface per amphiphile head (s) for the SDS/water mixture in the presence of additives [26]

Additive (M)	R (nm)	s (nm^2)
Ethanol		
0.685	1.82	0.632
1.712	1.66	0.662
3.082	1.50	0.756
n-Propanol		
0.535	1.81	0.801
1.338	1.70	1.032
n-Butanol		
0.873	1.78	1.225
1.091	1.87	1.127
2.182	2.59	1.247
tert-Butanol		
0.425	1.77	0.721
0.851	1.73	0.863
1.916	1.70	1.195
n-Pentanol		
0.461	1.83	1.206
0.923	2.62	1.307
1.108	2.82	1.407

tert-butanol, *n*-pentanol, *n*-hexanol, *n*-heptanol) has been shown to modify the value s of the surface per amphiphile head. Since s depends on the concentration of the additives, it does have influence on the spherical micellar size [26]. In Table 5.1, we give values of the micellar radius R and the surface per amphiphile head for SDS/water micelles in the presence of some additives [26].

Values of s typically increase with concentration of additives. However, depending on the additive, the micellar radius may decrease (ethanol, propanol, *tert*-butanol) or increase (*n*-butanol and *n*-pentanol) with the concentration of an additive. The behavior of R as a function of the concentration of additive can also be non-monotonic, decreasing up to a minimum value, and then increasing with concentration. This behavior seems to be related to the hydrophobic volume that is needed to accommodate the additive in the micellar medium.

5.2.2.2 *Interactions between micelles and local structures in the concentrated regime* In the semi-dilute and concentrated regimes, the interactions between micelles cannot be neglected. Indeed, these interactions should be responsible for the onset of ordered anisotropic liquid phases and even give rise to more complex amphiphilic non-micellar structures. As the concentration of amphiphile increases, the interactions between micelles are accounted for by electrostatic van der Waals attractions, as well as steric and hydration repulsive forces [11], depending on various physico-chemical parameters, as ionic strength and pH of the solution, type of solvent, and temperature. The competition among these interactions leads to the transition from the micellar isotropic phase, and defines the sequence of possible ordered structures.

Quasielastic light scattering experiments have been performed to measure micellar diffusivity of several surfactant/brine mixtures (e.g. CTAB, cetyltrimethyl ammonium bromide; MyTAB, myristyltrimethyl ammonium bromide; and DoTAB, dodecyltrimethyl ammonium bromide, and SDS), over a wide range of surfactant concentrations, but in a range of salt concentration such that the electrostatic interactions are still dominant [27]. It has been shown that the behavior of the diffusivity of micelles scales as the ratio of the surfactant to the salt concentrations.

Interactions between inverted micelles have been calculated by Monte Carlo simulations and then compared with experimental results of small-angle neutron scattering of a mixture of sodium di-2-ethylhexyl sufosuccinate (AOT) and water [28]. The dispersion force, coming from the correlated fluctuations in the ionic distribution inside micelles, is comparable to the classical van der Waals interaction.

In micelles of mixtures with two amphiphiles, with bilayer structures, there have been observations of a local pseudo-lamellar ordering in the micellar isotropic phase, in the neighborhood of the nematic phases.

5.2.2.3 *Direct and inverted isotropic micellar phases* We now discuss some properties of direct L_1 phases in the neighborhood of liquid crystalline regions. Examples of phase diagrams presenting micellar isotropic phases in this neighborhood are given in Figs 1.6, 1.9, 1.19, 3.8, and 3.10. In particular, there are several measurements of electric conductivity [29–31], density [32,33], resistance and resistivity [34], as a function of temperature and relative concentrations of the different components of the mixture, in the vicinity of nematic domains. In the nematic region, the principal electrical conductivities (parallel and perpendicular to the director) assume distinct values, but tend continuously to a unique limit at the nematic–L_1 transition. Measurements of mass density d as a function of temperature show a monotonic decrease in the nematic region, a small increase at the nematic–L_1 transition temperature T_{NI}, and small monotonic decrease for $T > T_{\mathrm{NI}}$. A similar behavior is displayed by the electric resistance through the nematic–L_1 transition. On the other hand, resistivity as a function of temperature, in both nematic and L_1 phases, decreases linearly with T, but with a different angular coefficient (which is smaller in the L_1 phase).

In the neighborhood of the discotic nematic phase, micelles of binary mixtures in the L_1 phase are expected to have disk-like shapes, with thickness of about twice the length of the amphiphile, $2l$, and diameter of about $4l$. In ternary mixtures, however, micelles were shown to present lower-symmetry shapes. A KL/DeOH/water mixture, in the L_1 phase, has been shown to display the same pseudo-lamellar local ordering of micelles in the neighboring nematic phases [35]. This indicates that micelles in the L_1 phase, in this particular region of the phase diagram, have a biaxial symmetry, as a flattened prolate ellipsoid, with three mutually orthogonal two-fold symmetry axes. Also, the isotropic characteristics of this phase are a consequence of orientational fluctuations of these micelles. The typical dimensions of these micelles were estimated as $A' = 8.5$ nm, $B' = 5.5$ nm, and $C' = 2.6$ nm, where C' refers to the potassium laurate bilayer and A' and B' are dimensions in the plane perpendicular to the bilayer.

Inverted isotropic micellar systems have been shown to present remarkable static and dynamic properties. In the L_2 phase, viscoelasticity is strongly dependent on the volume fraction of micelles ϕ (in other words, the volume fraction of the amphiphile) [36]. For example, in the range $0.05 \lesssim \phi \lesssim 0.5$, a mixture of sodium di-2-ethylhexyl sufosuccinate (AOT) and decane has inverted micelles of radius 1.5 nm and aggregation number of about 22. The dynamics of the interaction among micelles in the L_2 phase was investigated by time-of-flight ultrasonic measurements, for frequencies between 2 and 45 MHz, and Brillouin techniques, at higher frequencies. At low frequencies, micelles behave as isolated objects. At higher frequencies, however, there is a typical behavior of an instantaneous micellar network. Electrical conductivity depends on ϕ^2, for small values of ϕ, but deviates from this behavior at larger values of ϕ [37]. This deviation comes together with the appearance of a secondary slow dielectric relaxation process, which coexists with the typical faster relaxation of single particles. Sound velocity measurements for the AOT/decane mixture, with $\phi \lesssim 0.2$, give increasing values with ϕ, but are almost frequency independent (about 1230 m/s; increasing slightly with frequency).

An inverted micellar structure in a mixture of the *n*-decyl octaoxyethylene glycol monoether ($C_{10}E_8$) and dodecane was investigated by static light scattering measurements in the vicinity of the *consolute point*, defined by the critical solution temperature (28.05°C, in the case of this mixture) and critical composition (30 wt%, in this mixture) on a temperature–composition phase diagram [38]. This consolute point is an extremum if we draw a graph of temperature versus ϕ for a binary mixture; at this point, the homogeneous liquid begins to separate into two immiscible liquid volumes. The scaling behavior of the correlation length and the osmotic compressibility of this system near the consolute point were shown to be compatible with critical indices belonging to the universality class of the three-dimensional Ising model.

An interesting pressure-induced phenomenon was observed in infrared spectroscopy experiments for the micellar reversed phase of the 1,2-dioleoyl phosphatidylethanolamine/water mixture [39]. At about 9 kbar, the reversed micellar phase transforms into a lamellar phase.

The phase diagrams of some of the lyotropic mixtures, for instance, mixtures of sodium 4-(1′-heptylnonyl) benzenesulfonate (SHBS), isobutyl alcohol, *n*-dodecane, and a brine (with NaCl) [40], display a stable bicontinuous structure separating regions of L_1 and L_2 phases. On the basis of the analysis of small-angle X-ray scattering, viscosity, and electrical conductivity experiments, it has been suggested the existence of a continuous progression, from the bicontinuous structure towards the buildup of direct or inverse micelles, depending on the location in the phase diagram.

5.3 The sponge L_3 phase

Sponge or L_3 phases are one of the most interesting structures in the phase diagram of lyotropic mixtures. They are optically isotropic and do not display any long-range order. Experimental observations of sponge phases indicate the existence of a microstructure with a surfactant bilayer, with a bicontinuous structure, separating two solvent domains over macroscopic distances [41].

Sponge phases in multidimensional phase diagrams were observed in both water-rich and oil-rich regions, mostly in the vicinity of swollen lamellar phases. For example, there are sponge phases in the following mixtures: cetylpyridinium chloride (CPCl), *n*-hexanol, and brine (0.2 M NaCl) [42]; cetylpyridinium bromide, *n*-hexanol, and brine (0.2 M NaCl) [42]; sodium bis 2 ethylhexyl sulfosuccinate (AOT) and brine [43]; *n*-dodecylbetain, pentanol, and water [43]; AOT, dodecanol, and water [44]; sodium dodecylsulfate (SDS), 1-pentanol, cyclohexane, and water [45]; didodecyldimethylammonium bromide, tetradecane, and water [46]; Na-octylbenzene sulfonate, *n*-pentanol, and brine (0.5 M NaCl) [47]; and SDS, pentanol, dodecane, and water [48].

Different experimental methods and techniques have been used to investigate the structure and physico-chemical properties of sponge phases: freeze-fracture electron microscopy, measurements of transport properties [49], X-ray (SAXS) [50] and neutron (SANS) [51] scattering and diffraction, NMR [44], and light scattering [43]. In the following sections, we discuss some physico-chemical, dynamical, and structural properties of these phases.

5.3.1 *Light scattering experiments: Osmotic compressibility, diffusion, and relaxation times*

Elastic and quasi-elastic light scattering experiments give information on the correlation length ξ of the mean curvature fluctuations of the bicontinuous structure, the behavior of the osmotic compressibility, and the coefficient of cooperative diffusion D_c.

Consider the case of diluted samples. The scattered intensity as a function of the modulus of the scattering wave vector q may be written as [41]

$$I_l(q) = C_1 \left[C_2 + \frac{\tan^{-1}(q\xi/2)}{q\xi/2} \right], \tag{5.17}$$

where C_1 and C_2 depend on the form of the interactions. Fluctuations of the local volume fraction of amphiphiles in the lyotropic mixture have essentially two origins: (i) thermal fluctuations of the local cell size of the structure; (ii) thermal fluctuations of the mean curvature of the structural amphiphilic membrane locally averaged over a volume of dimensions comparable to the cell. However, it is important to note that there is no cell as in the usual case of solid state physics. To some extent, the structure of the L_3 phase can be sketched as a *melted cubic* structure of lattice parameter $\overline{d}$ [44,52]. In contrast to a cubic phase, in the L_3 phase there is no positional long-range order and periodicity.

We now use some thermodynamic and scaling arguments in order to write an expression for the free energy of a system of (non-intersecting) membranes [43]. First, let us recall that the bending elastic energy $\mathrm{d}E$ of an area element $\mathrm{d}A$ of a bilayer membrane is given by the quadratic form

$$\mathrm{d}E = \left[\frac{1}{2}K(c_1 + c_2)^2 + \overline{K}c_1c_2\right]\mathrm{d}A, \tag{5.18}$$

where c_1 and c_2 are two principal curvatures, K is the mean curvature modulus, and $\overline{K}$ is the Gaussian curvature modulus (which has no counterpart in one dimension). The rigidity moduli, $K/k_{\mathrm{B}}T$ and $\overline{K}/k_{\mathrm{B}}T$, are pure numbers; a dilation λ changes c_1 and c_2 into c_1/λ and c_2/λ, and $\mathrm{d}A$ into $\lambda^2\mathrm{d}A$, so that the elastic energy $\mathrm{d}E$ remains unchanged under the dilation. We now consider two membrane systems of areas A and $A' = \lambda^2 A$, within the volumes V and $V' = \lambda^3 V$, respectively. The ratio $V/A = V'/A'$ represents the characteristic distance $\overline{d}$ between membranes. Apart from small fluctuations, any membrane configuration of the first system should correspond to a (dual) membrane configuration of the second system. These dual configurations have the same elastic energy, and therefore contribute with the same weight to the expression of the free energy. This means that the contribution of these large-scale configurations to the free energy, as well as the elastic energy itself, are scale invariant under the dilation by any factor λ. The free energy, including the contribution of small ripples, given by a term of the form $\mu_{\mathrm{A}}A$, is written as

$$F = \mu_A A + B_A(K, \overline{K}, T)\frac{A^3}{V^2}, \tag{5.19}$$

where the second term is invariant under dilation. Using the definition of the volume fraction, $\phi = (A\delta)/V$, and taking into account that the partition function depends on K, $\overline{K}$, and T through the ratios K/T and $\overline{K}/T$ only, we have the free-energy density,

$$f = \frac{F}{V} = \mu_\varphi \phi + T\beta\left(\frac{K}{T}, \frac{\overline{K}}{T}\right)\phi^3. \tag{5.20}$$

Note that we are assuming non-intersecting membranes, and that these arguments do not account for the effects of disconnected subunits.

From Eq. (5.20), we obtain the osmotic pressure,

$$\pi = -f + \phi \frac{\partial f}{\partial \phi} \sim \phi^3, \tag{5.21}$$

so that the intensity scattered at zero wave vector is given by

$$I_l(q \to 0) \sim \phi \left(\frac{\partial \pi}{\partial \phi} \right)^{-1} \sim \phi^{-1}. \tag{5.22}$$

Using these equations in order to fit the experimental data, it is possible to obtain ξ and $\overline{d}$ as a function of ϕ [53]. For a mixture of AOT, D_2O, and NaCl, Skouri and coworkers found $\overline{d} \sim 50$ nm, and $\xi \sim 10^2$ nm, both depending on ϕ. However, the ratio $\xi/\overline{d} \simeq 3.2$ was shown to be independent of ϕ. From Eq. (5.22), the osmotic compressibility is expected to be proportional to ϕ^{-1} [41,43], although there are experimental results indicating deviations from this behavior [53]. The presence of small wavelength curvature fluctuations renormalizes both rigidity moduli of the structure, associated with the mean and Gaussian curvatures of the lipid membrane, and the parameter A [43]. In the particular case of the osmotic compressibility, the corrections introduced by this renormalization lead to

$$[I_l(q \to 0)\phi]^{-1} \propto \ln \phi. \tag{5.23}$$

This behavior was observed in several multicomponent mixtures with sponge phases, as AOT/brine, n-dodecylbetain/pentanol/water, and cetylpyridinium bromide/n-hexanol/brine. Although there is still some controversy [54], data for the pseudo-binary mixtures AOT/$NaNO_3$/water and AOT/NaCl/water seem to be well accounted for without the need of this logarithmic correction [55,56].

Quasi-elastic light scattering allows measurements of the relaxation of concentration fluctuations, which is a single exponential function with a typical time τ_D [43],

$$\tau_D^{-1}(q, \phi) = D_c(\phi) q^2, \tag{5.24}$$

where $D_c(\phi)$ is the diffusion coefficient, which was shown to be proportional to the volume fraction ϕ. Typical values of D_c in mixtures based on AOT range from $\sim 0.2 \times 10^{-7}$ to $\sim 1.1 \times 10^{-7}$ cm^2/s for $\phi \sim 0.05$ and $\phi \sim 0.38$, respectively.

Static light scattering experiments performed with samples from different regions of the multidimensional phase diagrams of lyotropic mixtures with a sponge phase show evidences of a second-order symmetric–asymmetric phase transition [57,58]. In the symmetric situation, the surfactant layer divides the space into two equal volumes (inside and outside); the asymmetric case corresponds to a division into two different volumes (e.g. less inside than outside). The usual expansion of the free energy, with only one (density) order parameter, does not account for both symmetric and asymmetric phases, and a phase transition. Coulon and coworkers [57] explained the problem by introducing an additional order parameter, which takes into account the difference between the inside and the outside volumes of the structure.

5.3.2 *Small-angle X-ray and neutron scattering experiments*

Small-angle X-ray (SAXS) and neutron (SANS) diffraction techniques give essential information on the local ordering and structure of the sponge phase. The analysis of these scattering experiments leads to the possibility of building structural models. X-ray experiments are sensitive to electronic density contrasts in the bicontinuous structure, coming from different electronic distributions of polar heads of the amphiphilic molecules, of the solvent, and of the carbonic chains. Neutron experiments are sensitive to the cross section of atoms in the lyotropic mixture, being particularly useful if the same mixture has both ^{2}H and ^{1}H isotopes in different parts of the bicontinuous structure. In a certain sense, both X-ray and neutron diffraction techniques give essentially complementary information about the structure at nanoscales, referring to different ranges of the modulus q of the scattering vector.

We now analyze the scattering patterns of the sponge phase at small and high scattering angles. In the small-q range, patterns present [44,49,50,53,59] a broad peak due to correlations between pieces of lamellae. In q space, the position of this peak can be written as a function of the inter-layer repeating distance $\overline{d}$ (assuming a locally lamellar-type structure) and the relative volume ϕ,

$$q_m = \frac{2\pi}{\overline{d}} = \frac{2\pi\phi}{\gamma_q\delta}, \tag{5.25}$$

where γ_q is a number larger than 1 (usually, $\gamma \sim 1.5$) [42] and δ is the thickness of the bilayer. Typical values of δ, obtained from SANS experiments, are 2.9 and 3.1 nm, for mixtures of SDS, pentanol, water, and NaCl, and of SDS, pentanol, dodecane, and water, respectively [49]. Comparing these values to the thickness α of the lamellae of the neighboring lamellar phases in the phase diagrams of the SDS/pentanol/water/NaCl and SDS/pentanol/dodecane/water mixtures, we have $\alpha = 1.9$ nm and $\alpha = 2.1$ nm, respectively. These results give $\delta/\alpha \sim 1.5$, which corresponds to the value of γ_q in the expression for q_m. In the neighboring lamellar phase, it was shown that $\alpha \propto \phi^{-1}$, although a more accurate analysis of the diffraction data suggests corrections to this dependence [60]. As in the case of the lamellar phases, this process of introducing solvents in the structure of the L$_3$ phase is called *swelling* of the bilayer structure. In the high-q range, patterns present the typical form factor of a flat bilayer. In mixtures with carbonic chain tails, patterns also present the typical broad band due to the distance between chains (of about 0.5 nm). At large values of q, from the point of view of local structures, L$_\alpha$ and L$_3$ phases present bilayers with the same morphology. In direct sponge phases, the "layer thickness" is about twice the length of the principal amphiphile. In inverse water-based sponge phases, the layer thickness is related to the water layer, which in general does not have a typical length scale and changes according to the relative concentrations of the components of the mixture. The scattering curves display a crossover from a behavior of the

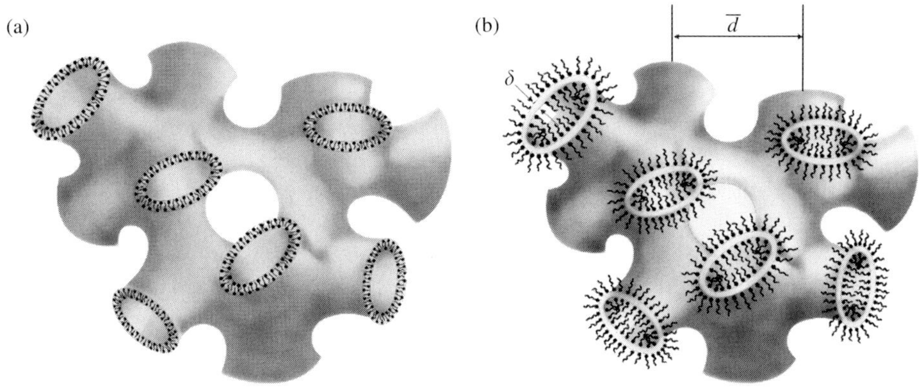

FIG. 5.4. Sketch of the sponge phase structure. Parameters δ and $\bar{d}$ represent the thickness of amphiphilic bilayer (water layer, for an inverted phase) and the mean distance between channels: (a) direct structure; (b) inverted structure.

form $1/q^2$, for $q < 2\pi/\bar{d}$, to a form $1/q^4$, for $q > q_m = 2\pi/\bar{d}$, which characterize the scattering of a flat bilayer with random orientation and a thin interface, respectively.

In a very dilute regime, the sponge phase of a mixture of Na-octylbenzene sulfonate, n-pentanol, and a brine (0.5 M NaCl) [47] presents an interesting behavior. Neutron scattering intensity is given by

$$I_n(q) \sim q^{-2} \exp(-\delta^2 q^2/12), \tag{5.26}$$

with $\delta = 1.5$ nm, which is a characteristic behavior of a random dispersion of a large lamellar-type structure of thickness δ. This mixture presents a micellar isotropic phase, L_1, in the vicinity of the L_3 region, with long cylindrical micelles of radius about 1.5 nm. The analysis of the scattered intensity as a function of q, along a path in the phase diagram from the L_3 to the L_1 phase, indicates that the transition should be mediated by the appearance of microscopic defects, probably edges, in the bilayer-type structure of the former sponge phase.

In summary, the available scattering results indicate that a direct L_3 phase structure can be modeled as a microstructure in which a surfactant bilayer, of thickness δ and multiply connected topology, separates two solvent regions, symmetric and asymmetric, over macroscopic distances [41] (see Fig. 5.4(a); as shown in Fig. 5.4(b), there have been observations of inverted structures).

5.3.3 *Electrical conductivity and viscosity measurements*

Electrical conductimetry, referring to measurements of the electrical conductivity σ of the mixture, is another experimental technique widely used to investigate the structural properties of the L_3 phase. Usually, this type of measurement is made with a Wayne Kerr Autobalance Universal Bridge operating at frequencies in the kilohertz range. The mobility of the ions in the mixture is expected to depend on the topology of the amphiphilic structure. Conductivity measurements can then be compared with theoretical calculations from different structural arrangements of amphiphilic molecules in the L_3 phase, which indicates the most probable structural model, compatible with the experimental observations.

We now discuss some examples of experimental results supporting the bicontinuous character of the sponge phase structure. Porte and coworkers [42] measured σ as a function of the volume fraction of brine, ϕ_w, in a mixture of cetylpyridinium bromide, n-hexanol, and brine (0.2 M NaCl), and found an almost constant value, slightly increasing as $(1 - \phi_w)$ increases from 0 to about 0.2. For $(1 - \phi_w) = \phi_a \to 0$, that is, if the volume fraction of surfactants goes to zero, we have

$$\sigma_s = \lim_{\phi_a \to 0} \sigma \sim \frac{2}{3}\sigma_0, \tag{5.27}$$

where σ_0 is the conductivity of the brine only. This indicates that the structure of the surfactant in the sponge phase, even in a very dilute regime, imposes a constant finite obstruction to the mobility of ions. This "obstruction factor," defined as $\sigma_r = \sigma_s/\sigma_0$, can be calculated from microscopic structural models; in the case of a flat bilayer with random orientation, Jonsson *et al.* [61] found $\sigma_r = 2/3$, in agreement with experimental observations. A similar result was obtained for a quasi-binary mixture of pentaethylenglychol n-dodecyl ether ($C_{12}E_5$) and water, with the addition of 0.1 wt% of NaCl [62]. It was shown that the relative (with respect to the brine) conductivity in the L_3 phase is smaller than in the L_1 phase. Defining the reduced conductivity, $\sigma' = \sigma/(\sigma_0\phi_w)$, it was observed that σ' decreases from about 0.8 to about 0.5 in the range $0 \lessapprox \phi < 0.18$. This decrease is more pronounced in the range $0 \lessapprox \phi \lesssim 0.02$; moreover, in the limit of large ϕ, the value of σ' agrees with theoretical calculations and experimental values for σ_r in the cubic bicontinuous phase. The abrupt increase of σ' at $\phi \lesssim 0.02$ could indicate the disruption of the bicontinuous structure, giving rise to fragments of bilayers or even to the formation of micelles in a very dilute regime of the mixture. This abrupt increase of the reduced conductivity as a function of solvent concentration in highly diluted L_3 phases, which was also observed in a mixture of SDS, pentanol, water, and NaCl [49], seems to be a general feature of the sponge phases.

As a general conclusion, conductivity measurements are consistent with the bicontinuous character of the L_3 phase structure. Measurements of the *relative viscosity*, defined as the ratio $\eta_r = \eta/\eta_s$, where η is the viscosity of the solution and η_s is the viscosity of the solvent, are also consistent with this picture. Porte and coworkers [42] measured η_r as a function of ϕ_a in the cetylpyridinium bromide/n-hexanol/brine mixture, along a path of the partial phase diagram from the micellar isotropic phase to the sponge phase. An abrupt increase of η_r, by almost a factor 3, was observed in the transition from the micellar phase to the L_3 phase, which is consistent with the bicontinuous character of the sponge phase structure. An interesting result was recently reported by Porcar and coworkers [63] when studying the behavior of the L_3 phase of the cetylpyridinium chloride/hexanol/brine mixture under shear in a Couette cell. The measurements of η_r as a function of the *shear rate* $\dot{\gamma}$ show three different regions: (i) at low values of $\dot{\gamma}$, a typical Newtonian behavior with constant η_r; (ii) at high values of $\dot{\gamma}$, another region with constant η_r, but a factor 3 smaller than in the previous region; (iii) an intermediate region in which η_r decreases continuously. These results indicate that shear can impose local modifications of the sponge-type structure, with a possible shear-induced phase transition.

5.3.4 *Flow-induced birefringence*

One of the most spectacular properties of the L_3 phase is flow-induced birefringence [43,63,64]. Under shear, or even under a gentle shaking, there appears a velocity gradient and a transient flow birefringence, which relax with a characteristic time, from some *seconds* up to *minutes*, depending on the system and relative concentrations. For example, the AOT/dodecanol/water mixture shows a relaxation time of about 250 s [44]. Induced birefringence was shown to be proportional to $\dot{\gamma}\phi^{-2}$ [43]; also, it increases with the water content of the mixture. Samples of the L_3 phase of a mixture of SDS, pentanol, dodecane, and water, placed in a Couette cell, show the existence of a *critical shear rate* $\dot{\gamma}_c$, as a function of ϕ, above which there is flow birefringence [64]. This critical shear rate was shown to behave as [65]

$$\dot{\gamma}_c \propto \frac{k_B T}{\eta_s}\left(\frac{\phi}{\delta}\right)^3. \tag{5.28}$$

Fusion of membranes has been suggested as a microscopic mechanism in the sponge structure in the presence of shear. Above the critical shear rate, the deformation imposed to membranes could lead to a *local fusion process*, forming anisometric structures responsible for the observed birefringence. Yamamoto and Tanaka [66] reported experimental evidences of a shear-induced sponge–lamellar first-order phase transition in the $C_{12}E_5$/water mixture. Shear seems to narrow, and finally suppress, the region of coexistence of lamellar and sponge phases. More recently, it has been shown that both dynamic and static processes, as wetting [45], can induce the appearance of an anisotropic, probably lamellar, phase.

Porte and coworkers proposed that shear-induced optical birefringence should be written as [43]

$$\Delta n = B_{\text{flow}}\dot{\gamma}, \tag{5.29}$$

where B_{flow}, with dimension of time, expresses a dynamical feature of the medium. This parameter can be positive or negative, depending on the sign of the induced birefringence. In terms of the volume fraction of amphiphile in the lyotropic mixture, it should behave as $B_{\text{flow}} \sim \phi^{-2}$. In the linear regime (small values of $\dot{\gamma}$), the optical axis of the shear-induced anisotropic structure is expected to orient at about 45° off the direction of the velocity gradient, on the plane of both velocity and velocity gradient. Measurements of $B_{\rm flow}$ as a function of ϕ for the CPCl/hexanol/brine mixture indicate the existence of corrections to this simple scaling law. Indeed, a better representation is given by

$$B_{\rm flow} \sim \Lambda_1 \phi^{-2} + \Lambda_2 \phi^{-1}, \tag{5.30}$$

where Λ_1 and Λ_2 are constants. In the range $0.03 \lesssim \phi \lesssim 0.3$, $B_{\rm flow}$ changes from approximately 720×10^{-10} to -5×10^{-10} s.

5.4 Calculations for the sponge phase

A random surface model was first proposed by Cates and coworkers [58,67], and later developed by Golubovic and Lubensky [68], for explaining a symmetry-breaking transition within the sponge phase, as well as the transition from the sponge to the neighboring swollen lamellar phase. The idea consists in considering the configurations of non-intersecting, defect free, random surfaces, defined by Ising spin variables on a lattice of (cubic) cells. It is difficult, however, to carry out any analysis of this random-surface model beyond the mean-field level, which makes it interesting to resort to a Landau phenomenological expansion of the free energy. In this section, we briefly review these two types of treatment.

5.4.1 *The lattice model of random surfaces*

Consider a cubic lattice of Ising (up or down) spin variables. A spin configuration of the lattice corresponds to a collection of up and down domains of spins. We now imagine the bilayers along the interfaces between domains of opposite spins (and smooth out any sharp corners that may arise). Removing the spins and keeping the bilayers, the resulting surface has no defects and divides the space into continuous regions of identical material, which may be called I (inside) and O (outside). Ising-type Hamiltonians are invariant under changes of up and down spins, which is equivalent to the symmetry between I and O domains of space. The spontaneous breaking of this I/O symmetry may lead to a (second-order) phase transition in the context of the random surface model.

According to this proposal [58], the space is coarse grained by cubic cells of size ξ, of the order of the distance d between bilayers, and each cell is associated with a two-state variable, $\phi_i = 0, 1$. A part of the surface is drawn between neighboring sites i and j with $\phi_i \neq \phi_j$. We then construct spin Hamiltonians in order to mimic bending and Gaussian elastic energy terms [68]. For example, in a crude approximation, we may use the spin Hamiltonian

$$H = a(K) \sum_{(i,j)} N_{ij}, \tag{5.31}$$

where $a(K)$ depends on the mean curvature modulus K, the sum is over nearest-neighbor pairs of cells, and we take into account pair interactions only,

$$N_{ij} = \phi_i(1-\phi_j) + (1-\phi_i)\phi_j. \tag{5.32}$$

Note that $N_{ij} = 1$ if $\phi_i \neq \phi_j$, and $N_{ij} = 0$ otherwise, such that it is equivalent to an "occupation number of broken bonds." The prefactor $a(K)$, which should be linearly dependent on K, comes from considerations of the local elastic energy.

A standard mean-field analysis can be developed on the basis of the well-known inequality

$$F \le F_0 + \langle H - H_0 \rangle_0 = F_{\rm MF}, \tag{5.33}$$

where F is the free energy, H_0 is a trial Hamiltonian,

$$H_0 = -\sum_i \eta\phi_i, \tag{5.34}$$

η is a variational parameter, $\langle \cdots \rangle_0$ is an average with respect to an ensemble defined by H_0, and the free energy in the mean-field approximation, $F_{\rm MF}$, is an upper bound to F. The trial free energy is given by

$$\frac{1}{V}F_0 = -\frac{k_{\rm B}T}{V}\ln {\rm Tr}\exp(-\beta H_0) = -\frac{k_{\rm B}T}{\xi^3}\ln[1+\exp(\beta\eta)],$$

where $k_{\rm B}T = 1/\beta$. Now it is convenient to introduce the definition

$$\psi = \langle \phi_i \rangle_0, \tag{5.35}$$

which may be interpreted as the probability of occupancy of a cell, and to write η in terms of ψ. It is thus straightforward to obtain the (coarse-grained) mean-field expression

$$\frac{1}{V}F_{\rm MF} = F_{\rm bend} - TS_{\rm mix}, \tag{5.36}$$

where

$$F_{\rm bend} = \frac{3}{\xi^3}a(K)\psi(1-\psi) = \frac{1}{\xi^3}8\pi K\psi(1-\psi) \tag{5.37}$$

is an elastic free-energy density, with a suitable choice of $a(K)$, and

$$S_{\rm mix} = -\frac{k_{\rm B}}{\xi^3}[(1-\psi)\ln(1-\psi) + \psi\ln\psi] \tag{5.38}$$

is the standard form of the entropy of a mixture. The mean-field solutions come from the minimization of $F_{\rm MF}$ with respect to ψ.

Calculations for this simple model already lead to some interesting results. Let us assume that the free energy of the swollen lamellar phase $\rm L_\alpha$ is given by

Helfrich's expression [58],

$$F_{L_\alpha} = \frac{3\pi^2}{128}\frac{\phi}{\delta}\frac{(k_B T)^2}{K\alpha^2}, \tag{5.39}$$

where α is the interlamellar spacing. Then we have:

(i) at large values of the ratio between the mean curvature modulus and the temperature, K/T, the lamellar swollen phase L_α is stable. Upon lowering K/T, there is a transition to a symmetric sponge L_3 phase (with $\psi = 1 - \psi = 1/2$);

(ii) upon further lowering the ratio K/T, the system undergoes a symmetry-breaking transition ($\psi \neq 1/2$).

As pointed out by different authors, a more complete representation of the elastic interactions requires the inclusion of at least four-spin interactions in the expression of the spin Hamiltonian [58,68]. For example, in the coarse-grained mean-field treatment, the free energy should include terms of the form

$$F_G = \frac{1}{\xi^3} 4\pi\overline{K}\psi(1-\psi)\{1 - 16\psi(1-\psi)[\psi^2 + (1-\psi)^2]\}, \tag{5.40}$$

where $\overline{K}$ is the Gaussian curvature modulus. At small values of the ratio K/T, we still have a symmetric sponge phase. Also, in this phase, it has been shown that the free energy obeys the thermodynamic scaling form, given by Eq. (5.20), with the addition of a logarithmic correction.

Although lattice models are fascinating, and may lead to a number of interesting results, we now turn to the much more general (yet phenomenological) Landau approach.

5.4.2 *Landau expansion and phase diagrams*

The Landau expansion for the free energy of a lyotropic mixture with a sponge phase is usually written in terms of two order parameters [58]. One of them is chosen as

$$\rho = \phi - \phi^*, \tag{5.41}$$

where ϕ is the volume fraction of surfactant and ϕ^* is a reference value (note that ϕ and ϕ^* play a similar role as the specific volume and the critical specific volume of an ordinary fluid near the critical point). The underlying I/O symmetry of the random-surface models suggests the introduction of the second order parameter, η, so that $\eta = 0$ in the symmetric phase, and $\eta \neq 0$ if the symmetry is spontaneously broken.

An expansion for the grand thermodynamic potential, $\Phi = f - \mu\rho$, where f is a density of Helmholtz free energy and μ is a chemical potential, may be written as

$$\Phi = -\mu\rho + \frac{1}{2}a\rho^2 + \frac{1}{4}\rho^4 + \frac{1}{2}A\eta^2 + \frac{1}{4}\eta^4 + \frac{1}{2}\rho\eta^2, \tag{5.42}$$

where we keep the most relevant, lowest-order, terms only. Note that, as in a usual fluid, we can always shift ρ, and eliminate a ρ^3 term. Also, the presence of even powers of η, and the form of the coupling, $\rho\eta^2$, are chosen to preserve the I/O symmetry (the coefficient of this coupling term can be conveniently set to a positive value without any loss of generality). Given the chemical potential μ, and the coefficients a and A, the thermodynamic solutions come from the minimization of this grand potential with respect to the order parameters ρ and η.

As pointed out by Roux and coworkers [58], the same form of Landau expansion for the thermodynamic potential also appears in the context of the spin-1 Blume–Emery–Griffiths model, which was originally proposed to describe the tricritical behavior in mixtures of ^{3}He and ^{4}He. Phase diagrams associated with this form of thermodynamic potential are known to display a number of distinct topologies, including tricritical and higher-order multicritical behavior. In Fig. 5.5(a) and (b), we sketch two typical phase diagrams, in terms of field variables (the coefficient a, usually associated with temperature, and the chemical potential μ) at fixed values of the coefficient A; in Fig. 5.5(c) and (d), the corresponding diagrams are drawn in terms of the coefficient a and the density ρ (which may be experimentally more relevant).

It is relatively easy to use Eq. (5.42) in order to derive analytic expressions for the second-order lines and the tricritical points. For example, we have

$$\frac{\partial\Phi}{\partial\rho} = -\mu + a\rho + \rho^3 + \frac{1}{2}\eta^2 = 0, \tag{5.43}$$

from which we can write the expansion of ρ in terms of even powers of η,

$$\rho = C_1 + C_2\eta^2 + C_4\eta^4 + \cdots, \tag{5.44}$$

where C_1 comes from the solution of the equation

$$-\mu + aC_1 + C_1^3 = 0, \tag{5.45}$$

$C_2 = -1/(2a + 6C_1^2)$, and so on. Inserting this expansion for ρ into the Landau expansion, we have

$$\Phi = \Phi_0 + \frac{1}{2}(A + C_1)\eta^2 + A_4\eta^4 + \cdots, \tag{5.46}$$

where it is not difficult to write an expression for the coefficient A_4. The line of second-order transitions is given by $A + C_1 = 0$, with $A_4 > 0$, which leads to

$$\mu + Aa + A^3 = 0, \tag{5.47}$$

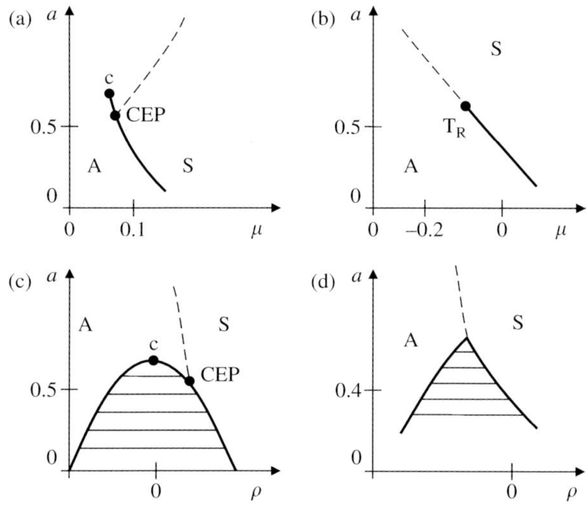

FIG. 5.5. Sketch of some phase diagrams associated with the Landau expansion for the free energy of lyotropic mixtures with a sponge phase: (a) and (c) correspond to $A = -0.2$; (b) and (d) correspond to $A = 0.2$. We indicate the symmetric (S) and ordered (A) sponge phases. In (a) and (b), heavy solid lines represent first-order transitions; continuous transitions are represented by dashed lines. C, $\mathrm{T_R}$, and CEP, label critical, tricritical, and critical end points, respectively. In (c) and (d), we draw the corresponding phase diagrams in terms of field and density variables, and indicate some tie lines of phase coexistence.

supplemented by the stability condition

$$a(2a-1)+6A^2(a-1)>0. \tag{5.48}$$

In the particular case $A = 0$, the second-order line is given by $\mu = 0$, and $a > 1/2$; if the expansion is carried out to the next-order term, it is possible to show that $\mu = 0$ and $a = 1/2$ is a double critical endpoint. Also, we can do some simple numerical calculations to draw the first-order lines sketched in Fig. 5.5.

We now turn to the calculation of correlations functions, which can be experimentally measured by small-angle light-scattering techniques. It is then necessary to add appropriate gradient terms to the Landau expansion. Let us include the simplest of these gradient terms only, and use a Gaussian approximation for the treatment of fluctuations. For a particular soluble case [58], we write

$$\Phi = -\mu\rho + \frac{1}{2}a\rho^2 + \frac{1}{2}\rho\eta^2 + g_0(\eta), \tag{5.49}$$

where

$$g_0(\eta) = \frac{1}{2}A\eta^2 + \frac{1}{4}\eta^4 + \frac{1}{2}\gamma_\eta|\nabla\eta|^2. \tag{5.50}$$

Note that there is no gradient coupling between the surfactant density $\rho(x)$ at different spatial positions, so that it is possible to integrate out this order parameter. We then have

$$\Phi(\eta) = \frac{1}{2}A'\eta^2 + \frac{1}{4}\lambda\eta^4 + \frac{1}{2}\gamma_\eta|\nabla\eta|^2, \tag{5.51}$$

where $A' = A + \mu/a$, and $\lambda = 1 - a/2$. In the Gaussian approximation, for small values of λ, we have the Ornstein–Zernicke form,

$$\langle\eta_0\eta_r\rangle = \frac{1}{4\pi\gamma_\eta r}\exp\left[-\frac{r}{\xi_\eta}\right], \tag{5.52}$$

where $\xi_\eta^2 = \gamma_\eta/A'$. As pointed out by Roux and coworkers [58], this behavior cannot be measured directly, since there is no contrast between I and O regions which contain an identical solvent. However, fluctuations of η affect the surfactant density–density correlation function,

$$g(r) = \langle\delta\rho(0)\delta\rho(r)\rangle, \tag{5.53}$$

where $\delta\rho(x) = \rho(x) - \rho(0)$. From Eqs. (5.49) and (5.50), we can see that the probability distribution for $\rho(x)$ is a Gaussian with mean value $\mu/a - \eta(x)^2/2a$ and variance $1/a$. Thus, we have

$$g(r) = \frac{1}{4a^2}\left[\langle\eta(r)^2\eta(0)^2\rangle - \langle\eta(r)^2\rangle^2\right] + \frac{1}{a}\delta(r). \tag{5.54}$$

Since we are assuming Gaussian distributions, we also have

$$\langle\eta(r)^2\eta(0)^2\rangle = \langle\eta(r)^2\rangle^2 + 2\langle\eta(r)\eta(0)\rangle^2, \tag{5.55}$$

from which we write the density–density correlation for the surfactant,

$$g(r) = \frac{1}{4a^2(4\pi\gamma_\eta r)^2}\exp\left[-\frac{2r}{\xi_\eta}\right] + \frac{1}{a}\delta(r), \tag{5.56}$$

whose Fourier transform, given by Eq. (5.17),

$$I(q) = C_1\left[C_2 + \frac{\tan^{-1}(q\xi_\eta/2)}{q\xi_\eta/2}\right], \tag{5.57}$$

where $C_1 = \xi_\eta/(16\pi a^2\gamma_\eta^2)$ and $C_2 = (16\pi a\gamma_\eta^2)/\xi_\eta$, can be measured by static small-angle light-scattering experiments. As we have already mentioned, this characteristic $1/q$ behavior has indeed been observed close to the transitions (in the symmetric phase).

References

[1] P. Ekwall (1975). In *Advances in Liquid Crystals*, Ed. G. H. Brown, Academic Press, New York, vol. 1.

[2] L. E. Scriven (1977). In *Micellization, Solubilization and Microemulsions*, Ed. K. L. Mittal, Plenum, New York.

[3] W. C. Preston (1948). *J. Phys. Colloid Chem.* **52**, 84.

[4] C. Tanford (1980). In *The Hydrophobic Effect: Formation of Micelles and Biological Membranes*, Wiley, New York.

[5] A. M. Figueiredo Neto (1998). *Phase Transitions in Complex Fluids*, Eds by P. Tolédano and A. M. Figueiredo Neto, World Scientific, Singapore.

[6] A. A. Aniansson, S. N. Wall, M. Almgren, H. Hoffmann, I. Kielmann, W. Ulbricht, R. Zana, J. Lang, and C. Tondre (1976). *J. Phys. Chem.* **80**, 905.

[7] F. K. Gottberg, K. A. Smith, and T. A. Hatton (1998). *J. Chem. Phys.* **108**, 2232.

[8] R. J. Hunter (1993). *Foundations of Colloid Science* , 5th edn, Oxford Science Publications, USA, vol. 1.

[9] W. U. Malik and A. K. Jain (1967). *J. Inorg. Nucl. Chem.* **29**, 2825.

[10] B. R. Roszak (1990). *J. Colloid Interface Sci.* **140**, 538.

[11] J. N. Israelachvili (1991). *Intermolecular and Surface Forces*, Academic Press, New York.

[12] C. S. Shida and V. B. Henriques (2001). *J. Chem. Phys.* **115**, 8655.

[13] K. Hamano, N. Kuwahara, T. Koyama, and S. Harada (1985). *Phys. Rev. A* **32**, 3168.

[14] K. Hamano, K. Fukuhara, N. Kuwahara, E. Ducros, M. Benseddik, J. Rouch, and P. Tartaglia (1995). *Phys. Rev. E* **52**, 746.

[15] M. Kahlweit (1981). *Pure Appl. Chem.* **53**, 2069.

[16] J. C. Desplat and C. M. Care (1996). *Mol. Phys.* **87**, 441.

[17] P. H. Nelson, G. C. Rutledge, and T. A. Hatton (1997). *J. Chem. Phys.* **107**, 10777.

[18] N. A. Mazer, G. B. Benedek, and M. C. Carey (1976). *J. Phys. Chem.* **80**, 1075.

[19] D. Stigter, R. J. Williams, and K. J. Mysels (1955). *J. Phys. Chem.* **59**, 330.

[20] K. Granath (1953). *Acta Chem. Scand.* **7**, 297.

[21] A. T. Bernardes, V. B. Henriques, and P. M. Bisch (1994). *J. Chem. Phys.* **101**, 645.

[22] L. Maibaum, A. Dinner, and D. Chandler (2004). *J. Phys. Chem. B* **108**, 6778.

[23] G. B. Benedek (1969). *Optical Mixing Spectroscopy with Applications to Problems in Physics, Chemistry, Biology and Engineering*, Presses Universitaire de France, Paris.

[24] P. J. Missel, N. A. Mazer, G. B. Benedek, C. Y. Young, and M. C. Carey (1980). *J. Phys. Chem.* **84**, 1044.

[25] V. K. Aswal and P. S. Goyal (2000). *Phys. Rev. E* **61**, 2947.

[26] M. Almgren and S. Swarup (1983). *J. Colloid Interface Sci.* **91**, 256.
[27] J. Briggs, R. B. Dorshow, C. A. Bunton, and D. F. Nicoli (1982). *J. Chem. Phys.* **76**, 775.
[28] A. Luzar and D. Bratko (1990). *J. Chem. Phys.* **92**, 642.
[29] P. J. Photinos and A. Saupe (1986). *J. Chem. Phys.* **84**, 517.
[30] P. J. Photinos, G. Melnik, and A. Saupe (1986). *J. Chem. Phys.* **84**, 6928.
[31] P. Photinos and A. Saupe (1991). *Phys. Rev. A* **43**, 2890.
[32] M. Stefanov and A. Saupe (1984). *Mol. Cryst. Liq. Cryst.* **108**, 309.
[33] P. Photinos, S. Y. Xu, and A. Saupe (1990). *Phys. Rev. A* **42**, 865.
[34] P. J. Photinos, L. J. Yu, and A. Saupe (1981). *Mol. Cryst. Liq. Cryst.* **67**, 277.
[35] A. M. Figueiredo Neto, L. Liébert, and Y. Galerne (1985). *J. Phys. Chem.* **89**, 3737.
[36] L. Ye, D. A. Weitz, P. Sheng, S. Bhattacharya, J. S. Huang, and M. J. Higgins (1989). *Phys. Rev. Lett.* **63**, 263.
[37] S. Bhattacharya, J. Sowa, and J. S. Huang (1990). *Phys. Rev. Lett.* **65**, 131.
[38] J. P. Wilcoxon (1991). *Phys. Rev. A* **43**, 1857.
[39] P. T. T. Wong, S. F. Weng, and H. H. Mantsch (1986). *J. Chem. Phys.* **85**, 2315.
[40] E. W. Kaler, H. T. Davis, and L. E. Scriven (1983). *J. Chem. Phys.* **79**, 5685.
[41] D. Roux, M. E. Cates, U. Olsson, R. C. Ball, F. Nallet, and A. M. Bellocq (1990). *Europhys. Lett.* **11**, 229.
[42] G. Porte, J. Marignan, P. Bassereau, and R. J. May (1988). *J. Phys. (France)* **49**, 511.
[43] G. Porte, M. Delsanti, I. Billard, M. Skouri, J. Appell, J. Marignan, and F. Debeauvais (1991). *J. Phys II (France)* **1**, 1101; G. Porte (1992). *J. Phys.: Condens. Matter* **4**, 8649.
[44] M. Magalhães, D. Pusiol, M. E. Ramia, and A. M. Figueiredo Neto (1998). *J. Chem. Phys.* **108**, 3835.
[45] M. Magalhães, A. M. Figueiredo Neto, and P. Tolédano (2000). *Phys. Rev. E* **62**, 5847.
[46] I. S. Barnes, P. J. Derian, S. T. Hyde, B. W. Ninham, and T. N. Zemb (1990). *J. Phys. (France)* **51**, 2605.
[47] M. Filali, G. Porte, J. Appell, and P. Pfeuty (1994). *J. Phys. II (France)* **4**, 349.
[48] N. Lei, C. R. Safinya, D. Roux, and K. S. Liang (1997). *Phys. Rev. E* **56**, 608.
[49] D. Gazeau, A. M. Bellocq, D. Roux, and T. Zemb (1989). *Europhys. Lett.* **9**, 447.
[50] A. Maldonado, W. Urbach, R. Ober, and D. Langevin (1996). *Phys. Rev. E* **54**, 1774.
[51] G. Porte, J. Appell, P. Bassereau, and J. Marignan (1989). *J. Phys. (France)* **50**, 1335.
[52] B. Balinov, U. Olsson, and O. Söderman (1991). *J. Phys. Chem.* **95**, 5931.

[53] M. Skouri, J. Marignam, J. Appell, and G. Porte (1991). *J. Phys. II (France)* **1**, 1121.

[54] G. Porte, J. Appell, and J. Marignan (1997). *Phys. Rev. E* **56**, 1276.

[55] J. Daicic, U. Olsson, H. Wennerström, G. Jerke, and P. Schurtenberger (1995). *J. Phys. II (France)* **5**, 199.

[56] J. Daicic, U. Olsson, H. Wennerström, G. Jerke, and P. Schurtenberger (1995). *Phys. Rev. E* **52**, 3266.

[57] C. Coulon, D. Roux, and A. M. Bellock (1991). *Phys. Rev. Lett.* **66**, 1709.

[58] D. Roux, C. Coulon, and M. E. Cates (1992). *J. Phys. Chem.* **96**, 4174.

[59] M. Filali, J. Appell, and G. Porte (1995). *J. Phys. II (France)* **5**, 657.

[60] J. Engblom and S. T. Hyde (1995). *J. Phys. II (France)* **5** , 171.

[61] B. Jonsson, H. Wennerstrom, P. G. Nilsson, and P. Linse (1986). *Colloid Polym. Sci.* **264**, 77.

[62] R. Strey, R. Schomäcker, D. Roux, F. Nallet, and U. Olsson (1990). *J. Chem. Soc. Faraday Trans.* **86**, 2253.

[63] L. Porcar, W. A. Hamilton, and P. D. Butler (2002). *Phys. Rev. Lett.* **89**, 168301.

[64] O. Diat and D. Roux (1995). *Langmuir* **11**, 1392.

[65] M. E. Cates and S. T. Miler (1989). *Phys. Rev. Lett.* **62** , 1856.

[66] J. Yamamoto and H. Tanaka (1996). *Phys. Rev. Lett.* **77** , 4390.

[67] M. E. Cates, D. Roux, D. Andelman, S. Milner, and S. Safran (1988). *Europhys. Lett.* **5**, 733.

[68] L. Golubovic and T. C. Lubensky (1990). *Phys. Rev. A* **41**, 4343.

6

THE NEMATIC AND CHOLESTERIC PHASES

6.1 Introduction

We now discuss nematic and cholesteric phases in lyotropic mixtures. Both uniaxial and biaxial phases have been characterized in the extensively investigated phase diagram of a mixture of potassium laurate (KL), decanol (DeOH), and water (W or D, for H_2O and D_2O), which we use as an example of the rich polymorphism of these lyotropic mixtures. Lyotropic cholesteric phases can be obtained by doping with chiral molecules a lyotropic mixture with nematic phases.

Modifications of the shape anisotropy of micelles as a function of temperature and relative concentrations of the components, observed by X-ray scattering measurements, are an essential ingredient to account for the experimental phase diagrams. The reentrant features of the topology of these phase diagrams can be explained by the introduction of an extra non-critical order parameter, besides the usual order parameter related to the breaking of uniaxial and biaxial symmetries.

6.2 The potassium laurate/decanol/water mixture

The existence of uniaxial nematic phases in a mixture of potassium laurate, 1-decanol, and water, KL/DeOH/W, in particular the presence of a discotic nematic N_D phase, is known since the beginning of the 1970s. Long [1] already reports the existence of the N_D phase in the mixture KL/potassium chloride/decanol-OD/D. The topology of the phase diagram of the KL/DeOH/D mixture was well established in the seminal work of Yu and Saupe [2]. Besides the identification of the calamitic nematic N_C and the discotic nematic N_D phases, in this work there is a clear characterization of a biaxial nematic N_B phase.

In the following sections, we mainly use this particular mixture to discuss several features of phase transitions in lyotropic nematics.

6.2.1 *Identification of the nematic structures by various experimental techniques*

Nematic phases were characterized by the complementary use of different experimental techniques, as light polarized optical microscopy (LPOM), laser conoscopy, nuclear magnetic resonance (NMR), X-ray and neutron scattering and diffraction.

In the LPOM experiments, we observe the temperature dependence of the phase textures of a liquid crystalline mixture located inside a transparent sample holder and forming a thin film of tenths of micrometers. Each phase is expected to display a typical texture characterized by the orientation of a symmetry axis with respect to the direction of the electric field of light, and affected by the presence of typical defects. This technique is particularly useful to determine the phase sequence of a mixture as a function of temperature. Also, with suitable conoscopic illumination, it is useful to carry out measurements of birefringence in oriented samples [3], which lead to unequivocal determinations of either uniaxial or biaxial phases. However, LPOM alone is not conclusive about the structure of the liquid crystalline sample.

Laser conoscopy [3, 4] is a powerful technique used to measure birefringences of lyotropic nematics in a more precise way. As birefringences of lyotropic mixtures are very small (of order 10^{-3}), we usually need thick samples, of the order of millimeters, in order to have precise measurements. Also, this technique requires good quality of the sample alignment, which is reflected in symmetrical and well-formed interference fringes.

NMR experiments can provide more detailed information about the structure, local ordering, and symmetry of the phases. Studies of the deuteron resonance splitting of D_2O, in particular if samples are rotated in the spectrometer, give important information on the characteristics of the samples in terms of the orientation with respect to the applied magnetic field.

Scattering and diffraction techniques (X-ray and neutrons) complete the analysis of the problem and lead to the determination of the structure of the samples.

6.2.2 *The calamitic, discotic, and biaxial phases*

Let us consider texture observations in KL-based mixtures inside flat glass capillaries (microslides), about 10 μm thick, in a polarizing light microscope, with the sample between crossed polarizers [5, 6].

Freshly prepared N_C samples present a typical schlieren texture. Applying a magnetic field $B \sim 10$ kG, during about 1 min, the texture presents inversion walls whose periodicity scales with the sample thickness [7]. This structure, however, is not stable; after a certain time in the field, there appear closed walls (ellipses); finally, it relaxes to a planar texture, with the director **n** (optical or symmetry axes) parallel to **B**.

An N_D sample placed inside a freshly filled flat glass capillary, between crossed polarizers, also presents a schlieren-type texture. Usually, the nematic director **n** orients perpendicular to the flat glass walls of the sample holder. This orientation gives rise to a pseudo-isotropic texture under the polarizing microscope, which has an isotropic appearance, due to the orientation of the director parallel to the direction of light propagation. This is typical of a homeotropic alignment of the director, which can be improved by the application of a magnetic field in the plane of the wall surface (that is, perpendicular to **n**).

An N_B sample placed in a freshly filled flat glass capillary presents a texture that resembles the texture of the N_C phase. After being oriented, according to the procedure described in Chapter 1, samples in the N_B phase present a planar-type texture. In this orientational condition, both optical axes of the phase are oriented in a plane perpendicular to the flat glass surface (which contains the light propagation direction, the vertical axis, and the vector **B**), each axis forming an angle θ with respect to the vertical axis. In the N_B region, this angle θ depends on temperature and relative concentrations.

Although it is possible to measure the optical birefringence of liquid crystalline samples in most of the modern optical microscopes with a conoscopic facility [3], laser conoscopy [4] is preferentially used to measure both birefringences of biaxial lyotropics in the same experiment. Since birefringences of lyotropic nematics are small, thick samples are usually required for providing measurements of both Δn and δn (where Δn and δn refer to the birefringences that vanish in the uniaxial N_D and N_C phases, respectively). Usually, liquid crystalline films of about 2 mm have to be used to allow the observation of a reasonable number of conoscopic fringes (about six, three along each direction in the plane perpendicular to the laser beam direction), which are needed in order to measure the birefringences with an accuracy of about 10^{-5}. Figure 6.1 shows typical birefringence curves of a lyotropic mixture (KL/DeOH/D) as a function of temperature [4]. Three different regions are observed: for $10.5 < T < 18°C$, there is an N_D phase ($\Delta n = 0$ and $\delta n \neq 0$); for $18 < T < 20°C$, there is an N_B phase ($\Delta n \neq 0$ and $\delta n \neq 0$); for $20 < T \lesssim 45°C$, there is an N_C phase ($\Delta n \neq 0$ and $\delta n = 0$). These measurements of the symmetry-breaking order parameter lead to an unambiguous characterization of the phases, to the determination of transition temperatures, and to the characterization of an eventual critical behavior across the uniaxial–biaxial phase transition.

NMR experiments are extremely useful for the characterization of distinct nematic mesophases. Deuteron NMR measurements of HDO in the N_C phase, with D standing for ^2H, show a typical quadrupolar splitting, depending on temperature and relative concentration of the components of the mixture. Samples of the KL/DeOH/D mixture in the N_C phase have an anisotropic diamagnetic susceptibility, $\Delta\chi > 0$. In a NMR experiment, if the sample rotates around an axis perpendicular to the magnetic field, the observed spectra are typical of

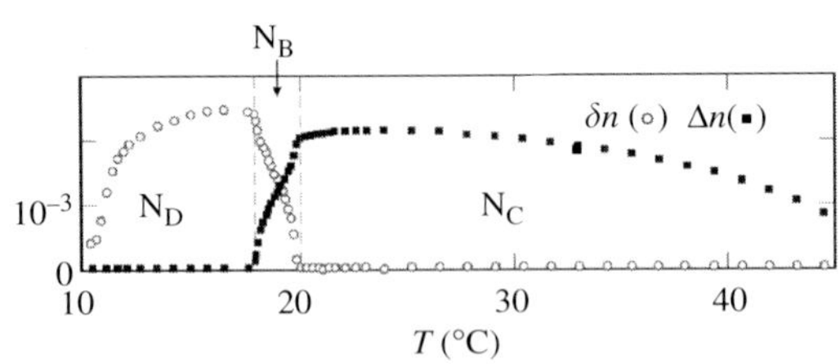

FIG. 6.1. Optical birefringences, Δn and δn, as a function of temperature, obtained from laser conoscopy measurements for a KL/DeOH/D mixture, with weight percentage composition 26/6.24/67.76 [4].

a two-dimensional sample (since the field leads to the spread of the director in the plane perpendicular to the spinning axis). The ^{2}H magnetic resonance of oriented HDO in the N_D phase, associated with $\Delta\chi < 0$, gives a typical quadrupolar splitting. In this case, if the sample rotates around an axis perpendicular to the magnetic field, **n** orients perpendicular to **B**, with the typical spectra of a homogeneously oriented sample. The temperature dependence of ^{2}H magnetic resonance splittings of oriented samples of HDO in the three nematic phases shows strong discontinuities (of the first derivative with respect to temperature) at the uniaxial–biaxial phase transitions [2, 6]. Values of the splitting increase from the borders of the uniaxial–biaxial phase transitions towards the middle of the region of the biaxial phase. In the uniaxial phases, the splitting curves have a smooth convex shape; in the N_B phase, this convex shape is even more pronounced.

Structures of the three nematic phases were determined by X-ray [8, 9] and neutron [10,11] diffraction experiments in oriented samples. The reciprocal space image of the N_C phase is a hollow cylindrical barrel, with the axis parallel to **n** and with more intense sides. There is a strong diffraction band along the direction perpendicular to **n**, which is attributed to the intermicellar positional correlations along this direction. A second-order band along the axis perpendicular to **n** is observed in the diffraction pattern. The reciprocal space image of the N_D phase is a hollow cylindrical barrel, with the axis parallel to **n** and with more intense caps. There is a strong diffraction band and a second-order band along the direction parallel to **n**. The reciprocal space image of the N_B phase is a hollow barrel of elliptical cross section in a plane perpendicular to its long axis, in other words, aflattened ellipsoid, with more intense caps. A strong diffraction band and a second-order band, related to intermicellar positional correlations, are observed along the direction perpendicular to the long ellipsoidal axis.

A model of *intrinsically biaxial micelles* was proposed on the basis of the detailed analysis of the profile of the X-ray diffraction bands obtained in all of the nematic phases [8, 9]. This model is consistent with the NMR results for the nematic mixtures and with the behavior of the optical birefringences as a function of temperature in the neighborhood of the nematic–nematic phase transitions. Available experimental results (diffraction and NMR measurements, behavior in the presence of a magnetic field, light scattering measurements) can be interpreted assuming that there are similar, direct, micelles in the three nematic phases. These micelles have a biaxial symmetry, as a flattened prolate ellipsoid, with three mutually orthogonal two-fold symmetry axes, and the three nematic phases can be regarded as consequences of orientational fluctuations of these micelles.

Estimates for typical dimensions of an orthorhombic micelle are of the order $A' = 8.5$ nm, $B' = 5.5$ nm, and $C' = 2.6$ nm, where C' represents the thickness of the potassium laurate bilayer, and A' and B' are dimensions in the plane perpendicular to the bilayer. The shape anisotropy of the micelles is about $A' : B' : C' = 3 : 2 : 1$. These dimensions change slightly with temperature, but no drastic variations are observed at the nematic–nematic phase transitions. Orientational fluctuations that degenerate the symmetry axis of the phase, perpendicular to the largest micellar surface, give rise to the N_D phase. Orientational fluctuations that degenerate the symmetry axis in the plane of the largest micellar surface, along the largest axis of the flattened ellipsoid, give rise to the N_C phase. Small amplitude orientational fluctuations along the three axes of symmetry of the micelles give rise to the N_B phase. The continuous modification of the shape anisotropy of micelles as a function of temperature is the driven mechanism of the observed phase transitions.

6.2.2.1 *The experimental phase diagram* Due to the presence of three components, the representation of the thermodynamic states of mixtures of potassium laurate, decanol and water, leads to a multidimensional phase diagram. It is usual to indicate the stability of different phases in a surface of this phase diagram, at constant alcohol concentration. In Fig. 6.2, we show a particular surface of the phase diagram of the KL/DeOH/D_2O mixture, at a fixed concentration of DeOH (6.24 wt%) [2, 12].

In this surface phase diagram, depending on temperature and concentration of KL, there is a low-temperature isotropic micellar phase $L_1^{(\ell)}$, a high-temperature isotropic micellar phase $L_1^{(h)}$, and three nematic phases, N_C, N_B, and N_D [2]. The N_B phase is located between the two uniaxial phases. Nematic phases are found at temperatures $10 \lesssim T \lesssim 45°C$. The temperature range of the N_B phase depends on the relative concentrations of the compounds of the mixture.

Nematic–isotropic and nematic–nematic transitions are of first and second order [4], respectively. In this particular lyotropic mixture, it is possible to

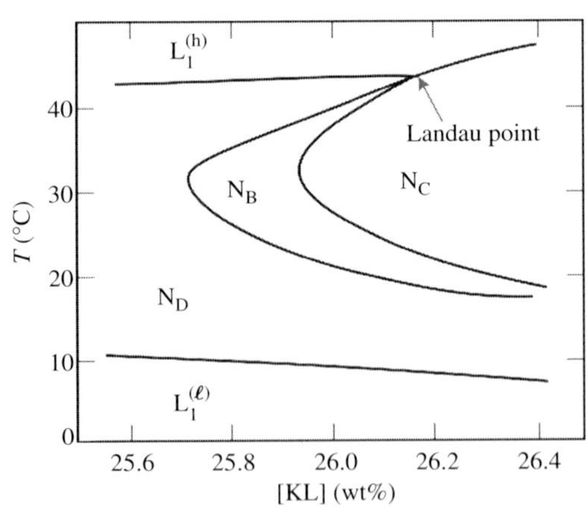

FIG. 6.2. Sketch of the surface of the phase diagram of the potassium laurate/1-decanol/D_2O mixture [2]. Concentration of DeOH is fixed at 6.24 wt%.

experimentally determine the Landau point, at which two critical lines meet a first-order border [13].

It is remarkable to observe the existence of a low-temperature nematic–isotropic phase transition. This transition is a consequence of the change of shape anisotropy of the micelles as a function of temperature. In this particular case, the parameter C' of the micelles increases for decreasing temperatures.

This polymorphism of the phase diagram, with three distinct nematic phases, has been observed in mixtures with at least two amphiphiles. Mixtures with only one amphiphile have uniaxial nematic phases only (discotic or calamitic, depending on the system), and no transitions mediated by a biaxial phase [14,15].

The experimentally observed topology of these phase diagrams is only partially described by the (mean-field) model developed by Freiser [16] and Alben [17]. The Landau point and its neighborhood are explained by the Freiser–Alben model, but it does not describe the low-temperature regions of the phase diagrams (in particular, the existence of an additional, low-temperature, Landau point).

6.2.2.2 *The uniaxial–biaxial phase transition* The uniaxial–biaxial phase transition is one of the most interesting features of lyotropic mixtures. Galerne and Marcerou [4] measured the temperature dependence of Δn and δn with a precision of about 0.02°C, and obtained the invariants of the dielectric tensor associated with the mixture of potassium laurate, decanol, and water (see Chapters 1 and 2). Since this dielectric tensor is traceless, one of the invariants is trivially zero. The other two invariants, I_1 and I_2, have a linear dependence on temperature in the neighborhood of the phase transition, in the uniaxial and in the biaxial regions. In Fig. 6.3, we show the invariants I_1 and I_2 as a function of temperature [4].

In Chapter 2, we have already discussed the Landau–de Gennes phenomenological theory of the uniaxial–biaxial phase transition. The free energy can be written in terms of the invariants of the traceless order parameter,

$$F = aI_1 + bI_2 + cI_1^2 + dI_2^2 + eI_1I_2 + \cdots, \tag{6.1}$$

where the coefficients a and b may be chosen as linear functions of temperature, and c, d, and e are fixed constants. The minimization of this free energy with respect to I_1 and I_2 leads to the equations of state in the neighborhood of the transitions. It is easy to see that the equilibrium values of I_1 and I_2 are linear functions of temperature, which is the expected critical behavior predicted by the mean-field approximation.

Saupe and coworkers [6, 13] used a mean-field approach in order to account for the experimental data. Measurements of one of the birefringences at the $\mathrm{N_B}$ and $\mathrm{N_C}$ phases, with a temperature precision of about 1 mK, indicate small deviations from the mean-field behavior at the uniaxial–biaxial phase transition. This critical behavior has indeed been claimed to belong the universality class of a three-dimensional XY model [18]. Deviations from the mean-field behavior were

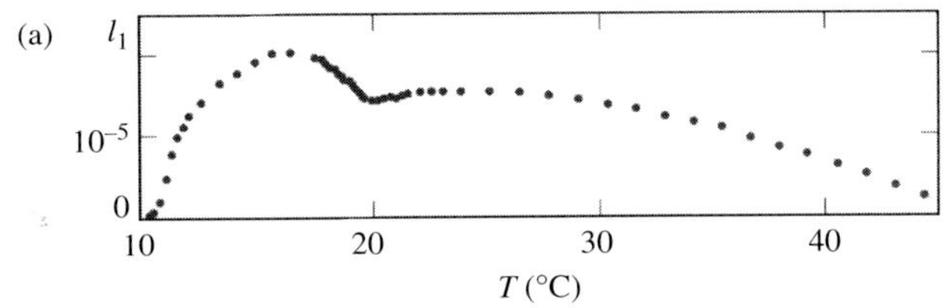

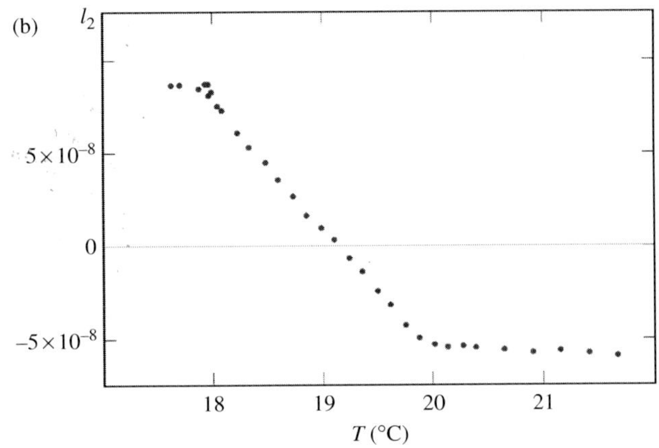

FIG. 6.3. Invariants of the symmetry-breaking order parameter (anisotropic part of the dielectric tensor) as a function of temperature, in the nematic region of the KL/DeOH/D mixture [4]: (a) I_1 invariant; b) I_2 invariant.

also observed [19] in measurements of the optical birefringence, in a temperature range $\Delta T/T_c \sim 10^{-4}$ from the uniaxial–biaxial transition temperature T_c. The critical exponents associated with the order-parameter, β, and the susceptibility, γ, of the KL mixture were found to be $\beta = 0.38 \pm 0.03$ and $\gamma = 1.29 \pm 0.06$, which can be compared with theoretical predictions [18] for the three-dimensional XY model, $\beta \approx 0.38$ and $\gamma \approx 1.35$.

The Z-scan technique [20], in the neighborhood of the nematic–nematic phase transition [21], has also been used to investigate the behavior of the nonlinear refraction indices and of the birefringence. In the vicinity of the transition temperature T_c, the nonlinear birefringence was shown to behave as $|T - T_c|^{-0.5}$ in both uniaxial and biaxial regions (see Fig. 6.4). The symmetry invariants of the order parameter were shown to be linear functions of temperature in the uniaxial nematic region, in good agreement with mean-field predictions.

6.2.3 *Phenomenological calculations for the nematic transitions*

The early mean-field calculations of Freiser [16] and Alben [17] have provided an explanation for the main features of the temperature-concentration phase diagrams of lyotropic mixtures, including the existence of a Landau multicritical

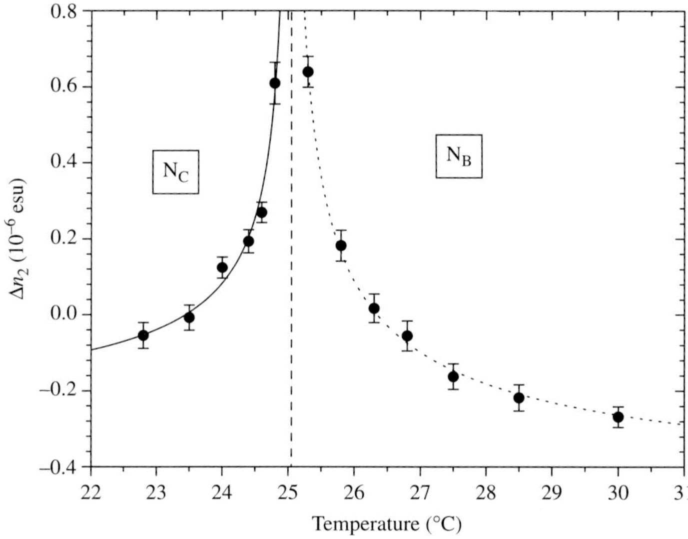

FIG. 6.4. Nonlinear birefringence, $\Delta n_2 = n_{2\|} - n_{2\perp}$, for a sample of potassium laurate, decanol, and water [21]. The subscripts $\|$ and $\perp$ refer to the configurations of the electric field $\vec{E}$ of the probe laser beam with respect to the magnetic field $\vec{B}$ used to orient the nematic N_C phase. The refraction index can be written as $n = n_L + n_2 I_0$, where n_L is the linear index of refraction, I_0 is the intensity of the incident light, and n_2 is the nonlinear index of refraction. The dashed line corresponds to $T_{c1} = 25.02°C$. The solid and dotted lines correspond to the fitting function $\Delta n_2 \propto |T - T_{c1}|^{-\beta}$, with $\beta_C = 0.55(3)$ and $\beta_B = 0.50(1)$ for the N_C and N_B phases, respectively.

point. The Freiser–Alben model is summarized by the Landau–de Gennes expansion of the free energy in terms of the invariants of the tensor order parameter, as discussed in considerable detail in Chapter 2.

We now recall the Landau–de Gennes expansion of the free energy,

$$F = F_0 + \frac{1}{2}AI_1 + \frac{1}{3}BI_2 + \frac{1}{4}CI_1^2 + \frac{1}{5}DI_1I_2 + \frac{1}{6}EI_1^3 + \frac{1}{6}E'I_2^2 + \cdots, \tag{6.2}$$

where F_0 is a regular function, $I_1 = \delta = \mathrm{Tr}\,\mathbf{Q}^2$ and $I_2 = \Delta = \mathrm{Tr}\,\mathbf{Q}^3$ are the invariants of the traceless tensor order parameter

$$\mathbf{Q} = \begin{pmatrix} -\frac{1}{2}(S+\eta) & 0 & 0 \\ 0 & -\frac{1}{2}(S-\eta) & 0 \\ 0 & 0 & S \end{pmatrix}, \tag{6.3}$$

and the coefficients depend on temperature and concentration. As we have pointed out in Chapter 2, a truncated expansion of the form

$$F_U = F_0 + \frac{1}{2}AI_1 + \frac{1}{3}BI_2 + \frac{1}{4}CI_1^2, \tag{6.4}$$

with $C > 0$, provides an explanation for the first-order isotropic–nematic transition (with $B \neq 0$), but it is unable to account for the biaxial phase.

We can still discard some of the higher-order coefficients in the full expansion of the free energy, given by Eq. (6.2). However, it is essential to assume $E' \neq 0$ in order to predict a stable biaxial phase. In Chapter 2, we have shown that the minimization of this free energy with respect to the components of the order parameter leads to some typical phase diagrams, with a Landau multicritical point, as sketched in Fig. 2.11.

Despite predicting a biaxial phase and a Landau multicritical point, the Freiser–Alben model (or the equivalent Landau–de Gennes approach) provides only a partial description of the experimentally observed topology of the nematic phase diagrams. It does not account for the low-temperature reentrant regions, in particular for the existence of an additional, low-temperature, Landau point. Tolédano and Figueiredo Neto [22,23] were then motivated to propose the inclusion of an extra non-critical order parameter associated with changes of shape anisotropy of the molecular aggregates. This scalar non-critical order parameter, which we call τ, is associated with the distribution of micellar shape anisotropies as a function of temperature and concentration.

Although it is possible to use symmetry arguments in order to make an attempt at deriving an expression for τ in the context of the model of intrinsically biaxial micelles [8,9], we just assume that τ is phenomenologically introduced in order to take into account the effects of the changes of shape anisotropy of the micellar aggregates. The parameter τ is supposed to describe the continuous configurational change of the micellar population as a function of temperature. In particular, it should reflect the tendency to a more isotropic shape at lower temperatures, which explains the reentrant low-temperature isotropic phase.

Consider the Landau–de Gennes free energy, given by Eq. (6.2), with $D = E = 0$, $C > 0$, and $E' \neq 0$, which is already sufficient for explaining the existence of a nematic biaxial phase,

$$F = F_0 + \frac{1}{2}AI_1 + \frac{1}{3}BI_2 + \frac{1}{4}CI_1^2 + \frac{1}{6}E'I_2^2 + \cdots . \tag{6.5}$$

The non-critical order parameter τ is introduced according to the following prescriptions: the lowest-order invariant of the original order parameter is redefined as $I_1 \to I_1 + \tau^2$ (but we keep terms up to order τ^2 only); the higher-order invariant I_2 is unchanged; new linear and quadratic terms in τ are added to the expansion. If we discard the regular term, the modified phenomenological free-energy is given by

$$F_\mathrm{m} = a_1\left(I_1 + \tau^2\right) + b_1 I_2 + a_2\left(I_1^2 + 2I_1\tau^2\right) + b_2 I_2^2 + c_1\tau + c_2\tau^2 + \cdots , \tag{6.6}$$

where the coefficients are functions of temperature and concentration. Note that we are changing the notation ($a_1 = A/2$, $b_1 = B/3$, $a_2 = C/4$, $b_2 = E'/6$) in order to make closer contact with the calculations of Tolédano *et al.* [23]. Using the expressions of the tensor invariants, $I_1 = (3S^2+1)/2$ and $I_2 = 3S(S^2-\eta^2)/4$, we may write

$$F_{\mathrm{m}} = F_{\mathrm{m}0} + F_{\mathrm{m}\tau}, \tag{6.7}$$

where

$$F_{\mathrm{m}0} = \alpha(S) + \beta(S)\eta^2 + \frac{1}{2}\gamma(S)\eta^4, \tag{6.8}$$

and

$$F_{\mathrm{m}\tau} = c_1\tau + [a_1 + c_2 + a_2(3S^2 + \eta^2)]\tau^2. \tag{6.9}$$

Note that the polynomial functions $\alpha(S)$, $\beta(S)$, and $\gamma(S)$ are given by the same expressions already obtained in Chapter 2. The τ-independent part of the free energy, $F_{\mathrm{m}0}$, leads to the well-known results of the Freiser–Alben model, including a stable biaxial phase and a Landau multicritical point. The minimization of the total free energy with respect to the parameter τ leads to the relation

$$\tau = -\frac{1}{2}c_1 \left[a_1 + c_2 + a_2\left(3S^2 + \eta^2\right)\right]^{-1}, \tag{6.10}$$

which shows that the equilibrium value of τ is linearly dependent on the coefficient c_1. As we have pointed out in Chapter 2, since τ is proportional to c_1, the transition lines are invariant under the change $c_1 \to -c_1$, which explains the reentrant low-temperature topology of the nematic phase diagrams [23]. The phenomenon of reconstruction of the Freiser–Alben model is represented in a graph of c_1 versus b_1, as sketched in Fig. 2.15 (in this figure, $c_1 \equiv C_1$ and $b \equiv B$), with regions of high- and low-temperature isotropic phases. From a more formal point of view, it is important to remark that Tolédano and Figueiredo Neto have shown that Eq. (6.10) may be obtained from a much more general scheme of minimization of a quadratic form, as the free energy F_{m}, given by Eq. (6.6).

Although this discussion has been restricted to the context of the Landau phenomenology, it should be mentioned that there are some recent attempts at devising microscopic statistical models to account for the nematic transitions. In particular, we mention the work of Henriques and Henriques [24], based in part on a previous proposal of Oliveira and Figueiredo Neto [15]. As we have shown in Chapter 2, at the mean-field level the model of a random distribution of interacting prolate and oblate micelles accounts for the biaxial phase and the Landau multicritical point. The explanation of reentrance phenomena, however, still needs some extra ingredient, as the phenomenological non-critical order parameter τ.

6.3 Nematic phases in other lyotropic mixtures

There are two main groups of lyotropic mixtures with nematic phases: (a) mixtures with uniaxial phases (N_C, N_D) only; (b) mixtures with the three nematic phases (N_C, N_B, and N_D). Particular surfaces of the phase diagrams of some of these mixtures can be found in the literature. In many cases, however, experimental investigations as a function of the relative concentrations of the components have been carried out at room temperature only.

From the experimental point of view, it is important to remark that the N_B phase has never been found alone, without the vicinity of N_C and N_D phases. Moreover, as we have already pointed out, all mixtures presenting the N_B phase have at least two amphiphiles (a surfactant and a cosurfactant).

We now review and list the main lyotropic mixtures with nematic phases. An extensive review of mixtures with lyotropic nematics can be found in the work of Forrest and Reeves [25].

6.3.1 *Binary mixtures*

As we have already mentioned, binary mixtures present only one type of uniaxial nematic phase, N_C or N_D, depending on temperature and relative concentrations. The components are necessarily an amphiphilic molecule and (heavy or light) water. In Table 6.1, we list some common binary lyotropic mixtures, and give some references about investigations of their physico-chemical properties.

Let us present some examples of typical phase diagrams. In Fig. 6.5, we sketch the phase diagram of the mixture CsPFO/water [26]. There is a nematic N_D phase between lamellar (L) and isotropic (ISO) phases, and a coexistence region (POL). In Fig. 6.6, we sketch the phase diagram of the TP6EO2M/water mixture [27]. This diagram displays a nematic N_C phase, columnar phases C_H, and coexistence regions. The N_C phase is located near the region of a columnar hexagonal phase (in Fig. 6.5, note the presence of a lamellar phase in the neighborhood of the N_D region).

Table 6.1 Binary mixtures presenting uniaxial nematic (N_C or N_D) phases. The label *water* means that both heavy (D) and light (W) water were used to prepare the lyotropic mixture

Mixture	Reference
Cesium perfluoro-octanoate (CsPFO)/water	[26]
2,3,6,7,10,11-Hexa-(1,4,7-trioxaoctyl)-triphenylene/water	[27]
Disodium cromoglycate/water	[28, 29]
Tetradecyltrimethylammonium bromide (MTAB)/D	[30]
Ammonium perfluoro-nonanoate/W	[26]

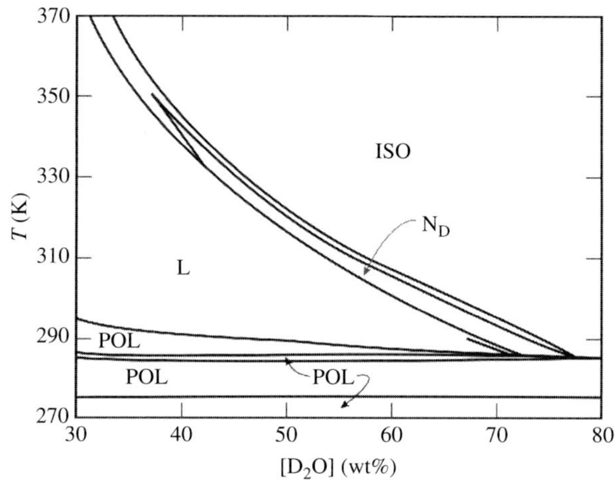

FIG. 6.5. Sketch of the phase diagram of a mixture of cesium perfluoro-octanoate and water [26].

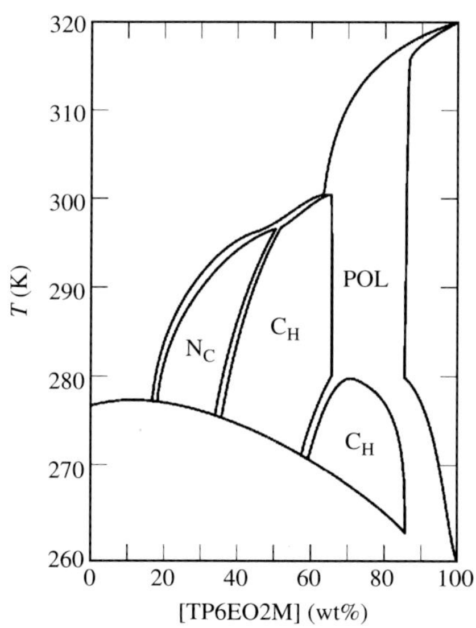

FIG. 6.6. Sketch of the phase diagram of a mixture of 2,3,6,7,10,11-hexa-(1,4,7-trioxaoctyl)-triphenylene and water [27].

6.3.2 *Multicomponent mixtures*

Nematic lyomesophases have also been found in multicomponent mixtures. The presence of one amphiphilic molecule and water is a necessary condition to observe nematic structures. The other components may be another amphiphile (a cosurfactant) or a salt. We have already mentioned the example of the KL/decanol/water ternary mixture. The three nematic phases have also been clearly identified in multicomponent lyotropic mixtures of sodium decylsulfate-(SdS)/1-decanol/water [31], and KL/decylammonium chloride-DaCl/water [32]. In Table 6.2, we list some multicomponent lyotropic mixtures, and give some references about investigations of their physico-chemical properties.

The multidimensional phase diagrams of these mixtures are generally complicated and not completely known. The characterization of an equilibrium state, at given values of temperature and pressure, requires the determination of the amount of each component of the mixture. In the best cases, some particular surfaces of these phase diagrams have been investigated. Also, it is not an easy task to clearly determine regions of different nematic phases and the associated phase boundaries. Detailed measurements of the temperature dependence of suitable order parameters are always essential for constructing these phase diagrams.

Some of the phase diagrams in the literature were drawn on the basis of optical microscopic observations of textures and X-ray diffraction patterns of some samples in a particular phase domain. These procedures, however, just give a coarse view of a particular surface of the phase diagram; they are not

Table 6.2 Multicomponent (from three up to six components) mixtures presenting nematic phases. The label *water* means that both heavy (D) and light (W) water were used to prepare the lyotropic mixture. The three nematic phases, N_C, N_B and N_D, have been found in mixtures marked with the tag #

Mixture	Reference
SdS/1-decanol/water (#)	[31]
KL/DaCl/water (#)	[32]
Sodium dodecylsulfate (SLS)/DeOH/W	[33]
C_8 or C_{10} alkyl sulfates/corresponding alcohol/Na_2SO_4/water	[34–36]
Sodium decylsulfate (SdS)/DeOH water	[10, 37]
DaCl/NH_4Cl/water	[38, 39]
Disodium cromoglycate/NaCl/water	[29]
Cetyltrimethylammonium bromide-CTAB/NaBr/KL/water	[40]
CTAB/DeOH/NaBr/KL/water	[40]
CTAB/decanol/NaBr/water	[40]
Decyltrimethylammonium bromide-DTAB/NaBr/KL/water	[40]
DTAB/CTAB/DeOH/NaBr/KL/water	[40]
DTAB/DeOH/NaBr/KL/water	[40]
MTAB/DeOH/water	[41]
KL/KCl/water	[42]
SdS/Na_2SO_4/water	[42]
Cesium decylsulfate (CsdS)/Na_2SO_4/water	[42]

sufficient for the determination of complete phase sequences. Taking into account these limitations, we now describe some phase diagrams of these multicomponent lyotropic mixtures.

In Fig. 6.7, we sketch a surface of the phase diagram of the SLS/DeOH/W mixture [33]. This surface was obtained by optical observations, without measurements of the order parameter. Therefore, phase boundaries are approximately marked only. Two uniaxial phases were identified as a function of DeOH content (at fixed concentrations of SLS). No biaxial phase was observed. However, since the order parameter was not measured, the existence of a biaxial phase between the two uniaxial phases cannot be ruled out. The labels POL, ISO, COA refer to polyphasic, isotropic and coagel domains, respectively. Complementary experiments are still needed to answer a number of questions about this phase diagram.

In Fig. 6.8, we show a surface of the phase diagram of the MTAB/DeOH/water mixture [41]. As in the previous case, there seems to be a transition between two uniaxial nematic phases.

In Fig. 6.9, we show a surface of the phase diagram of the SdS/1-decanol/water mixture [31]. In this case, there were optical birefringence measurements. The N_B phase was clearly characterized between the two uniaxial nematic phases. It is interesting to note the existence of hexagonal and lamellar regions in the neighborhood of the N_C and N_D phases, respectively.

In Fig. 6.10, we show a surface of the phase diagram of the KL/DaCl/water mixture [32]. Different and complementary techniques (optical microscopy, laser conoscopy, and X-ray diffraction) have been used to draw this phase diagram. The N_B region was clearly identified between the two uniaxial nematic phases.

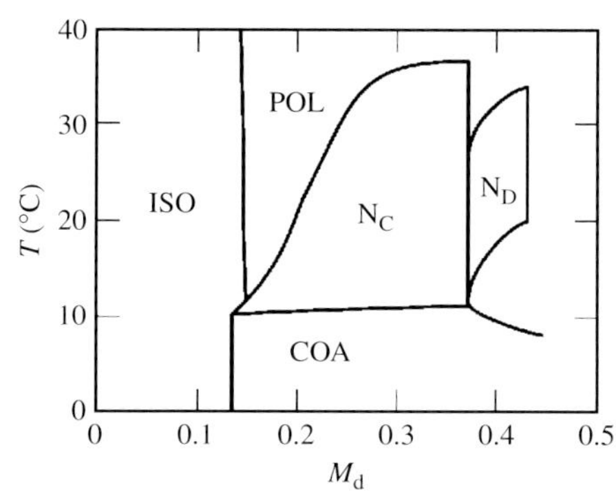

FIG. 6.7. Sketch of the phase diagram of the SLS/DeOH/W mixture [33]. Labels POL, ISO and COA refer to polyphasic, isotropic and coagel regions, respectively. In the horizontal axis, the parameter M_d represents the relative molar ratio of DeOH with respect to SLS.

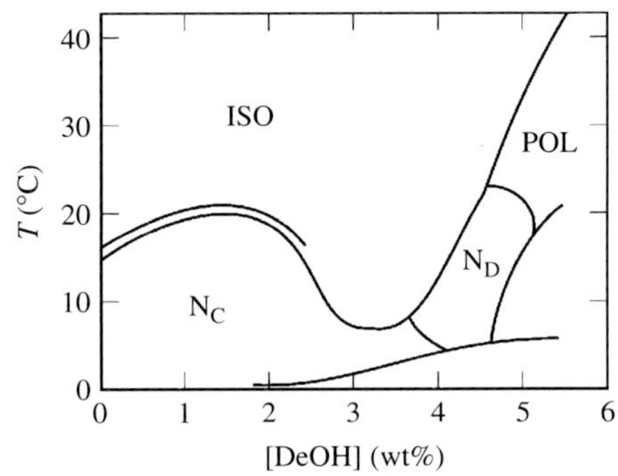

FIG. 6.8. Sketch of a surface of the phase diagram of the MTAB/DeOH/water mixture [41], at the molar ratio [MTAB + DeOH]/[D] = 0.032. Labels POL, L, and ISO refer to polyphasic, lamellar, and isotropic regions, respectively.

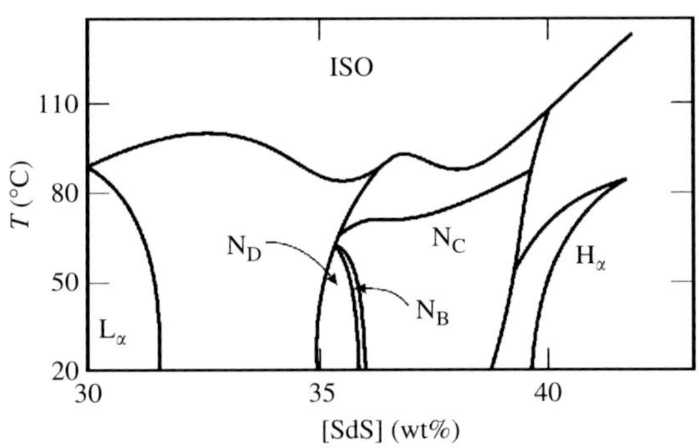

FIG. 6.9. Sketch of a surface of the phase diagram of the SdS/1-decanol/water mixture [31]. Labels ISO, L_α, and H_α refer to isotropic, lamellar, and hexagonal phases, respectively. The width of the N_B domain is restricted to about 0.1 wt% of the concentration of SdS. The water concentration is fixed at 57 wt%.

There is a puzzling result from NMR spectroscopy measurements for the sodium dodecylsulphate/decanol/water mixture. The ^{2}H resonance data were explained with the assumption of two types of biaxial nematic phases, N^+_{Bx} and N^-_{Bx}, in the partial isothermal representation, between the N_C and N_D phases [43]. In order to characterize the behavior in the presence of an external magnetic field, we write the anisotropy of the diamagnetic susceptibility,

$$\Delta\chi = \chi_{33} - \frac{1}{2}(\chi_{11} - \chi_{22}), \tag{6.11}$$

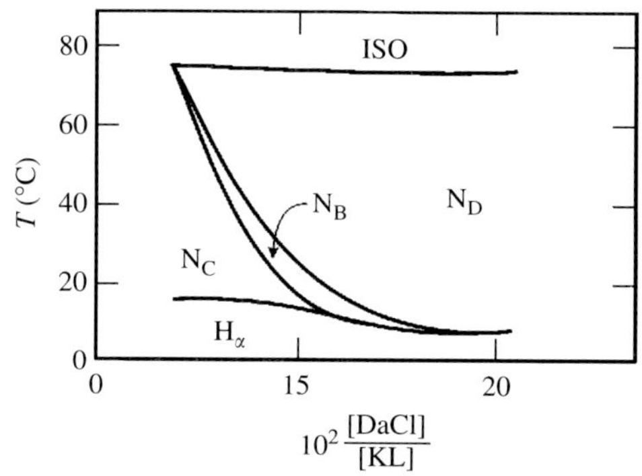

FIG. 6.10. Sketch of a surface of the phase diagram of the KL/DaCl/water mixture [32]. Labels ISO and H_α refer to isotropic and hexagonal phases, respectively. The horizontal axis represents the relative molar ratio of DaCl with respect to KL. The molar ratio $[\mathrm{DaCl} + \mathrm{KL}]/[\mathrm{water}] = 5 \times 10^{-2}$ is kept fixed.

where the subscripts refer to the three orthogonal directions in space. If $\Delta\chi > 0$, a biaxial phase aligns along the axis of the largest diamagnetic susceptibility parallel to the magnetic field (N^+_{Bx}). If $\Delta\chi < 0$, a biaxial phase aligns along the axis with the smallest diamagnetic susceptibility perpendicular to the magnetic field (N^-_{Bx}). In the context of the NMR techniques, we say that phases N^+_{Bx} and N^-_{Bx} align the largest residual *electric field gradient* component along and perpendicular to the magnetic field, respectively. Based on NMR quadrupole splitting measurements, Quist [43] concluded that the transitions between the nematic phases of this system are first order. A similar study was recently performed for another lyotropic mixture composed by tetradecyltrymethylammonium bromide (TTAB), *n*-decanol, and water [44]. An isothermal surface of the phase diagram of this mixture is sketched in Fig. 6.11.

6.4 Lyotropic cholesteric mixtures

Cholesteric lyomesophases can be obtained by adding a chiral molecule to a lyotropic mixture with nematic phases. The addition of an amphiphilic chiral molecule gives rise to an *intrinsic* cholesteric phase. Alternatively, the addition of a non-amphiphilic chiral molecule gives rise to an *extrinsic* cholesteric.

Three types of lyotropic cholesterics are known, Ch_C, Ch_D, and Ch_B. [45]–[47]. The subscripts C, D, and B refer to the original nematic phases, calamitic, discotic and biaxial, which have been cholesterized by the addition of the chiral dopant. As in the case of nematics, mixtures with only one amphiphile give rise to only one of the cholesteric phases (Ch_C or Ch_D). Mixtures with at least two amphiphiles can give rise to the three cholesteric phases.

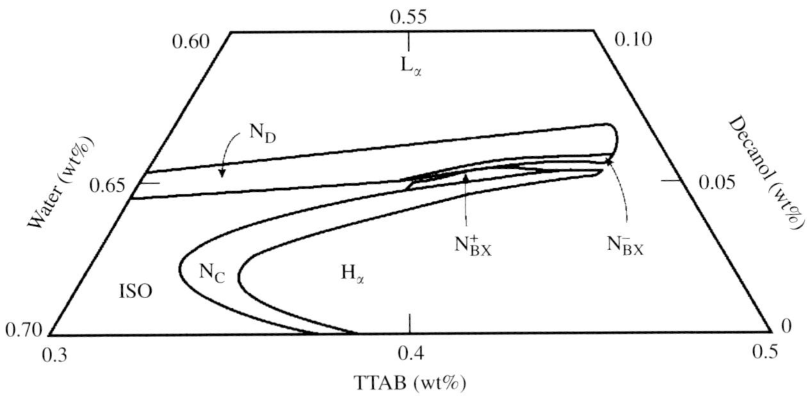

FIG. 6.11. Sketch of the surface of the phase diagram of a mixture of TTAB (tetradecyltrymethylammonium bromide), *n*-decanol and water (at 10 wt% of D_2O), at room temperature (22°C) [44].

6.4.1 *An introductory example*

As a typical example of a cholesteric lyotropic, we now discuss the behavior of a mixture of potassium laurate, decanol, brucine sulfate, and water, KL/DeOH/BS/D_2O [48]. Since the molecules of brucine sulfate (BS) are not amphiphilic, this is an example of an *extrinsic* cholesteric. Typical ratios of molar concentrations of brucine sulfate and potassium laurate molecules are given by $[BS]/[KL] \sim 10^{-3}$.

6.4.1.1 *The uniaxial and biaxial cholesteric phases* Observations of textures of cholesteric lyotropic phases in an optical polarized microscope show a typical fingerprint pattern. It is possible to identify a sequence of parallel stripes, with different colors, for high birefringent samples and white light illumination, and with different gray tonalities, for samples with small birefringence. In the middle of the stripes of the helicoidal fingerprint textures, Ch_C and Ch_D mesophases show a pseudo-isotropic region, with the director oriented in a homeotropic configuration (in other words, at these positions, the director that is twisting along the helical axis becomes parallel to the direction of light propagation). In the Ch_B phase, however, there is no pseudo-isotropic region [47]. The pitch P of the helicoidal structure, with a typical length in the range of micrometers, depends on temperature, concentration c_m of the chiral dopant [49,50], and shape anisotropy of the micelles [51]. Ch_D, Ch_B, and Ch_C cholesteric phases can be recognized by observing the behavior of the fingerprint textures when the sample is subjected to a magnetic field. In the presence of strong enough magnetic fields (or small fields, if the mixtures are doped with ferrofluids), the helical axis of Ch_D and Ch_B phases aligns along the magnetic field. Ch_C phases under the same conditions untwist in the presence of a field, giving rise to a planar alignment of the director. The director of the former nematic phase, from which the cholesteric phase was originated, aligns perpendicular to the helicoidal axis in both Ch_C and Ch_D phases. In the Ch_B phase, the largest dimension of the micelles aligns along the helicoidal axis.

General features of the X-ray diffraction patterns of nematics and cholesterics are essentially similar. However, in the particular case of the mixture with BS, the shape of the strong first-order band is slightly different with respect to nematics. The edges of the diffraction band are bent towards small values of the modulus of the scattering vector [52], which indicates a micellar deformation. The presence of large BS molecules leads to a swelling of the pseudo-lamellar structure near their edges.

6.4.1.2 *The experimental phase diagrams* In Fig. 6.12, we show a particular surface of the phase diagram of the KL/DeOH/BS/D_2O mixture, at fixed concentrations of KL, DeOH and D_2O [48].

The topology of this surface of the phase diagram has some features that are different from lyotropic nematics. No Landau points were clearly observed and large polyphase regions surround the cholesteric domains. Cholesteric phases appear in the temperature range $10 \lesssim T \lesssim 35°C$. The increase of the relative concentration of BS (with respect to KL) favors the stabilization of the Ch_D mesophase. A different topology, closer to lyotropic nematics, was found by Melnik and Saupe [53] for another surface of the same phase diagram, as sketched in Fig. 6.13.

6.4.1.3 *The uniaxial–biaxial cholesteric phase transition* The inspection of the temperature evolution of the texture of a lyocholesteric sample under crossed polarizers indicates the continuous nature of the Ch–Ch phase transitions.

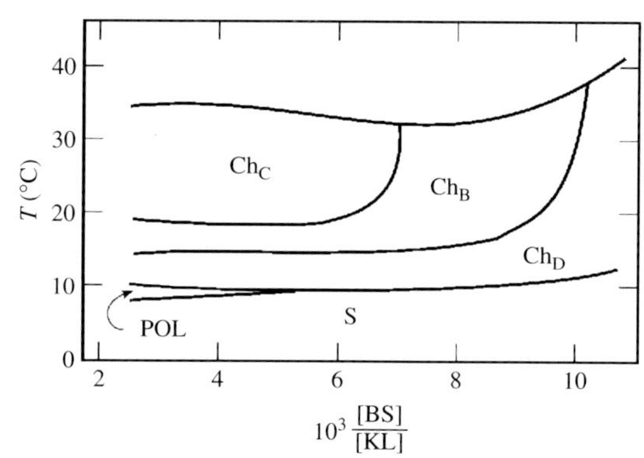

FIG. 6.12. Sketch of the surface of the phase diagram of the lyotropic cholesteric mixture of KL/DeOH/D/BS [48]. Labels POL and S refer to polyphase and gel regions, respectively. The horizontal axis represents the relative molar ratio of BS with respect to KL.

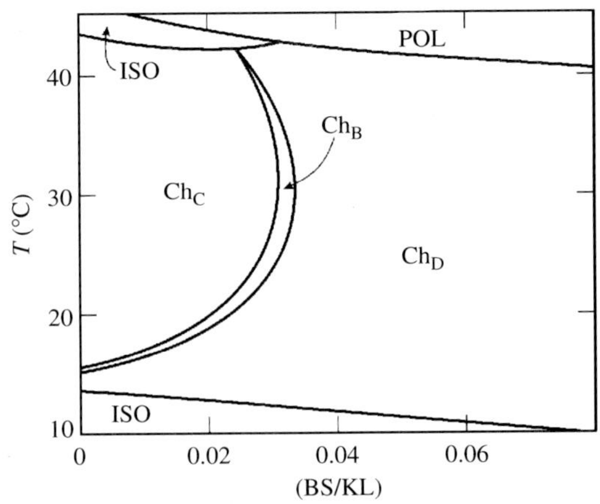

FIG. 6.13. Sketch of the surface of the phase diagram of the lyotropic cholesteric mixture KL/DeOH/D/BS [53]. Labels POL and ISO refer to polyphase and isotropic regions, respectively. The horizontal axis represents the relative weight percentage of BS with respect to KL.

The elastic chiral field is responsible for the chirality-induced biaxiality of the Ch–Ch phase transitions [54]. This is equivalent to the magnetic field-induced biaxiality at the uniaxial–biaxial phase transitions in nematic mixtures, but with a smaller amplitude. In the vicinity of the uniaxial–biaxial phase transition, the free-energy density in terms of an induced biaxial order parameter ζ should include uniform terms,

$$f_{\mathrm{u}} = \frac{1}{2}a(T)\zeta^2 + \frac{1}{4}b\zeta^4, \tag{6.12}$$

where $a(T) = a_0(T - T_{\mathrm{c}})$, and a_0 and b are positive coefficients, and an elastic gradient term, which can be written in the simple form [54]

$$f_{\mathrm{g}} = -cq_0^2\zeta, \tag{6.13}$$

where $q_0 = 2\pi/P$ is the helical wave number and c is a positive parameter. The dependence on the square of the helical wave number comes from the square of the first derivatives of the components of a suitable (biaxial) traceless tensor order parameter [54]. The minimization of the total free-energy density, $f = f_{\mathrm{u}} + f_{\mathrm{g}}$, with respect to ζ leads to the mean-field expression,

$$\zeta \simeq \frac{cq^2}{a_0(T - T_{\mathrm{c}})}. \tag{6.14}$$

The biaxial order parameter ζ is assumed to be proportional to the ratio $\delta n/\Delta n$, which can be determined from measurements of the optical birefringences, δn and Δn. Interferometric measurements performed in the neighborhood of the $\mathrm{Ch_D}$–$\mathrm{Ch_B}$ transition for different cholesteric mixtures [54] indicate that, for concentrations of chiral molecules of about 10^{-2} M%, close to the transition, ζ^2 is of order 10^{-2}. Experimental results for ζ versus T in the neighborhood of the transition show that the Ch–Ch phase transition is not of second order, although it is possible to go continuously from one phase to the other by a suitable manipulation of the chiral field.

Phase diagrams of cholesteric mixtures display similar phenomena of reentrance as the nematic counterparts. Again, this behavior can be accounted for by the introduction of a novel scalar non-critical order parameter τ related to changes in shape anisotropy of the molecular aggregates [23]. The expansion of the free energy should include phenomenological coupling terms involving τ and the helical wave number, which will lead to topological transformations of the phase diagram [55].

6.4.2 *Phenomenological theory of the cholesteric transitions*

The description of cholesteric structures requires the introduction of a space-dependent tensor order parameter, $\mathbf{Q} = \mathbf{Q}(\vec{r})$, and the inclusion of gradient terms in the expansion of the free energy. According to de Gennes and Prost [50], the free energy should be written as the sum of bulk and gradient terms,

$$F = F_{\text{bulk}} + F_{\text{grad}}. \tag{6.15}$$

The bulk term is given by a volume integral of the usual nematic free-energy density,

$$F_{\text{bulk}} = \int \mathrm{d}^3 r \left[\frac{1}{2} A I_1 + \frac{1}{3} B I_2 + \frac{1}{4} C I_1^2 \right], \tag{6.16}$$

where $I_1 = I_1(\vec{r}) = \operatorname{Tr} \mathbf{Q}^2$ and $I_2 = I_2(\vec{r}) = \operatorname{Tr} \mathbf{Q}^3$ are the invariants of a local traceless tensor order parameter. Note that we are not including higher-order terms. Also, note that $C > 0$, in order to ensure thermodynamic stability, and $B \neq 0$, for accounting to the first-order transition between isotropic and uniaxial nematic phases. As we have discussed in Chapter 2, the minimization of this bulk free energy favors a uniaxial form of the symmetric traceless tensor order parameter $\mathbf{Q}$.

The simplest expression for the gradient term includes derivatives of the symmetric traceless tensor $\mathbf{Q}$. With the restriction to second order in gradients and second order in $\mathbf{Q}$, there are just four rotationally invariant quantities [56]

$$\text{(i) } (\nabla_i Q_{jk})(\nabla_i Q_{jk}) = (\nabla_i Q_{jk})^2; \quad \text{(ii) } (\nabla_i Q_{ik})(\nabla_j Q_{jk}) = (\nabla_i Q_{jk})^2; \tag{6.17}$$

$$\text{(iii) } (\nabla_i Q_{jk})(\nabla_j Q_{ik}); \quad \text{(iv) } \varepsilon_{ijk}(\nabla_i Q_{js}) Q_{ks}; \tag{6.18}$$

where we are using the convention of summation over repeated indices, and $\varepsilon_{\alpha\gamma\delta}$ is a Levi–Civita symbol. Note that the antisymmetric third-rank tensor ε_{ijk}, which changes sign under spatial inversion, is a typical feature of cholesterics. As the third quantity differs from the second by a total derivative, the integral form of the gradient free energy is usually written as

$$F_{\text{grad}} = \int \mathrm{d}^3 r \left[\frac{1}{2} L_2 \left[(\nabla \times \mathbf{Q})_{ij} + 2q_0 Q_{ij} \right]^2 + \frac{1}{2} L_1 [(\nabla \cdot \mathbf{Q})_i]^2 \right], \tag{6.19}$$

where

$$(\nabla \times \mathbf{Q})_{ij} = \varepsilon_{ist} \nabla_s Q_{tj}, \quad (\nabla \cdot \mathbf{Q})_i = \nabla_j Q_{ji}. \tag{6.20}$$

The coefficients L_1 and L_2 are elastic positive constants, and q_0 is a (helical) wave number.

We now recall that the position-dependent director of a cholesteric helical phase may be written as

$$\vec{n} = \cos(q_0 z) \vec{i} + \sin(q_0 z) \vec{j},$$

where the pitch axis is along the z direction, $\vec{i}$ and $\vec{j}$ are unit vectors along the x and y axes, and the pitch is $P = 2\pi/q_0$ (ordinary nematics can be regarded as the limiting case $q_0 = 0$). This director can be used to construct the (uniaxial) order parameter

$$Q_{ij}^{\text{uh}} = \gamma \left(n_i n_j - \frac{1}{3} \delta_{ij} \right),$$

which may be written in the explicit form

$$\mathbf{Q}^{\text{uh}} = \gamma \begin{pmatrix} 1/6 & 0 & 0 \\ 0 & 1/6 & 0 \\ 0 & 0 & -1/3 \end{pmatrix} + \frac{1}{2}\gamma \begin{pmatrix} \cos 2q_0 z & \sin 2q_0 z & 0 \\ \sin 2q_0 z & -\cos 2q_0 z & 0 \\ 0 & 0 & 0 \end{pmatrix}, \tag{6.21}$$

where the superscript uh stands for uniaxial helix.

The main problem of the minimization of the total free energy is the competition between bulk and gradient terms. The bulk free energy (6.16) favors a uniaxial form of the tensor $\mathbf{Q}$. Therefore, $\mathbf{Q}^{\text{uh}}$, given by Eq. (6.21), minimizes the bulk free energy. On the other hand, it is not difficult to see that the gradient free energy is minimized for a tensor order parameter that satisfies the rotational condition

$$\nabla \times \mathbf{Q} = -2q_0 \mathbf{Q}. \tag{6.22}$$

With an additional algebraic work, we can show that this condition is satisfied by the second term of Eq. (6.21),

$$\mathbf{Q}^{\mathrm{bh}} = \begin{pmatrix} \cos 2q_0 z & \sin 2q_0 z & 0 \\ \sin 2q_0 z & -\cos 2q_0 z & 0 \\ 0 & 0 & 0 \end{pmatrix}, \tag{6.23}$$

which is called a biaxial helix (note that $\mathbf{Q}^{\mathrm{bh}}$ is associated with the maximal biaxial eigenvalues, -1, 0, and $+1$). Far below the transition from the isotropic phases, the bulk free energy is supposed to dominate, favoring the appearance of a phase associated with a helical uniaxial order parameter (which is the typical feature of a standard cholesteric phase). However, a dominant gradient free energy favors a strongly biaxial form of the tensor order parameter. The competition between bulk and gradient terms leads to intricate problems and the phenomena of the blue phases [56].

6.4.2.1 *The general helical order parameter* The minimization of the total mean-field free energy (6.15) is an intractable theoretical problem. It is then usual to resort to a trial order parameter for describing biaxial effects in a helical structure.

We now rewrite the free energy according to the more convenient notation of the pedagogical review of Wright and Mermin [56],

$$\phi = \int \mathrm{d}^3 r \left[t\chi_{ij}\chi_{ji} - \sqrt{6}\chi_{ij}\chi_{jk}\chi_{ki} + (\chi_{ij}\chi_{ij})^2 \right] + \int \mathrm{d}^3 r \left[\kappa^2 \left(\varepsilon_{ikl} \frac{\partial \chi_{\delta j}}{\partial x_k} + \chi_{ij} \right)^2 + \kappa^2 \eta \left(\frac{\partial \chi_{ij}}{\partial x_j} \right)^2 \right], \tag{6.24}$$

where

$$\phi = \frac{3^6 C^3}{2^4 B^4}(F - F_0), \quad \chi_{ij} = -\frac{3\sqrt{6}C}{4B} Q_{ij}, \tag{6.25}$$

with the coefficients

$$t = \frac{3CA}{4B^2}, \quad \kappa = \left(\frac{3^3 C L_2 q_0^2}{B^2} \right)^{1/2}, \quad \eta = \frac{L_1}{L_2}, \tag{6.26}$$

and we measure lengths in units of $1/2q_0$. Note that $C > 0$ and $B \neq 0$ in the neighborhood of the underlying uniaxial nematic (first-order) transition, so the parameter t may be viewed as a reduced temperature. Also, note that $\kappa = \xi q_0$, where ξ is a correlation length, gauges the degree of chirality. The remaining coefficient, $\eta = L_1/L_2$, is a ratio of elastic moduli that plays no essential role and may be discarded.

The general helical order parameter is usually written as

$$\chi = \sqrt{6}\lambda\cos\theta \begin{pmatrix} -1/6 & 0 & 0 \\ 0 & -1/6 & 0 \\ 0 & 0 & 1/3 \end{pmatrix} + \frac{\lambda}{\sqrt{2}}\sin\theta \begin{pmatrix} \cos z & \sin z & 0 \\ \sin z & -\cos z & 0 \\ 0 & 0 & 0 \end{pmatrix}, \tag{6.27}$$

where λ is related to the amount of angular ordering and θ represents the degree of (induced) biaxiality. We can distinguish several particular cases:

(i) $\theta = 0$ corresponds to a uniaxial nematic system;

(ii) $\theta = 2\pi/3$ corresponds to a uniaxial cholesteric system (with a simple uniaxial helix). We regain $\mathbf{Q}^{\mathrm{uh}}$, given by Eq. (6.21), with $\gamma = \lambda\sqrt{3/2}$ and the rescaling of z.

(iii) $\theta = \pi/2$ corresponds to the strongly biaxial case of Eq. (6.23).

Inserting this trial order parameter into the expression of the free energy and performing the integration over a region of volume v, we have

$$\frac{1}{v}\phi = \left(\frac{1}{2}\kappa^2 + t\right)\lambda^2 + \lambda^4 - \lambda^3\cos 3\theta + \frac{1}{2}\kappa^2\lambda^2\cos 2\theta. \tag{6.28}$$

Note that large values of λ favor $\cos 3\theta = 1$ (recall that $\theta = 2\pi/3$ corresponds to a uniaxial cholesteric system). On the other hand, large values of κ favor $\cos 2\theta = -1$ (recall that $\theta = \pi/2$ corresponds to a strongly biaxial cholesteric). In the limit $\kappa = 0$, we recover the behavior of a uniaxial nematic system.

There are a few analytical results that can be obtained from this expression for the free energy. For example, if we parametrize the amplitude λ by

$$\lambda = \frac{1}{6}\kappa^2\sinh\alpha, \tag{6.29}$$

and minimize with respect to θ, we have

$$\cos\theta = -\frac{1}{2}\tanh\frac{\alpha}{2} \tag{6.30}$$

and

$$\frac{1}{v}\phi = \frac{1}{9}\kappa^4 tu + \frac{1}{9}\kappa^4\left[t + \frac{1}{9}\kappa^2\left(\kappa^2 - \frac{9}{4}\right)\right]u^2 + \frac{2}{81}\kappa^6\left(\kappa^2 - \frac{3}{2}\right)u^3 + \frac{1}{81}\kappa^8 u^4, \tag{6.31}$$

where

$$u = \sinh^2\frac{\alpha}{2}. \tag{6.32}$$

From Eq. (6.31), if $\kappa > 3/2$, we see that ϕ is positive for $t > 0$, and can become negative for $t < 0$ and u sufficiently small. Therefore, there is a second-order transition at $t = 0$. Since $u = 0$ at the transition, α is also zero, and we have $\theta = \pi/2$, which characterizes a biaxial helix.

It is peculiar to have a second-order transition in the presence of a cubic term in the expression for the free energy. Let us look at this situation from a different point of view. We may write $\theta = \pi/2 + \epsilon$, where $\epsilon \ll 1$. We then have

$$\frac{1}{v}\phi = t\lambda^2 + \lambda^4 - 3\lambda^3\epsilon + \kappa^2\lambda^2\epsilon^2 + O(\epsilon^3). \tag{6.33}$$

Minimizing this free energy with respect to ϵ, we have $\epsilon = 3\lambda/(2\kappa^2)$, and

$$\frac{1}{v}\phi = t\lambda^2 + \left(1 - \frac{9}{4\kappa^2}\right)\lambda^4, \tag{6.34}$$

which gives rise to a second-order transition, at $t = 0$, provided that $\kappa > 3/2$.

We may now look at the first-order transition if $\kappa < 3/2$. Consider again the free energy given by Eq. (6.31), in terms of the variable u. The first-order transition occurs at $u_c \neq 0$, such that $\phi'(u_c) = 0$ and $\phi(u_c) = 0$. With some algebraic effort, it is not difficult to show the existence of a first-order border, given by

$$t_c = \frac{1}{8}\left[1 - 4\kappa^2 + \left(1 + \frac{4}{3}\kappa^2\right)^{3/2}\right]. \tag{6.35}$$

Along this border, we have

$$\cos\theta = -\frac{1}{2}\left(1 + \frac{4}{3}\kappa^2\right)^{1/2}. \tag{6.36}$$

According to this picture, the character of the helical cholesteric phase depends crucially on the parameter $\kappa = \xi q_0$. For small values of κ, there is a first-order transition to a cholesteric phase with an order parameter very close to a conventional uniaxial helix. For large values of κ, the transition turns into second order, with an order parameter close to a biaxial helix. However, for $\kappa > 3/2$, it has been shown the occurrence of much more interesting phenomena, related to the onset of the blue phases [56].

6.4.2.2 *Reentrance phenomena* The usual form of the phenomenological theory of Landau–de Gennes does not explain a number of features of the experimental phase diagrams of lyotropic cholesteric mixtures. For example, consider the temperature–concentration phase diagram of Fig. 6.13 for a mixture of potassium laurate, decanol, BS, and water. Besides the high-temperature isotropic–cholesteric transition, it should be interesting to have an explanation for the transition from the cholesteric to a low-temperature isotropic phase.

In close analogy with the treatment of the Freiser–Alben model, Tolédano and Figueiredo Neto have also invoked changes of the shape anisotropy of micelles in order to account for the reentrance phenomena in cholesteric mixtures. As we pointed out before, the idea consists in the introduction of a phenomenological non-critical order parameter τ that represents the effects of changes in the shape anisotropy of micellar aggregates with temperature and concentration. As a first approximation, the phenomenological free energy is still given by Eq. (6.6), with the choice $a_2 < 0$, which is supposed to mimic some of the features of cholesteric systems. Again, the reconstruction of the phase diagram leads to the explanation of the low-temperature isotropic phase.

The existence of two distinct biaxial regions in the phase diagrams of cholesteric systems suggests the phenomenological introduction of an additional non-critical parameter associated with chirality. According to Tolédano and collaborators [55], we then write $\nu = (\nu_0 P_0)/P$, where P is the helical pitch and ν_0 and P_0 are suitable normalization constants. If we redefine the lowest-order invariant of the Landau–de Gennes order parameter, $I_1 \rightarrow I_1 + \tau^2 + (\nu/\nu_0)^2$, and include terms up to second order in the non-critical parameters τ and ν, Eq. (6.6) for the free energy can be written as

$$F_m = a_1(I_1 + \tau^2 + \nu^2) + b_1 I_2 + a_2\left(I_1^2 + 2I_1\tau^2 + 2I_1\nu^2\right) + b_2 I_2^2 + c_1\tau + c_2\tau^2 + d_1\nu + d_2\nu^2 + e\nu\tau + \ldots \tag{6.37}$$

Standard minimization with respect to these variables yields phase diagrams with a wealth of lyotropic cholesteric phases as well as two independent isotropic regions [55].

The temperature dependence of the pitch is an effect that is not accounted for in the usual Landau–de Gennes phenomenological theory of cholesteric systems. However, it is possible to show [57] that the (very laborious) inclusion of higher-order elastic terms in the expansion of the free energy leads to a helical wave number of the form

$$q = \frac{bQ_0 + \kappa}{1 + gQ_0 + aQ_0^2 + hQ_2^2}, \tag{6.38}$$

where $Q_0 = \lambda\cos\theta$ and $Q_2 = \lambda\sin\theta$ should assume (temperature-dependent) equilibrium values, and the coefficients a, b, g, and h, are related to the extra q-dependent terms. Using the ideas of Tolédano and Figueiredo Neto, Longa and collaborators [58] introduced the non-critical parameter τ in the standard Landau–de Gennes treatment of cholesteric structures. The free energy, given by Eq. (6.28), is then supplemented with a linear and some quadratic extra terms,

$$\frac{1}{v}\phi_{\text{extra}} = c_1\tau + c_2\tau^2 + \left(1 + \frac{1}{2}\kappa^2 c_3\cos 2\theta\right)\tau^2\lambda^2, \tag{6.39}$$

where the coefficients c_1, c_2, and c_3 depend on temperature and concentration. According to this work, the main modification in the expression for the wave number q, given by Eq. (6.38), is the inclusion in the denominator of an extra term, proportional to τ^2.

6.4.3 *Cholesteric phases in other lyotropic mixtures*

In general, with the addition of a suitable chiral agent, all of the lyotropic mixtures with nematic phases can also present cholesteric phases. Some of these mixtures have been studied in a more extensive way, and surfaces of their phase diagrams can be found in the published literature. However, many of them were investigated at room temperature only, for a few values of the relative concentrations of their components (in particular, for changing concentrations of the chiral agent). From the experimental point of view, it should be noted that the Ch_B phase has never been found alone, without the vicinity of Ch_C and Ch_D phases.

In Table 6.3, we list the main lyotropic mixtures displaying cholesteric phases.

In general, as in the case of multicomponent mixtures with nematic phases, we do not know the complete structure of the associated multidimensional phase diagrams. In the best cases, we just know some particular surfaces of these phase diagrams.

In Fig. 6.14, we sketch a surface of the phase diagram of a potassium alaninate mixture [60], KL/DeOH/W/L-*N*-lauroyl potassium alaninate, determined by optical microscopy and X-ray diffraction observations. Besides the three cholesteric phases, a coagel (or gel) and a long-range ordered phase have also been characterized (labelled S in the figure). At higher temperatures, we have a polyphase region.

Table 6.3 Multicomponent mixtures presenting cholesteric phases. The label *water* refers to both heavy (D) and light (W) water used to prepare lyotropic mixtures. The three cholesteric phases, Ch_C, Ch, and Ch_D, have been found in mixtures marked with the tag #

Mixture	Reference
SdS/1-decanol/W/BS (#)	[47, 59]
KL/DeOH/W/L-*N*-lauroyl potassium alaninate (#)	[60]
KL/DaCl/W/BS (#)	[61]
L-*N*-lauroyl potassium serinate/KCl/*n*-DeOH/water/KOH	[62]
L-Di-sodium *N*-lauroyl-aspartate/NH_4Cl /DeOH/water	[63]
L-Di-sodium *N*-lauroyl-aspartate/Na_2 SO_4/DeOH/water	[63]
KL/KCl/water/cholesterol	[64]
KL/KCl/KOH/water/cholesterol	[64]
KL/KCl/DeOH/water/cholesterol	[64]
DaCl/NH_4 Cl/D/cholesterol	[65]
α-Alanine hydrochloride decylester/Na_2SO_4/*D*	[46]
K lauroyl-L-alaninate/K_2SO_4/DeOH/D	[25, 66]
CsdS/DeOH/water/tartaric acid	[45]
NH_4dS/DeOH/water/BS	[45]
DaCl/water/cholesterol	[45]

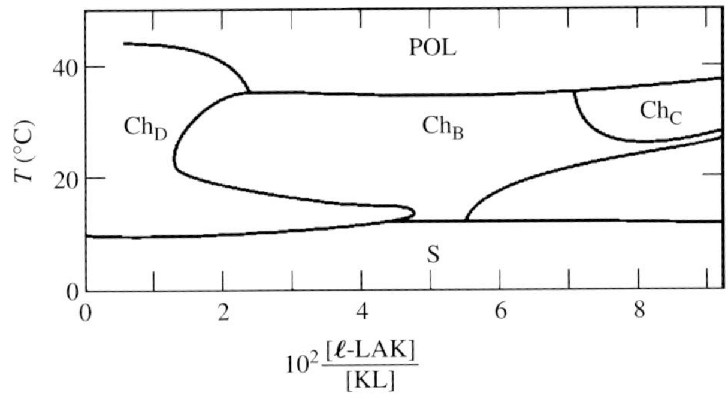

FIG. 6.14. Sketch of a surface of the phase diagram of the KL/DeOH/W/L-*N*-lauroyl potassium alaninate mixture [60]. The horizontal axis represents the relative molar ratio of L-*N*-lauroyl potassium alaninate with respect to KL.

References

[1] R. C. Long (1973). *J. Magn. Reson.* **12**, 216.
[2] L. J. Yu and A. Saupe (1980). *Phys. Rev. Lett.* **45**, 1000.
[3] M. Born and E. Wolf (1980). *The Principles of Optics*, Pergamon Press, Oxford.
[4] Y. Galerne and J. P. Marcerou (1983). *Phys. Rev. Lett.* **51**, 2109.
[5] A. Saupe (1977). *J. Colloid Interface Sci.* **58**, 549.
[6] A. Saupe, P. Boonbrahm, and L. J. Yu (1983) *J. Chim. Phys. (Paris).* **80**, 7.
[7] A. M. Figueiredo Neto, Ph. Martinot-Lagarde, and G. Durand (1984). *J. Phys. (Paris) Lett.* **45**, L-793.
[8] A. M. Figueiredo Neto, Y. Galerne, A. M. Levelut and L. Liébert (1985). *J. Phys. (Paris) Lett.* **46**, L-499.
[9] Y. Galerne, A. M. Figueiredo Neto, and L. Liébert (1987). *J. Chem. Phys.* **87**, 1851.
[10] Y. Hendrikx, J. Charvolin, M. Rawiso, L. Liébert, and M. C. Holmes (1983). *J. Phys. Chem.* **87**, 3991.
[11] Y. Hendrikx, J. Charvolin, and M. Rawiso (1986). *Phys. Rev. B* **33**, 3534.
[12] A. M. Figueiredo Neto, L. Liébert, and Y. Galerne (1985). *J. Phys. Chem.* **89**, 3737.
[13] G. Melnik, P. Photinos, and A. Saupe (1989). *Phys. Rev. A* **39**, 1597.
[14] Y. Galerne, A. M. Figueiredo Neto, and L. Liébert (1985). *Phys. Rev. A* **31**, 4047.
[15] M. J. de Oliveira and A. M. Figueiredo Neto (1986). *Phys. Rev. A* **34**, 3481.
[16] M. J. Freiser (1970). *Phys. Rev. Lett.* **24**, 1041.
[17] R. Alben (1973). *Phys. Rev. Lett.* **30**, 778.
[18] J. C. le Guillou and J. Zinn-Justin (1977). *Phys. Rev. Lett.* **39**, 95.
[19] G. Melnik, P. Photinos, and A. Saupe (1988). *J. Chem. Phys.* **88**, 4046.

[20] M. Sheik-Bahae, A. A. Said, T. H. Wei, D. J. Hagan, and E. W. Van Stryland (1990). *IEEE J. Quantum Electron.* **26**, 760.
[21] F. Cuppo, S. Gomez, and A. M. Figueiredo Neto (2003). *Phys. Rev. E* **67**, 051711.
[22] P. Tolédano, and A. M. Figueiredo Neto (1994). *Phys. Rev. Lett.* **73**, 2216.
[23] P. Tolédano, A. M. Figueiredo Neto, V. Lorman, B. Mettout, and V. Dmitriev (1995). *Phys. Rev. E* **52**, 5040.
[24] E. F. Henriques and V. B. Henriques (1997). *J. Chem. Phys.* **107**, 8036.
[25] B. J. Forrest and L. W. Reeves (1981). *Chem. Rev.* **81**, 1.
[26] N. Boden, P. H. Jackson, K. McMullen, and M. C. Holmes (1979). *Chem. Phys. Lett.* **65**, 476.
[27] N. Boden, R. J. Bushby, K. W. Jolley, and M. C. Holmes (1987). *Mol. Cryst. Liq. Cryst.* **152**, 37.
[28] N. H. Hartshorne and G. D. Woodard (1973). *Mol. Cryst. Liq. Cryst.* **23**, 343.
[29] L. J. Yu and A. Saupe (1982). *Mol. Cryst. Liq. Cryst.* **80**, 129.
[30] P. Photinos, S. Y. Xu, and A. Saupe (1990). *Phys. Rev. A* **42**, 865.
[31] R. Bartolino, T. Chiaranza, M. Meuti, and R. Compagnoni (1982). *Phys. Rev. A* **26**, 1116.
[32] E. A. Oliveira, L. Liébert, and A. M. Figueiredo Neto (1989). *Liq. Cryst.* **5**, 1669.
[33] L. Q. Amaral, and M. E. Marcondes Helene (1988). *J. Phys. Chem.* **92**, 6094; L. Q. Amaral (2002). *Braz. J. Phys.* **32**, 540.
[34] K. D. Lawson and T. J. Flautt (1967). *J. Am. Chem. Soc.* **89**, 5489.
[35] J. Charvolin, A. M. Levelut, and E. T. Samulski (1979). *J. Phys. (Paris) Lett.* **40**, L-587.
[36] L. Q. Amaral, C. A. Pimentel, M. R. Tavares, and J. A. Vanin (1979). *J. Chem. Phys.* **71**, 2940.
[37] Y. Hendrikx and J. Charvolin (1981). *J. Phys. (Paris)* **42**, 1427.
[38] D. M. Chen, F. Fujiwara, and L. W. Reeves (1977). *Can. J. Chem.* **55**, 2396.
[39] M. C. Holmes and J. Charvolin (1984). *J. Phys. Chem.* **88**, 810.
[40] P. Boonbrahm and A. Saupe (1984). *Mol. Cryst. Liq. Cryst.* **109**, 225.
[41] A. Saupe, S. Y. Xu, S. Plumley, Y. K. Zhu, and P. Photinos (1991). *Physica A* **174**, 195.
[42] F. Y. Fujiwara and L. W. Reeves (1980). *J. Phys. Chem.* **84**, 653.
[43] P. O. Quist (1995). *Liq. Cryst.* **18**, 623.
[44] A. A. Melo Filho, A. Laverde Jr., and F. Y. Fujiwara (2003). *Langmuir* **19**, 1127.
[45] K. Radley and A. Saupe (1978). *Mol. Phys.* **35**, 1405.
[46] M. Acimis and L. W. Reeves (1980). *Can. J. Chem.* **58**, 1533.
[47] A. M. Figueiredo Neto, Y. Galerne, and L. Liébert (1985). *J. Phys. Chem.* **89**, 3939.

[48] A. M. Figueiredo Neto and M. E. Marcondes Helene (1987). *J. Phys. Chem.* **91**, 1466.
[49] A. M. Figueiredo Neto, L. Liébert, and A. M. Levelut (1984). *J. Phys. (Paris)* **45**, 1505.
[50] P. G. de Gennes and J. Prost (1993). *The Physics of Liquid Crystals*, Clarendon Press, Oxford, 2nd edn.
[51] M. C. Valente Lopes and A.M. Figueiredo (1988). *Phys. Rev. A* **38**, 1101.
[52] A. M. Figueiredo Neto, A. M. Levelut, Y. Galerne, and L. Liébert (1988). *J. Phys. (Paris)* **49**, 1301.
[53] G. Melnik and A. Saupe (1987). *Mol. Cryst. Liq. Cryst.* **145**, 95.
[54] T. Kroin, A. M. Figueiredo Neto, L. Liébert, and Y. Galerne (1989). *Phys. Rev. A* **40**, 4647.
[55] P. Tolédano, A. M. Figueiredo Neto, and Z. A. de Sant'Ana (2000). *Phys. Rev. E* **61**, 486.
[56] D. C. Wright and N. D. Mermin (1989). *Rev. Mod. Phys.* **61**, 385.
[57] L. Longa and H.-R. Trebin (1989). *Phys. Rev. A* **39**, 2160; (1990). *Phys. Rev. A* **42**, 3453.
[58] L. Longa, M. Bachteler, A. M. Figueiredo Neto, G. Cholewiak, and F. A. Oliveira (2000). *Liq. Cryst.* **27**, 1669.
[59] L. J. Yu and A. Saupe (1980). *J. Am. Chem. Soc.* **102**, 4879.
[60] M. E. Marcondes Helene and A. M. Figueiredo Neto (1988). *Mol. Cryst. Liq. Cryst.* **162B**, 127.
[61] Z. A. de Sant'Ana and A. M. Figueiredo Neto (1992). *Phys. Rev. A* **46**, 7630.
[62] M. R. Alcantara, M. V. M. C. Melo, V. R. Paoli, and J. A. Vanin (1984). *Mol. Cryst. Liq. Cryst.* **107**, 359.
[63] M. R. Alcantara, M. V. M. C. de Melo, V. R. Paoli, and J. A. Vanin (1983). *Mol. Cryst. Liq. Cryst.* **90**, 335.
[64] M. R. Alcantara, M. V. M. C. de Melo, V. R. Paoli, and J. A. Vanin (1983). *Mol. Cryst. Liq. Cryst.* **95**, 299.
[65] F. Y. Fujiwara and L. W. Reeves (1976). *J. Am. Chem. Soc.* **98**, 6790.
[66] B. J. Forrest, L. W. Reeves, and M. R. Vist (1981). *J. Am. Chem. Soc.* **103**, 690.

7

THE LYOTROPIC ONE-, TWO- AND THREE-DIMENSIONALLY ORDERED PHASES

7.1 Lamellar phases

7.1.1 *Introduction*

Several types of lamellar phases, $L_\alpha, L_\beta, L_\gamma, L_{\beta'}, L_\delta, L_s$, and $P_{\beta'}$, have been observed in lyotropic phase diagrams [1,2]. In general, they are characterized by the presence of one-dimensional packings of lamellae. As discussed in Chapter 1, the topological organization of amphiphiles in the lamellar phases resembles a cellular membrane skeleton.

The L_α phase, also known as *neat soap*, is usually found in the region of large amphiphilic concentration of the phase diagrams [3]. In the L_α phase, the carbonic chains are in a liquid-like state. The L_β phase differs from the L_α phase by the state of the carbonic chains. In the L_β phase, the carbonic chains remain stiff, perpendicular to the plane of the lamellae, with the axes organized as a two-dimensional hexagonal lattice. There are, however, orientational fluctuations of the carbonic chains with respect to their long axes. The viscosity of the L_β phase ranges from approximately 1 to 10 P.

The L_γ phase is formed by a sequence of layers with L_α and L_β structures. The $L_{\beta'}$ phase differs from the L_β phase by the tilt of the axis of carbonic chains with respect to the normal to the lamellae. Tilt angles from 15 to 40° were observed in lipid–water mixtures [2]. In the L_δ phase, the carbonic chains remain stiff, perpendicular to the plane of the lamellae, and the axes are organized as a two-dimensional square lattice. As in the L_β phase, there are orientational fluctuations of the chains with respect to the long axes. The $P_{\beta'}$ phase displays rippled lamellae with the characteristics of the $L_{\beta'}$ phase.

Single-layered L_s phases were also observed in some lyotropic mixtures (e.g. in mixtures of Na caprylate, octan-1,8-diol, and water [3], almost in the center of the triangular phase diagram). Another lamellar-type phase found in ternary lyotropic mixtures, in the water-rich region of the phase diagram, is the so-called *mucous woven phase*. It is made of a double-layered structure, as in the usual L_α phase, but with a smaller optical anisotropy.

Mineral lyotropic lamellar phases [4] are mixtures of solid acids and water (e.g. a mixture of phosphatoantimonic acid and brine [5], $H_3Sb_3P_2O_{14}$/brine, where the brine may be NaCl in water). The structure of this phase resembles an aqueous dispersion of planar solid-like sheets of covalently bonded atoms.

7.1.2 *An introductory example: sodium dodecylsulfate-based mixtures*

We now give some examples of lamellar phases and their neighboring structures. Consider lyotropic mixtures based on molecules of sodium dodecylsulfate, $C_{12}H_{25}NaSO_4$, also known as SDS or SLS. These molecules form extensively investigated mixtures, with many examples of lamellar phases in binary, ternary, and quaternary systems.

7.1.2.1 *The* L_α *phase in binary and multicomponent phase diagrams*

Binary mixtures of SDS and water were investigated by Luzzati and coworkers [1,6]. At 75°C, for concentrations of SDS between 70 and 85 wt%, there is a well-characterized L_α phase. From X-ray diffraction experiments, the structural lattice parameter and the bilayer thickness are about 3 and 2.4 nm, respectively. The H_α phase is stable at the same temperature, but for smaller SDS concentrations (less than about 60 wt%).

Ternary mixtures of SDS, decanol (DeOH), and water also display lamellar phases. A partial isothermal, at 25°C, is shown in Fig. 7.1 [3]. The L_α phase occupies a large region in the geometrical center of the triangular representation. In the upper vertex of the triangle, corresponding to large DeOH concentrations (larger than 85 wt%), there is a stable inverted micellar phase, L_2. A direct micellar phase, L_1, is located near the left-down vertex of the triangle, at large water and small DeOH concentrations (smaller than 10 wt%). The hexagonal phase H_α is also present in this region of small DeOH concentration. This should be anticipated, since the presence of alcohol in the structure tends to reduce the

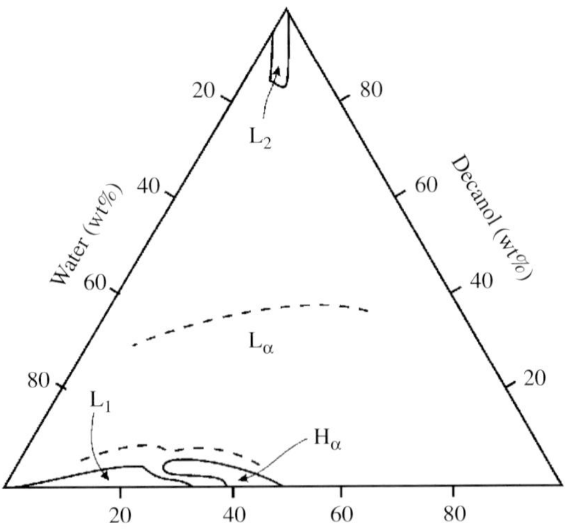

FIG. 7.1. Sketch of a partial isotherm, at 25°C, for a mixture of sodium dodecylsulfate (SDS), decanol, and water [3].

mean curvature of the amphiphile–solvent interface. If the DeOH concentration increases from 10 to about 40 wt%, and the concentration range of SDS extends from 10 to 60 wt%, which is also the approximate range of concentrations of water, there appears an L_α phase. The region of this L_α phase extends towards the large water concentration vertex of the partial isotherm, and is surrounded by large coexistence phase regions. These coexistence regions separate the L_α region from the regions of H_α and micellar isotropic L_1 phases.

The L_α phase was also found in a quaternary mixture of SDS, pentanol (POH), cyclohexane, and water. A particular partial isothermal representation (at room temperature and relative concentration [water wt%]/[SDS wt%] = 1.2) is shown in Fig. 3.12(b) of Chapter 3. Also, note a large, almost horizontal, region of L_α phase. As in the previous cases, the lamellar phase is stabilized at larger alcohol concentrations (with respect to the concentrations at which there is a hexagonal phase). The dodecane is preferentially located inside the lamellae, producing the swelling and increasing the bilayer thickness. If the concentration of dodecane increases, the region of the L_α phase becomes thinner and finally vanishes.

7.1.2.2 *Light scattering studies* As pointed out in different sections of this book, light scattering is one of the most powerful experimental techniques to study different aspects of the physics of lyotropics [7]. Typical scattering wave numbers ($q = 4\pi \sin\theta/\lambda$, where λ and 2θ are the laser wavelength and the scattering angle, respectively) range from 10^3 to $10^4\,\mathrm{cm}^{-1}$.

Let us increase the concentration of the solvent in a lyotropic mixture with a lamellar phase. From the experimental point of view, the increasing of the concentration of a non-polar solvent may lead to either the swelling of the lamellae, keeping the same lamellar structure, or to a phase separation between lamellar and isotropic phases. The dynamics of phase separation between L_α and isotropic phases in the quaternary SDS/POH/water/dodecane mixture, upon increasing the solvent concentration, can be investigated, e.g., by light scattering experimental techniques [8]. At the early stages of dilution, measurements as a function of time of the typical size of the isotropic phase regions growing out of a lamellar phase have been shown to follow a power law of the form $t^{1/3}$. Lattice parameters in the diluted lamellar phase reach values of about 60 nm. The growth of the isotropic phase domains in the lamellar phase is anisotropic, that is, it depends on the particular direction. Growth is faster in the in-plane direction, which is partially due to the anisotropy of the diffusion constants in the direction parallel and perpendicular to the lamellae. The anisotropy of the elastic constants, however, is also an important factor for explaining this kind of behavior.

Dynamic light scattering techniques lead to measurements of the layer compressibility modulus, $\overline{B}$, of lamellar phases. The hydrodynamics of lamellar lyotropic mesophases is quite complex. Besides the usual modes of thermotropic smectics, there is an additional mode, known as the *slip mode* [9], which is due to the coupling between amphiphile concentration and layer displacement fluctuations. In particular, the behavior of $\overline{B}$, calculated from the anisotropic dispersion relation of the slip mode, depends on the particular solvent used to swell the lamellar structure [10]. In lyotropic mixtures under dilution, regardless the polar or non-polar type of the solvent, the compressibility modulus was shown to be inversely proportional to d^3, where d is the interlayer spacing distance. Helfrich [11] proposed a mechanism of interaction between membranes, in which a long-range repulsive entropically driven interaction is due to the steric hindrance of the thermally excited undulations of the layers. The interlayer interactions can be studied by investigating the relaxation frequency of the slip mode. For SDS-based mixtures, undulation forces dominate in the case of dodecane or brine (0.4 M of a NaCl solution in water) dilutions. Electrostatic interactions dominate in pure water dilution (mixtures of SDS, pentanol, and water). The values of $\overline{B}$ range from 10^4 Pa (for less diluted oil SDS/pentanol/NaCl/water mixtures) to about 10 Pa (for most diluted brine SDS/dodecane/pentanol/water mixtures).

There is an interesting behavior if the lamellar structure is subjected to a shear. By varying the shear rate, closed-compact multilamellar vesicles can be formed [12], giving rise to a structure that has been called *onion texture* [13]. The process of formation of these vesicles can be studied by using small-angle light scattering techniques. The onset of undulation (buckling) instabilities in the lamellar structure has been suggested as the mechanism in order to account for the formation of these vesicles.

7.1.3 *The* L_α *phase in some lyotropic mixtures*

The L_α phase is found in most of the binary and ternary lyotropic mixtures [3]. For example, it has been found in binary mixtures of potassium laurate and water, sodium myristate and water, and dimethyldodecylamine oxide and water; also, in the ternary mixtures of potassium laurate, decanol and water, sodium caprylate, nonanol and water, and sodium octylsulfate, decanol and water.

In binary mixtures, the L_α region is usually found at concentrations of amphiphilic molecules larger than about 50 wt%, and at temperatures from 20 up to 300°C, depending on particular amphiphiles and relative concentrations. For example, in Fig. 7.2 we sketch a partial phase diagram of a mixture of potassium oleate and water [14]. Note the neighboring H_α and isotropic phases. Also, note large phase coexistence regions separating different phases.

In ternary mixtures, the L_α region is located almost at the middle of the triangular phase diagram [3], at about 30 wt% of solvent and 60 wt% of the principal amphiphile, as described in the introductory example of the preceding section. The cosurfactant (e.g. an alcohol), with a carbonic chain of smaller length than the main amphiphile, works in order to increase the radius of the surface curvature of the supermolecular aggregates, which favors flat surfaces. In the triangular phase diagrams, the lamellar phase is located in regions with increasing amounts of the cosurfactant with respect to cosurfactant concentrations at the

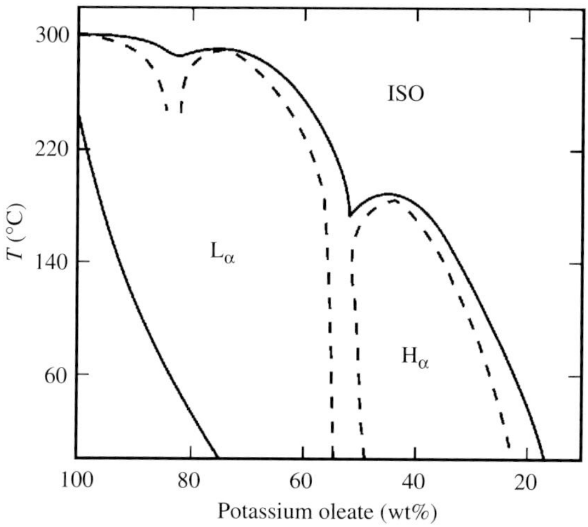

FIG. 7.2. Sketch of the temperature versus concentration phase diagram of a mixture of potassium oleate and water [14]. The label ISO refers to an isotropic phase.

location of the hexagonal phase. The L_α phase has the smallest surface per polar head with respect to other lyotropic structures.

A nematic structure is one of the neighboring phases of the lamellar structures. Quasielastic light scattering techniques are used to study the temperature behavior of the orientational fluctuations of nematics in the vicinity of the nematic–lamellar phase transition. For example, we have the phase diagram of a mixture of decylammonium chloride (DaCl), NH_4Cl, and water. Measurements of the elastic constants in the nematic phase indicate a divergence, according to the behavior of a three-dimensional XY model, as the temperature approaches the transition to the lamellar phase [15]. X-ray and nuclear magnetic resonance (NMR) experiments for the DaCl/NH_4Cl/water mixture suggest that the temperature-driven transition from the lamellar to the nematic phase occurs through the appearance of water holes in the lamellae. In other words, the lamellae are pierced by regions of water that increase as the temperature approaches the transition [16]. A similar interpretation was proposed to explain the neutron scattering data for the sodium decylsulfate/1-decanol/water mixture in the lamellar phase, in the neighborhood of the phase transition [17].

The topology of the phase diagram and the sequences of phases depend on different parameters related to the shape of the amphiphilic molecule, the interaction between these molecules and the solvent, and the interaction between the amphiphilic aggregates.

7.1.3.1 *The sodium dodecanoate/water mixture* The phase diagram of a mixture of sodium dodecanoate (or laurate) and water has already been sketched in Fig. 3.18 of Chapter 3. Lamellar L_α and hexagonal H_α phases are located between T_c (crystallization line) and the high-temperature isotropic phase. A large lamellar region is stabilized at larger amphiphile concentrations and extends to higher temperatures. Transitions between different liquid crystalline phases and between them and the isotropic phase are characterized by large regions of phase coexistence.

7.1.3.2 *The AOT/water mixture* A mixture of sodium bis-2-ethylhexyl sulfosuccinate (AOT) and water is another example of a lyotropic system with a large lamellar phase region [18]. The phase diagram of this mixture, in terms of temperature and concentration of AOT, is sketched in Fig. 7.3. Note that there are some different features in the phase diagrams of Figs 7.2 and 7.3. For example, the locus of the lamellar phase occurs at AOT concentrations *smaller* than those of the hexagonal phase (in other words, at a fixed temperature, there is a hexagonal–lamellar transition if we decrease the AOT concentration).

If a soap is one of the components of a lyotropic binary mixture, there appears a hexagonal–lamellar transition in the phase diagram, at a given temperature, if we increase the soap concentration. The shape of the amphiphilic molecule strongly influences the topology of these phase diagrams. In particular, the topology of the phase diagram of Fig. 7.3 is strongly dependent on the shape anisotropy of the AOT molecule and the volumes occupied by the

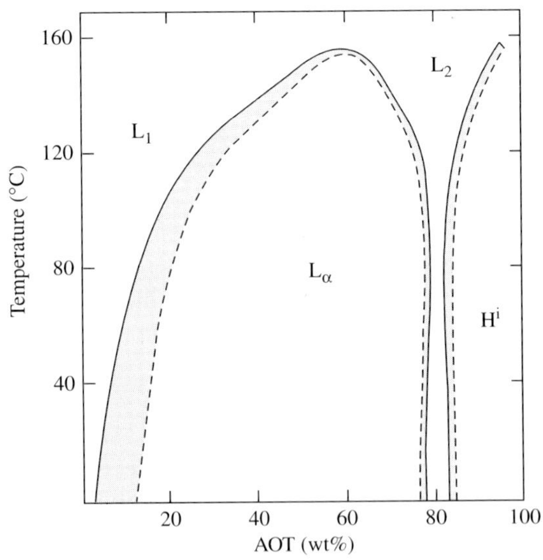

FIG. 7.3. Sketch of the phase diagram of a mixture of sodium bis-2-ethylhexyl sulfosuccinate (AOT) and water [18]. The gray regions correspond to phase coexistence.

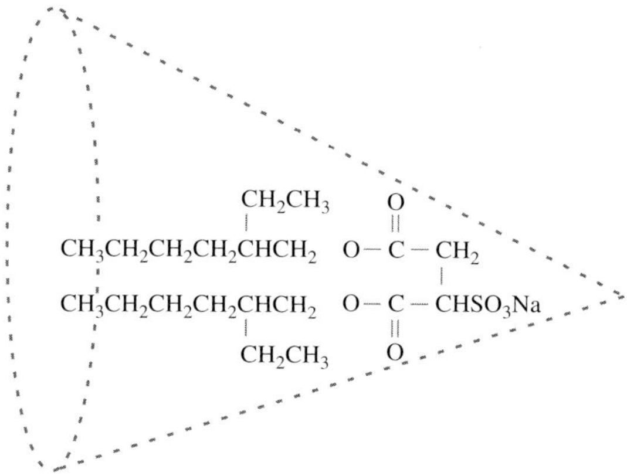

FIG. 7.4. Structure of the AOT ($C_{20}H_{27}NaO_7S$) molecule. O and S atoms are in the polar heads. Note that the Na atom is placed in the neighborhood of S and O atoms.

polar and non-polar parts of the molecule. Figure 7.4 shows the structure of the AOT molecule. It can be involved by a cone of circular base, with the polar part of the molecule placed near the apex of the cone and the non-polar part extending towards the base. This topology clearly favors inverted-type amphiphilic aggregates. Thus, at large AOT concentrations, inverted hexagonal (H^i) and micellar isotropic (L_2) phases become stable. If we decrease the AOT concentration, there appear lamellar and direct micellar isotropic (L_1) phases.

7.1.3.3 *The* $C_{12}E_5$*/water mixture* A mixture of pentaethyleneglycol *n*-dodecyl ether ($C_{12}E_5$) and water [19] presents lamellar L_α, direct (L_1) and reverse (L_2) isotropic micellar, sponge (L_3), hexagonal (H_α), and cubic (Q) phases, as a function of temperature and relative concentrations of the compounds (see Fig. 3.8 in Chapter 3). The topology of the phase diagram shows that, at a given temperature (e.g. 19°C), the lamellar phase is located between cubic and micellar inverted phases. As in the case of soap-based mixtures (e.g. potassium laurate/water), the phase diagrams display a hexagonal phase at amphiphilic concentrations smaller than those where the lamellar phase is stable. The stabilization of the lamellar phase, even at small amphiphilic concentration but at higher temperatures (larger than about 55°C), is an interesting feature of the phase diagram. In this system, particularly in the lamellar phase, the steric and entropic repulsion between membranes is the dominant interaction for explaining the stabilization of the structure. It can be highly swollen (e.g. with a layer separation of about 300 nm, and bilayers of about 3.8 nm) and still keep the lamellar structure. Dynamic light scattering experiments carried out on highly swollen lamellar phases [20] indicate that the bilayer elasticity is characterized by a curvature, or bending, elastic constant of the order of $k_B T$.

7.1.4 *Structures of the lamellar phases*

Besides the usual structure of the L_α phase, as sketched in Fig. 1.13 of Chapter 1, other structures have been found and characterized in different lyotropic systems. These structures have different organizations of the carbon chains, distinct topology of the surfaces (flat or undulated), and various values of thickness of the (single or double) layer.

7.1.4.1 *Phospholipid/water membrane structures* Phospholipids are important structural components of living tissues in plants, animals, and microorganisms. If they are mixed with water, around room temperature, these amphiphilic molecules may form lamellar phases, which are called $L_\beta, L_\gamma, L_{\beta'}, L_\delta$, and $P_{\beta'}$, with different structures as compared with the usual L_α phase.

In the L_β phase, carbonic chains remain stiff, perpendicular to the plane of the lamellae. The long axes are organized on a two-dimensional hexagonal lattice. There are orientational fluctuations about these long chain axes. The L_γ phase is formed by a sequence of layers with L_α and L_β structures.

In the $L_{\beta'}$ phase (see Fig. 1.14(b) of Chapter 1), the axes of the carbonic chains are tilted, by about 15–40° with respect to the normal to the lamellae [21].

In the L_δ phase, carbonic chains remain stiff, perpendicular to the plane of the lamellae. The long axes are organized on a two-dimensional square lattice. As in the L_β phase, there are orientational fluctuations about the long chain axes.

The structure of the $P_{\beta'}$ phase displays rippled lamellae with the characteristics of the $L_{\beta'}$ phase (see Fig. 1.14(a) of Chapter 1) [21,22]. The surface undulations of the lamellae, suggested by X-ray diffraction measurements, were observed by transmission electron microscopy (TEM) in a freeze-fracture replica of a DMPC/water sample [23]. The amplitude and wavelength of undulations were estimated as 4.5 and 16 nm, respectively. This wavelength seems to be weakly dependent on temperature and water content of the sample [24].

7.1.4.2 *Single-layered* L_s *structures* Lamellar single-layered structures are not so common, but they were also observed in some lyotropic phase diagrams (e.g. in a mixture of sodium caprylate, octan-1,8-diol (also called 1,8 octandiol) and water, as shown in Fig. 7.5 [25,26]).

In terms of spacing distances, these structures are characterized by a layer thickness of about the same typical length of the cosurfactant. In the example of Fig. 7.5, the locus of the L_s phase corresponds to approximate concentrations 40/15/45 wt% of sodium caprylate/octan-1,8-diol/water. The ratio between the molar fraction of the principal amphiphile and octandiol is about 2. The neighboring phases are L_1 and H_α. The presence of 1,8-octandiol [$C_8H_{18}O_2$, or HO–$(CH_2)_8$–OH] seems to be the responsible for the existence of the single-layered L_s phase. In other lyotropic mixtures with the same principal amphiphile

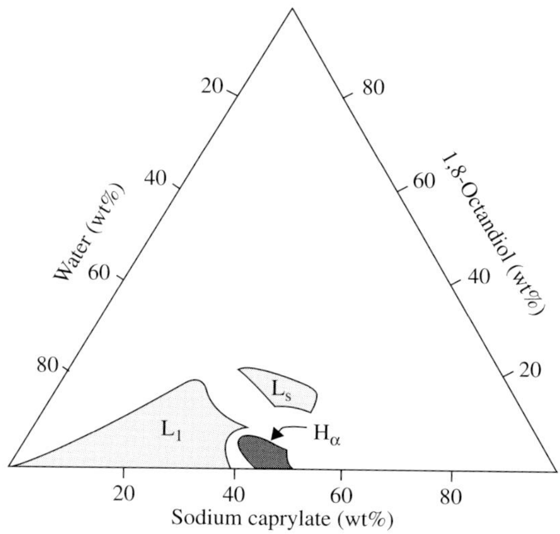

FIG. 7.5. Sketch of a partial isotherm of a mixture of sodium caprylate, 1,8-octandiol, and water, at 20°C [3].

and a different alcohol, this phase is not observed (e.g. in mixtures of sodium caprylate, pentanol, and water [27], and sodium caprylate, decanol, and water [28]). Different *diol* groups are located on the surfaces of the lamellae, defining the layer thickness. The L_s phase is not observed in phase diagrams of mixtures in which octandiol is replaced by smaller *diol* chains (as ethylene glycol and glycerol).

7.1.4.3 *Swollen lamellar* L_α *phases* We now consider diluted L_α phases, for increasing solvent (water) concentrations at fixed values of temperature and pressure. Two types of behavior were observed in lamellar phases under dilution: (i) in one case, the lattice parameter d (i.e. the repeating distance between layers, measured by X-ray diffraction) increases with the solvent concentration ϕ_s; (ii) in other cases, d remains constant as ϕ_s increases. In the first case, d and the amphiphile concentration ϕ_a behave according to the relation

$$\ln(d/d_0) \propto \ln(1/\phi_a),$$

where d_0 is a constant; this is a phase of *expanding* (*swelling*) type. In the *non-expanding* (*non-swelling*) phases, the ratio d/d_0 is constant as ϕ_s increases. For example [3]: (i) AOT/water, at 20°C, and monolaurin/water, at 29°C, present expanding phases; (ii) potassium laurate/water, at 86°C, presents a non-expanding phase.

In expanding phases, the surface per polar head of the amphiphilic molecules in the interface between the amphiphilic aggregates and the solvent, as well as the bilayer thickness, are not significantly affected if we increase ϕ_s. Water molecules essentially occupy the interlayer space, leading to a one-dimensional type of swelling. On the other hand, in non-expanding phases, since d remains constant, the increase of water concentration brings about an increase of the amphiphile surface per polar head and even a decrease of layer thickness. The type and characteristics of the hydrophilic part of the amphiphilic molecules, specially those related to the head hydration phenomenon, are important parameters to be considered in the swelling process. For example, amphiphile heads (in this discussion, counterions and ionized groups are taken into account) that can bind more water molecules are able to increase the surface per polar head, incorporating more water molecules in the interface. This "excess"of water in the interface is supposed to keep the interlayer distance almost constant. These water molecules, under some conditions, may even be introduced in the amphiphilic aggregates, in the neighborhood of the first carbon atoms of the chain. In soaps, Gallot and Skoulios [29] showed that the surface per polar head (S) and the ratio between the number of moles of water and soap (Z) behave according to the relation

$$\log(S/S_1) = q \log(Z),$$

where S_1 is a constant representing the surface per polar head if $Z = 1$, and q is a parameter depending on the particular soap molecule (e.g. $S_1 = 0.241\,\text{nm}^2$ and $q = 0.24$, for Na at 86°C; $S_1 = 0.256\,\text{nm}^2$ and $q = 0.21$ for K at 86°C [3]).

Mineral lyotropic lamellar phases have a remarkable behavior under dilution [4]. Let us call ϕ_m the mineral volume fraction. At a very dilute regime ($0.033\% < \phi < 0.18\%$), the structure determined from X-ray scattering is formed by large ($\gtrsim 300\,\text{nm}$) flat sheets. Even at the maximum swelling, the large number of Bragg peaks (of the order of 7) indicates the continuing existence of the lamellar order. The behavior of the lamellar spacing distance d as a function of $1/\phi_\text{m}$ shows a crossover from an almost linear increasing (for $1/\phi_\text{m} \lesssim 150, d \propto 1/\phi_\text{m}$) to a plateau (for $1/\phi_\text{m} \sim 500, d \sim 200\,\text{nm}$). The thickness of the layers, of about 1.1 nm, is independent of ϕ_m.

There is, however, another possibility to swell a lamellar structure, different from the situations that we have been discussing in this section. A non-polar solvent enters into the lamellae and may increase the bilayer thickness. In Fig. 7.6(a), we show a sketch of a usual lamellar L_α^1 phase; layers of the polar solvent are thicker than the lamellae. In Fig. 7.6(c), the non-polar solvent swells the lamellar structure (L_α^2); the layers of the polar solvent are thinner than the lamellae. From the theoretical point of view, it has been shown [30,31] that the phase diagram obtained from a free energy with a two-component order parameter (associated with molecular segregation and relative concentration) presents two independent stability regions (of L_α^1 and L_α^2 phases). To some extent, the structure of the L_α^2 phase can be regarded as an *inverted* lamellar phase. Between these two structures, there is a symmetric phase, in which the layers of polar solvent and the lamellae have the same thickness (see Fig. 7.6(b)).

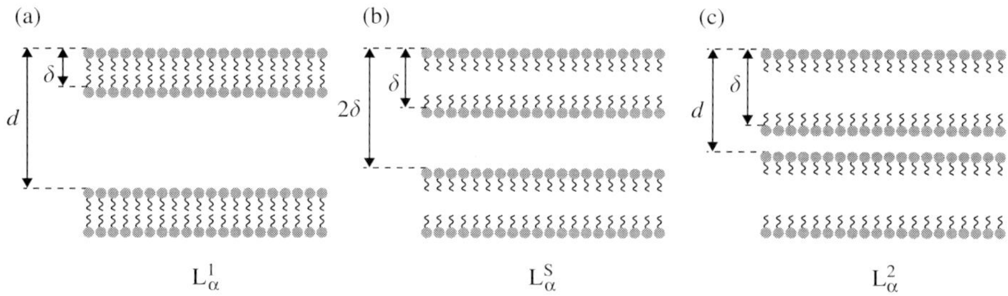

FIG. 7.6. Sketch of the lamellar structure under dilution with a non-polar solvent (polar solvent is placed between lamellae and non-polar solvent is placed inside lamellae): (a) usual lamellar structure (L^1_α); (b) symmetric case (L^s_α); polar and non-polar layers have the same thickness; (c) *inverted* lamellar structure (L^2_α).

7.1.4.4 *Undulation and flexibility of the* L_α *phase* Different experimental techniques have been used to investigate undulations in lamellar phases (in particular, X-ray and dynamic light scattering [32], static light-scattering [33], neutron scattering [34], and NMR [35] measurements). The counterion screening strongly affects the undulation behavior of lamellar structures with charged layers. In unscreened layers, the electrostatic interactions lead to small amplitude undulations; however, this amplitude increases if we increase the concentration of monovalent electrolytes. It is interesting to emphasize that undulations contribute to the stabilization of the lamellar phase.

Elastic properties of a lamellar phase can be characterized by three elastic constants, the layer compression modulus $\overline{B}$, and two smectic curvature moduli, K_1 and $\overline{K}$, which are related to the lamellar periodicity d. These two curvature moduli are related to the mean bending modulus κ and the bilayer Gaussian bending modulus $\overline{\kappa}$ by the relations $K_1 = \kappa/d$ and $\overline{K} = \overline{\kappa}/d$. In particular, $\overline{\kappa}$ is responsible for the bilayer topology. In lyotropic lamellar phases, with a bending elastic constant of the order k_BT, the static and dynamic structures of the phase are strongly influenced by thermal fluctuations of the undulations of layers. From the study of the hydrodynamic baroclinic–undulation mode [36], it is possible to determine the bending rigidity K_1 and the compressibility modulus $\overline{B}$, which are macroscopic characteristics of the lamellar system. We then determine the rigidity K, or flexibility, of the membrane, which is a property of the system at a local scale. We have the following examples of values of the rigidity [32]: $K = 0.5\,k_BT$, for sodium octylbenzene sulfonate/pentanol/brine and n-dodecylcarboxybetain/pentanol/D_2O mixtures; $K = 1.4\,k_BT$, for a cetylpyridinium chloride/hexanol/brine mixture. As discussed above, undulation fluctuations give rise to long-range steric repulsions [11,37].

The flexible structures are strongly affected by shear, which can promote reconstructive phase transitions to an onionlike phase made of multilamellar vesicles, e.g.. Shear experiments, under *shear stress* $0.03 < \sigma < 0.12\,\mathrm{dyn/cm^2}$,

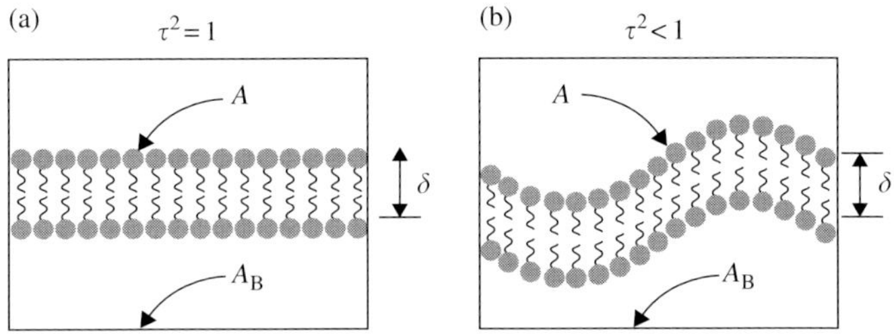

FIG. 7.7. Definition of the *crumpling parameter* $\tau^2 = A_B/A$, where A is the real area of the membrane and A_B is the projection of the area: (a) $\tau^2 = 1$; (b) $\tau^2 < 1$.

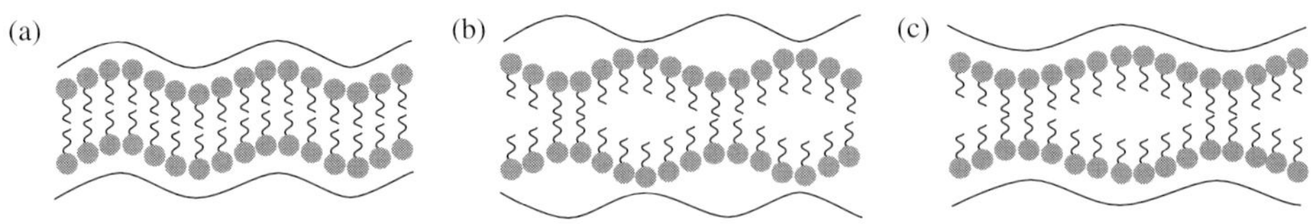

FIG. 7.8. Sketch of the low-frequency hydrodynamic modes of a lamellar structure [9,36]: (a) transverse shear mode; (b) peristaltic mode; (c) undulation–baroclinic mode.

and with *shear rate* $1.5 < \dot{\gamma} < 1\,\mathrm{s}^{-1}$, performed in a mixture of $C_{12}E_5$ and water, show that the behavior of the flow of the lamellar phase is non-Newtonian, and that the average spacing distance decreases whereas the out-of-plane layer displacement fluctuations increase as a function of shear [38].

A useful parameter that accounts for the undulation of the lamellae is the *crumpling* τ^2 [39,40]. It is defined as the ratio between the projection of the area of the membrane, A_B, and the real value of the area, A, as indicated in Fig. 7.7. If $\tau^2 = 1$, the membrane is flat; if $\tau^2 < 1$, it is undulated. In the particular case of swollen lamellar phases, if we assume small compressibility and small heat capacity, there are three low-frequency hydrodynamic modes [9,36] (see Figs 7.8)(a)–(c)): (a) a transverse shear mode; (b) a peristaltic mode; (c) an undulation–baroclinic mode, which comes from the coupling between concentration and layer displacement fluctuations. Brochard and de Gennes [9] called slip modes the cases (b) and (c). Quasi-elastic light scattering experiments, in the frequency range from 10^2 to 10^6 Hz, are expected to be sensitive to this last mode, since the other modes have larger characteristic frequencies (with a typical value of order 10^2 MHz).

From the point of view of elastic properties, it is interesting to investigate the behavior of lamellar phases with the addition of a certain amount of a water-soluble polymer [41]. The introduction of polyvinylpyrrolidone in a mixture of cetylpiridinium chloride, hexanol, and water, with weight concentrations of polymer in water of about 20 wt%, has been shown to produce no noticeable changes of values of κ with respect to the undoped lyotropic mixture. On the other hand, the presence of the polymer decreases the value of $\overline{\kappa}$ with respect to the undoped sample, which is as yet a difficult result to be explained.

There may be high-amplitude fluctuations of the lamellar structures, in particular in the vicinity of phase transitions. In the case of the transition between the L_α and a bicontinuous gyroid-type phase structure, these fluctuations lead to a *perforating* lamellar structure, with the formation of holes in the lamellae [42], giving rise to a different topology of the amphiphilic aggregates.

7.1.4.5 *The lamellar–inverse hexagonal transition* Due to symmetry considerations, the lamellar–hexagonal phase transition is expected to be first order. Regions of different phases are separated by regions of phase coexistence, as it can be seen for L_α–H_α transitions, in Figs 3.7 and 3.10 of Chapter 3, for a binary and a ternary mixture. In the following sections, we discuss in more detail the intermediate phases that may be present between the lamellar and hexagonal phases.

The L_α–H^i_α transition is observed as a function of either temperature [43,44] or pressure [45]. Lyotropic mixtures presenting this transition include 1,2-dioleoyl-sn-glycero-3-phosphoethanolamine (DOPE) and water, and 1-palmitoyl-2-oleoyl-sn-phosphatidylethanolamine (POPE) and water. Rappolt and coworkers [44] have recently proposed a mechanism to account for the formation of the inverted rods in the lamellar structure. It is assumed that there appears a line of defects (e.g. a water core) between consecutive lamellae. This mechanism, sketched in Fig. 7.9, produces a region of unfavorable curvatures in the lamellae, which leads to the formation of the first rod (whose axis is perpendicular to the plane of Fig. 7.9). The minimization of the volume of the interstitial region induces the formation of new cores and, consequently,

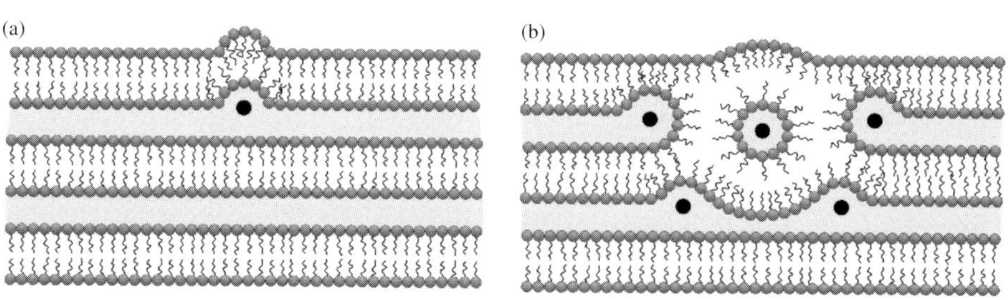

FIG. 7.9. Sketch of a possible mechanism for the lamellar–inverse hexagonal phase transition [44]. Black dots represent the water cores: (a) beginning of the process with the creation of a line defect; (b) formation of the first cylinder, with an axis perpendicular to the plane of the figure.

new rods. The number of water molecules per amphiphile has been observed to increase from 14 in the L_α phase to 18 in the H^i_α phase.

Besides temperature, pressure was shown to induce a phase transition between the lamellar and the inverted hexagonal phases (e.g. in a mixture of DOPE and water [45]). The domain of the lamellar phase increases with pressure (until about 2 kbar). In the H^i phase, the lattice parameter of the hexagonal structure also increases with pressure. The experimental results obtained so far seem to indicate that an intermediate cubic phase may be stabilized between lamellar and hexagonal phases.

7.2 Two- and three-dimensionally ordered phases

7.2.1 *Introduction*

In conventional solid crystals, discrete symmetry groups reflect the translational invariance, along three independent directions, of the mean molecular positions in equilibrium. In a liquid crystal, molecules or molecular aggregates are usually positionally disordered. The discrete groups of translation are then associated with the spatial pattern of orientation of molecules or molecular aggregates.

In two-dimensional ordered structures, amphiphilic molecules self-organize in aggregates of large anisometry, as cylinders and ribbons, at least 100 nm long. Three-dimensional lyotropic structures can be formed by micellar or bicontinuous molecular aggregates. The presence of these high-dimensionally ordered structures in a particular phase diagram has no connection with the number of components of the lyotropic mixture. In fact, two- and three-dimensional structures have been observed in some binary mixtures. Although we have two- or three-dimensional structures, which can be characterized by typical scattering patterns, in general we do not have long-range positional order (as in a three-dimensional solid). Some authors even say that in lyotropic liquid crystals there is short or medium-ranged positional order.

We pay special attention to the structural characterization of different phases. X-ray and neutron scattering and diffraction are the basic techniques for the determination of these structures. Complementary techniques, as nuclear magnetic resonance, may give key information on local ordering, which is fundamental for a more complete description of the amphiphilic aggregates and their space organization.

7.2.2 *Two-dimensional phases*

Hexagonal (H), monoclinic (M), rectangular (R), and tetragonal (C, direct; K, inverted) two-dimensional phases have been observed in lyotropic mixtures. All of these structures are optically anisotropic.

7.2.2.1 *Direct and inverted hexagonal phases* There are three types of lyotropic hexagonal phases [3]: direct (H_α), inverted (H^i_α), and complex (H_c)

hexagonal phases. In these structures, amphiphilic molecules self-assemble as long cylindrical aggregates. The carbonic chains remain in a liquid-like state. If hexagonal and isotropic micellar phases occupy neighboring regions in a phase diagram, the transition from the micellar to the hexagonal phase can be regarded as a process of increasing the length of the micellar cylindrical aggregates. This mechanism, taking into account shape anisotropy and the concentration of cylindrical stiff objects, sometimes called "particles," has been theoretically treated by Flory [46]. The approach is more complicated for lyotropic mixtures, since the micelles, in the isotropic micellar phase, and the long cylinders, in the hexagonal phase, are flexible and deformable objects, changing shape anisotropy as a function of external parameters (as temperature and relative concentrations of the components).

Location of the phases in the phase diagrams The H_α phase (see Fig. 1.15(a)) is present in phase diagrams of binary mixtures of an amphiphile and water at amphiphilic concentrations of about 50 wt% (e.g. see Figs 1.6, 3.7, and 3.8). A micellar isotropic phase, at increasing temperatures or water concentrations, and a lamellar phase, at increasing amphiphile concentrations, are in the neighborhood of this H_α phase. Since the transitions between the hexagonal and their neighboring phases are of first order, there are large characteristic regions of phase coexistence. The H_α phase is also found in the phase diagrams of ternary and quaternary mixtures (e.g. see Figs 3.10 and 3.13). In partial isothermal representations of ternary mixtures, the H_α phase is located at low co-surfactant concentrations, and at about 50 wt% of the principal surfactant.

The H^i_α phase (Fig. 1.15(b)) is also found in phase diagrams of binary [18] and multicomponent [3] lyotropic mixtures. Amphiphiles that combine the geometrical characteristics of bulky paraffinic regions and a small polar head are potential candidates to present inverted structures. In a binary mixture of sodium bis-2-ethylhexyl sulfosuccinate (AOT) and water (see Fig. 7.3), the H^i_α phase is located in a range of large amphiphilic concentrations ($\gtrsim$82 wt%), and high temperatures (higher than 100°C). The stabilization of an inverted phase in this phase diagram is strongly dependent on the molecular topology of AOT. Binary mixtures of lecithin and water [47] also present a reversed hexagonal phase, at high temperatures (above 100°C), surrounded by an inverted micellar isotropic phase at lower temperatures. In the partial isotherm of phase diagrams of ternary mixtures, the H^i_α phase is usually located in the upper vertex region of the triangular representation (mixture of Na caprylate/1-decanol/water at 20°C, [3]). However, depending on the particular amphiphile, it can be located in a different region. For example, in the partial isothermal representation of a mixture of AOT, decanol, and water, the H^i_α phase is located in the region of right vertex of the triangular representation (in other words, in a region with concentrations of AOT larger than 60 wt% and concentrations of decanol less than 30 wt%).

The complex hexagonal phase has been found in several binary lyotropic mixtures; e.g., in Na myristate ($C_{14}H_{27}NaO_2$) and water, K palmitate ($C_{16}H_{31}KO_2$)

and water, and Na laurylsulfate ($C_{12}H_{25}NaO_4S$) and water [48,49]. It is located in different regions of the phase diagrams between L_α and H_α phases.

Determination of the structures The determination of the structure of a hexagonal phase is mainly achieved by X-ray and neutron diffraction and scattering techniques. In neutron diffraction and scattering experiments, samples have to be prepared with deuterated materials in order to give the necessary contrast between different parts of the structure, mainly between the solvent and the alkyl chain [50,51]. The coherent scattering of neutrons by deuterons is much stronger than by protons. Coherent scattering lengths of H and D are about -0.37×10^{-12} and 0.67×10^{-12} cm, respectively [52]. The contrast technique can be used to determine the molecular distribution of a particular species in the structure. It is used for studying the deformation of the circular cross section of cylindrical amphiphilic aggregates into a flattened cylinder (which is also called a *ribbon*).

Experimentalists always face the important, and difficult, task of achieving a good degree of orientation of samples. In most of the situations, the diffraction peaks of the lyotropic structures appear at small scattering angles, which correspond to large spacing distances, of the order of 1–10 nm.

If X-rays are used as probes, the diffraction intensity depends on the electronic density contrast between the amphiphile head and the surrounding medium (e.g. between the paraffinic chains and water). The analysis of diffraction patterns of lyotropic phases is essentially performed in two regions of reciprocal space: (i) in the high-angle region, with $s = 2\sin\theta/\lambda_x \sim 2.5\,\text{nm}^{-1}$, where 2θ and λ_x are the scattering angle and the X-ray wavelength, respectively; (ii) in the small-angle region, with $s \sim 0.1\,\text{nm}^{-1}$. At high angles, water-based lyotropic mesophases display two broad bands, at $s \sim 2.22\,\text{nm}^{-1}$ and $s \sim 3.13\,\text{nm}^{-1}$, due to the mean distances between carbons of paraffinic chains and water molecules, respectively. These bands do not have an oriented profile, even if oriented phases are used to obtain the diffraction patterns, which is an indication of the liquid-like state of the paraffinic chains (and water) in these mesophases. In the range of small angles, diffraction patterns of H_α, H^i_α, and H_c phases display several Bragg peaks, with characteristic distances according to ratios $1:\sqrt{3}:\sqrt{4}:\sqrt{7}:\cdots$, which are typical of a two-dimensional hexagonal structure in direct space. These patterns are consistent with the picture of long cylindrical amphiphilic aggregates, organized as a hexagonal two-dimensional lattice. In crystallographic terms, the reciprocal spacings s_{hkl} of the diffraction, with Miller indices [53] h, k and l, are written as $s^2_{hkl} = a^{*2}(h^2+k^2-hk)$, where a^* is the lattice parameter of the reciprocal unit cell. Figures 1.15(a) and (b) show sketches of the H_α and H^i_α structures. In the H_α phase, the diameter of the cylinders is about twice the extended length of the principal amphiphile.

Typical lattice parameters d of selected amphiphiles in water, in the H_α phase, are given in Table 7.1. The diameter of the cylinders, d_a, and the mean surface per polar head, S, can be calculated if we know the concentrations of

Table 7.1 Lattice parameters of phases H_α and H_c of selected amphiphiles in water; d corresponds to H_α, and d_c to H_c ([49])

Amphiphile	d (nm)	d_c (nm)
$\mathbf{C_{18}H_{35}NaO_2}$	6.3	12.4
$\mathbf{C_{18}H_{33}NaO_2}$	5.4	10.7
$\mathbf{C_{16}H_{31}KO_2}$	5.0	10.5
$\mathbf{C_{14}H_{27}KO_2}$	4.6	9.3

the different components of the mixture, their partial specific volumes, $\overline{v}_s$ for the solvent and $\overline{v}_a$ for the amphiphile, and the weight concentration of amphiphile per gram of the mixture, c_w. Thus, we have

$$d_a = d\left[\frac{\pi}{2\sqrt{3}}\left(1+\frac{\overline{v}_s}{\overline{v}_a}\frac{1+c_w}{c_w}\right)\right]^{-1/2}.$$

It should be noted that the partial specific volume of component 1 of a binary mixture is given by $\overline{v}_1 = (\partial V/\partial g_1)_{T,P,g_2}$, where V, g_i, T and P are the total volume of the mixture, mass of component i, temperature, and pressure.

In the H_α phase of water-based mixtures, the shape of the cross section of the cylindrical amphiphilic aggregates depends on the water concentration. This result seems to be related to the hydration of the heads of the amphiphiles in the structure [54]. Let us call ϕ_a the volume fraction of the amphiphile in the mixture. In binary mixtures of fatty acid soaps, the lattice parameter d behaves according to the relation

$$\log(d/d_0) \propto \log(1/\phi_a),$$

where d_0 is a constant, and the slope is usually small [29,54]. In mixtures with small water concentrations, the lattice parameter becomes almost independent of ϕ_a, which indicates a possible modification of the shape of the amphiphilic aggregate (e.g. a modification of the cross section perpendicular to the cylindrical axis). The diameter of the cylindrical aggregates slightly increases with ϕ_a. The surface per polar head, at the interface between aggregates and water, increases with the water concentration in the mixture. Gallot and Skoulios [29] showed that, in the case of soaps, the surface per polar head S and the ratio between the number of moles of water and soap, Z, behave according to the same relation for the lamellar phase,

$$\log(S/S_1) = q\log(Z),$$

where S_1 is a constant, representing the surface per polar head if $Z = 1$, and q is a parameter that depends on the particular soap molecule (e.g. we have $S_1 = 0.397\,\text{nm}^2$ and $q = 0.09$, for a sodium soap at 86°C; and $S_1 = 0.396\,\text{nm}^2$ and $q = 0.10$ for a potassium soap at the same temperature [3]). In the

H_α phase, the surface per polar head has intermediate values between those found in lamellar and micellar isotropic phases. In a mixture of potassium laurate and water, $S \sim 0.53\,\text{nm}^2$ in the H_α phase, $S \sim 0.35\,\text{nm}^2$ in the lamellar phase, and $S \sim 0.65\,\text{nm}^2$ in the micellar isotropic phase [55].

Quantities d and S have distinct behavior in the H^i_α structure and the corresponding direct phase; d is almost constant for increasing (small) amounts of water, and slightly increases with ϕ_w (where we recall that ϕ_w is the volume fraction of water in the mixture); S increases with ϕ_w. A logarithmic plot of S/S_1 as a function of Z shows different linear regions, with different angular coefficients [56].

In the H_c phase, lattice parameter d_c obtained from the analysis of X-ray diffraction patterns is about twice the lattice parameter of the H_α phase for the same amphiphile (see Table 7.1). In this structure, it has been assumed that long cylinders are formed by closed bent double layers, as sketched in Fig. 7.10 [48,49].

The local configuration: Growth and shape anisotropy of aggregates The local configuration of cylindrical-type amphiphilic aggregates is affected by several parameters, as temperature, pH, and ionic strength of the mixture. Let us discuss some quasielastic light scattering experiments performed in a mixture of sodium dodecylsulfate (SDS), NaCl, and water, in the temperature range $10 < T < 85°\text{C}$, for NaCl concentrations $0.15 < c_n < 0.8\,\text{M}$, and SDS concentrations $1.7 \times 10^{-2} < c_s < 6.9 \times 10^{-2}\,\text{M}$ [57,58]. For $c_n > 0.3\,\text{M}$, amphiphilic aggregates present a prolate ellipsoidal shape, with a minor axis of about 5.0 nm and a major axis increasing with T and decreasing with c_n. The aggregation number $\widetilde{n}$, defined as the mean number of amphiphilic molecules per aggregate, at

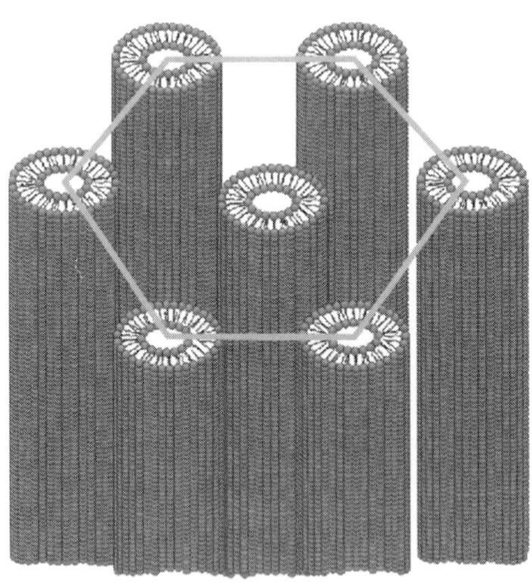

FIG. 7.10. Sketch of the hexagonal complex phase H_c.

fixed SDS and NaCl concentrations, $c_s = 6.9 \times 10^{-2}$ M and $c_n = 0.6$ M, increases from $\widetilde{n} \sim 60$, at $T = 85°$C, corresponding to an almost spherical micelle, to $\widetilde{n} \sim 1600$, at $T = 18°$C, corresponding to a major axis of about 135 nm. Fixing the temperature at 25°C, and with the same value of the NaCl concentration $c_s, \widetilde{n}$ increases from $\widetilde{n} \sim 80$, at $c_n = 0.15$ M, to $\widetilde{n} \sim 1000$, at $c_n = 0.6$ M. In these cylindrical-type aggregates, it has been observed that $\widetilde{n} \sim \sqrt{c_s}$, at fixed T and c_n. The width of the distribution of aggregation numbers is ±70% of the mean value, which indicates a large polydispersity of the major axis (length of the cylinders) of the molecular aggregates (since the minor axis is almost independent of T, c_n and c_s).

The growth of the major axis of the cylindrical aggregates, keeping the minor axis approximately unchanged, is a consequence of what we may call the *ending effect*. For example, consider a spherocylindrical aggregate, formed by a cylindrical core and two semispherical caps. The molecular packing is expected to be more compact in the cylindrical core than in the caps. Variable parameters as T, c_n, and c_s seem to affect these caps preferentially, increasing or decreasing the core length.

NMR experiments indicate that the local behavior of amphiphilic molecules is practically unchanged in all of the lyotropic phases (hexagonal as well as lamellar, cubic, etc). Fast molecular translational diffusion and conformational motions of paraffinic chains have been detected in hexagonal and lamellar phases [59]. However, the surface per polar head is larger in the direct hexagonal than in the lamellar phase. In order to account for deuteron magnetic resonance results, the packing of the heads of amphiphilic molecules in the cylindrical surface was proposed to be direction-dependent. Along the generators of the cylinders, heads of the amphiphile molecules pack as in a lamellar phase; along the circumference, the heads are more distant from each other [60].

Deformation of the cylinders In the hexagonal phases of binary mixtures, cylindrical aggregates of amphiphilic molecules are expected to display a simple circular cross section. However, in particular if there is a cosurfactant in the mixture, there may appear stable fattened cylinders, which are also known as *ribbons*.

Hendrikx and coworkers [61,62] performed contrast neutron scattering experiments in the ternary mixtures of potassium laurate (KL), decanol, and water, and sodium decylsulfate, decanol, and water, with convenient changes of H_2O by D_2O, using deuterated KL and perdeuterated decanol. The sections of the cylindrical amphiphile aggregates depend on the molar ratio, $R_{d/a}$, between decanol and the main amphiphile. The deviations of the cylinders from a circular shape increase with the ratio $R_{d/a}$. This deformation produces cylindrical cross sections of elliptical shape, and degrees of ellipticity increasing with $R_{d/a}$, until the formation of ribbons. It is interesting to point out that this deformation occurs before the system enters into another phase. A phase transition (to a rectangular phase) takes place at the moment the cylindrical cross sections of the aggregates are sufficiently far from a circular shape. Contrast experiments also lead to the preferential location of alcohol molecules, in the *flat part* of the amphiphilic aggregates.

Shear effects The peculiar behavior of the direct hexagonal phase under shear is strikingly different from the usual behavior of nematic and lamellar phases. There are rheological, as well as combined rheological and small-angle light-scattering, rheological-NMR and rheological-neutron scattering experiments in order to investigate dynamic aspects of the hexagonal phases.

In a rheological experiment, a *creep curve* is a plot of the time dependence of the compliance, $J(t) = \gamma(t)/\sigma$, where $\gamma(t)$ is the measured time-dependent shear strain and σ is the (constant) shear stress. The *flow curve* is the plot of the shear stress as a function of the shear rate $\dot{\gamma}$. The SI units of σ and $\dot{\gamma}$ are Pa and s^{-1}, respectively, while γ is dimensionless. Among other effects, in rheological experiments we observe shear-induced phase transitions and shear-orientation of amphiphilic aggregates.

A lyotropic mixture of $H_3C(CH_2)_{13}OCH[CH_2O(CH_2CH_2O)_4CH_3]_2$, also called $C_{14}G(E_4M)_2$, and water [63] displays a hexagonal phase, which has been investigated under shear using the rheo-SALS (rheological-small angle light scattering) technique [64]. At constant shear stress or shear rate, the long axes of the cylindrical aggregates become oriented either parallel or perpendicular to the flow direction. At smaller shear rates, or at short creep times, there are observations of orientational correlations of the molecular aggregates along a direction perpendicular to the flow. After long creep times, or at large shear rates, the axes of the cylindrical aggregates become oriented parallel to the direction of the flow. More recent experiments performed in the same mixture with complementary techniques, as optical microscopy, SALS, SANS (small-angle neutron scattering), birefringence, and NMR [65], show that, at nanoscopic length scales, the 1 0 plane (where $h = 1$ and $k = 0$ are the Miller indices) of the hexagonal structure is oriented parallel to the shear plane. The six-fold symmetry is lost under shear, but it is recovered if the shear ceases. The average orientations of the director were shown to be always parallel to the direction of the flow. Also, there are observations of orientational correlations, associated with an undulation of the director in a direction perpendicular to the flow direction, at a larger length scale, of the order of micrometers, as compared with the typical diameter of a cylinder.

A cone–plate rheo-NMR experiment performed in a mixture of hexa (ethylene glycol) monododecyl ether ($C_{12}E_6$) and D_2O, in the region of the hexagonal phase, has shown that [66]: a) shear macroscopically orients a previously non-oriented sample (at large shear rates, cylindrical axes are aligned parallel to the flow); (b) the aligned state shows no relaxation after removing shear; (c) the apparent shear viscosity decreases with the shear-alignment of the cylinders; (d) reorientation processes *do not depend* on the shear rate (but depend on the shear strain).

Shear melting was observed in the hexagonal phase of a mixture of SDS, pentanol, cyclohexane, and a brine (NaCl) [67]. Two states with different viscosities were identified as a function of shear. In the small shear regime ($\dot{\gamma} \leq 100\,\mathrm{s}^{-1}$), the axes of the cylindrical aggregates are aligned parallel to the shear flow, keeping a two-dimensional hexagonal structure. At larger shear rates ($\dot{\gamma} \geq 1000\,\mathrm{s}^{-1}$), however, the two-dimensional long-range arrangement is destroyed, leading to a two-dimensional liquid-like structure of amphiphilic aggregates, with the cylindrical axes along the flow. The six-fold symmetry is recovered a few seconds after removing the shear. It has been proposed that this mechanism of shear melting involves an activation energy, E_A, related to the end-cap energy of the cylindrical aggregates [68]. This energy, estimated as $E_\mathrm{A} \sim 30k_\mathrm{B}T$, should account for the energy cost for breaking a cylinder into two pieces. In the range of temperatures $285 < \mathrm{T} < 309\,\mathrm{K}$, the linear elastic modulus G_0, also known as *shear modulus*, does not display a monotonous behavior as a function of temperature: it increases to a maximum value, $G_0 \sim 230\,\mathrm{Pa}$, at $T \sim 291\,\mathrm{K}$, and then decreases to $G_0 \sim 120\,\mathrm{Pa}$, at 309 K. This decreasing of G_0 at higher temperatures, after reaching a maximum value, was attributed to the increasing number of end caps due to the activated breaking of the cylinders.

7.2.2.2 *Monoclinic, orthorhombic, and tetragonal phases* In this section, we discuss monoclinic (M), orthorhombic (also known as *rectangular*, R), and tetragonal (C for direct, and K for inverted) phases. Let us recall some crystallographic concepts. In Fig. 7.11, we sketch a unit cell in real space with lattice parameters a, b, and c, angles α, β, and γ, and lattice vectors $\vec{a}$, $\vec{b}$, and $\vec{c}$.

Since we are dealing with two-dimensional lattices, and amphiphilic aggregates that are long cylinders or ribbons, one of the dimensions is assumed to be much larger than the other two. Relations pertaining to axial lengths and angles,

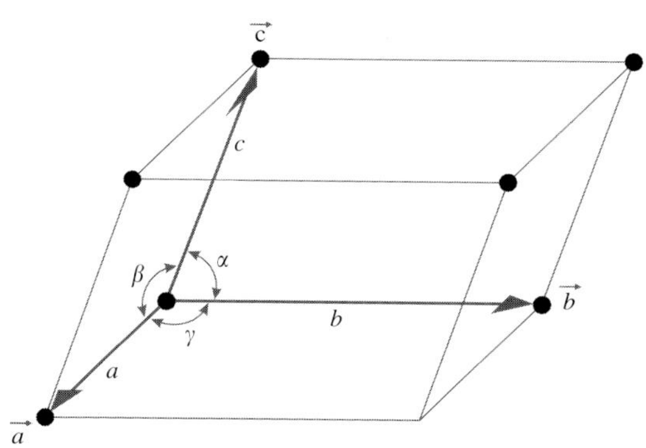

FIG. 7.11. Sketch of a unit cell.

Table 7.2 Crystallographic systems with the respective axial lengths and angles

System	Axial lengths	Angles
Tetragonal	$a = b \neq c$	$\alpha = \beta = \gamma = \pi/2$
Orthorhombic	$a \neq b \neq c$	$\alpha = \beta = \gamma = \pi/2$
Monoclinic	$a \neq b \neq c$	$\alpha = \gamma = \pi/2 \neq \beta$
Hexagonal	$a = b \neq c$	$\alpha = \beta = \pi/2, \gamma = 2\pi/3$

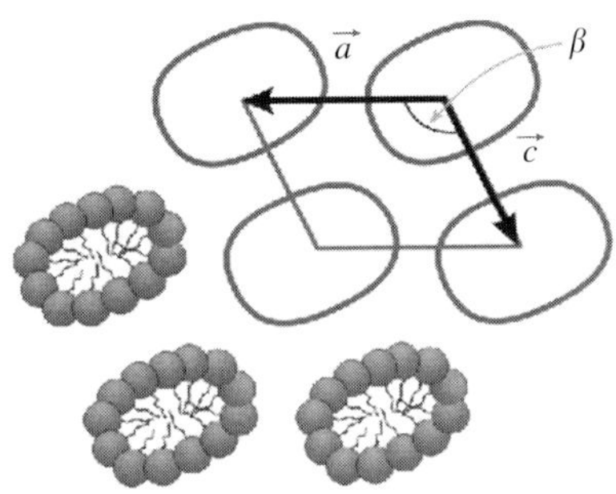

FIG. 7.12. Sketch of a cross section of the monoclinic M_α phase structure; the amphiphilic aggregates have a ribbon-type shape.

in the monoclinic, orthorhombic, tetragonal and hexagonal crystallographic systems, are shown in Table 7.2.

A monoclinic two-dimensional phase M_α has been identified in a mixture of SDS and water, between the hexagonal and the lamellar phases [69,70]. In real space, the structure of this phase was associated with a "deformed parallelogram,"originated from the hexagonal phase. One of the edges becomes longer than the other one, and the angle between them decreases continuously. This phase, which has been called "deformed hexagonal" by Luzzati and coworkers [49], was observed in soap/water mixtures, between H and H_c phases. In the SDS/water mixture, the lattice parameters and the angle of the M_α phase, where the subscript α represents the liquid-like state of the paraffinic chains, were estimated as $a = 4.88\,\text{nm}$, $c = 3.93\,\text{nm}$, $b \gg a, c$, and $\beta \simeq 1.99$ ($\simeq$114.1°). In Fig. 7.12, we show the sketch of a cross section of the M_α phase, in which the amphiphilic aggregates have a ribbon-type shape. Diffraction peaks observed in the M_α phase of the SDS/water mixture correspond to the following $h\,k$ pairs of Miller indices: $1\,0$, $0\,1$, $1\,1$, $2\,0$, $1\,\bar{2}, 2\,\bar{2}$ or $0\,2$, and $3\,0$ or $3\,\bar{2}$.

A direct rectangular phase R_α of a two-dimensional orthorhombic system, formed by long cylinders with a ribbon-type cross section, has been observed between lamellar and hexagonal phases of some binary ($C_{18}H_{33}NaO_2$ and water [6]) and ternary (sodium decylsulfate, decanol, and water [71]) mixtures. The subscript α represents the liquid-like state of the carbonic chains.

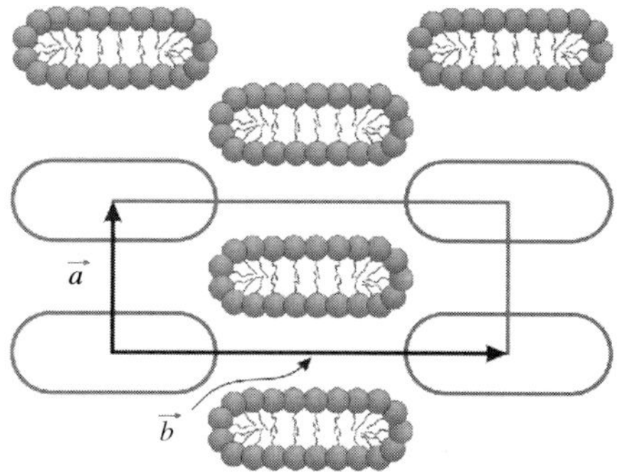

FIG. 7.13. Sketch of a cross section of the direct rectangular R_α phase structure; the amphiphilic aggregates have a ribbon-type shape.

In Fig. 7.13, we show the sketch of a cross section of this phase structure. In crystallographic terms, the reciprocal spacings are written as $s_{hk}^2 = h^2a^{*2} + k^2b^{*2}$, where $a^*$and b^* are the lattice parameters of the reciprocal unit cell. The positions of the Bragg peaks of the X-ray diffraction spectra of the R_α phase are distributed according to the ratios $1:1/2:1/3:\cdots$.

Square phases (C for direct, and K for the inverted structures) associated with a two-dimensional tetragonal system, formed by long cylinders with non-circular, almost square, cross sections, also known as "white phases," have been observed in some ternary lyotropic mixtures. Consider the C phase. It is located between L_1 and L_α regions. X-ray diffraction patterns of the C phase present Bragg peaks with positions according to the ratios $1:1/2:1/3:\cdots$. Lyotropic mixtures of sodium caprylate ($C_8H_{15}NaO_2$), nonanol and water, and of decaoxyethylene glycol monolauryl ether, oleic acid and water, are known to present the C phase [26]. In Fig. 7.14(a), we show a cross section of this phase structure. In a mixture of sodium caprylate, decanol and water, in the C phase, Fontel and coworkers [54] have shown that the lattice parameter a depends on the volume fraction of amphiphiles ϕ_a according to the relation

$$\log(a/a_0) \propto \log(1/\phi_a),$$

where a_0 is a constant. However, the surface per polar head remains constant, of order $0.41\,\text{nm}^2$, as a function of the amount of water in the mixture. The proposal of a square shape for the cross sections of the cylinders was not based on the X-ray diffraction patterns only. It also comes from considerations about the molar concentrations of different components of the mixture [3]. The C phase is stable in alcohol and soap-rich regions of the phase diagrams, mostly in the alcohol-rich region, with a water content of about 50 wt%.

We now discuss the inverted K phase, which is sketched in Fig. 7.14(b). It was also observed in ternary mixtures, as $CH_3(CH_2)_8COOK$ (potassium caprate),

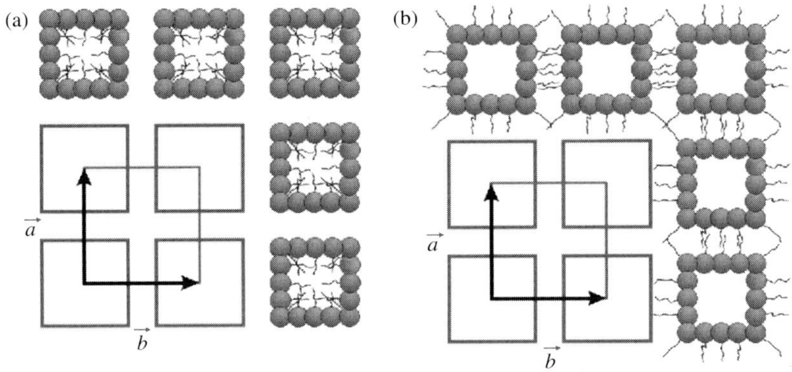

FIG. 7.14. Sketch of a cross section of the square phase structure; the amphiphilic aggregates have a square-type shape, $a = b$; (a) direct, C; (b) inverted phase, K.

octanol, and water [72], with small amounts of water. X-ray diffraction experiments display Bragg peaks in positions according to the ratios $1:1/2:1/4:1/8$. The lattice parameter seems to be weakly sensitive to the water content of the mixture.

7.2.3 *Three-dimensionally ordered phases*

Three-dimensional lyotropic structures can be formed by micellar (direct and inverted) or bicontinuous molecular aggregates. Several distinct three-dimensional structures have been experimentally observed in lyotropic mixtures: cubic micellar ($\mathrm{Q_m}$), hexagonal micellar ($\mathrm{H_m}$), rhombohedral micellar (Rh), tetragonal micellar (T_α), bicontinuous cubic ($\mathrm{Q_b}$), bicontinuous sponge ($\mathrm{L_3}$), and bicontinuous mesh. We now discuss some features of these structures.

7.2.3.1 *Direct and inverted cubic micellar phases* From the optical point of view, direct and inverted cubic micellar phases are isotropic. Two types of lyotropic micellar cubic phases were identified, $\mathrm{Q_m^{223}}$, with a primitive unit cell and a homogeneous lipid composition, and $\mathrm{Q_m^{227}}$, with a fcc unit cell [73] of $Fd3m$ space group and heterogeneous lipid composition. $\mathrm{Q_m^{223}}$ was observed in chemically pure mixtures of lipids and solvents, and $\mathrm{Q_m^{227}}$ requires mixtures of water-miscible and water-immiscible lipids and a solvent [73]. Binariy mixtures of dodecyltrimethylammonium chloride (DTAC) and water, *N*,*N*,*N*-trimethylamino dodecanoimide and water, and palmitoyllyso phosphatidyl choline (PLPC) and water, are known to present cubic phases. Also, a ternary mixture of dioleoylphosphatidylcholine, dioleoylglycerol and water presents a cubic phase. X-ray diffraction patterns of fcc phases display Bragg peaks with characteristic distances according to the ratios $1:\sqrt{4/3}:\sqrt{8/3}:\sqrt{11/3}$.

These phases may have spherical or ellipsoidal micelles [74]. The primitive cell of the Q_m^{223} phase has six direct, disk-like, micelles of symmetry $\bar{4}2\,m$, and two spherical micelles of symmetry $m3$. The ratio between area and volume of these micelles is about the same. The values of the lattice parameters of the Q_m^{223} phase, in mixtures of DTAC and water and PLPC and water, were estimated as 8.5 and 13.7 nm, respectively, always larger (about twice) than the values of the lattice parameters of the neighboring hexagonal phase.

In the inverted cubic phase Q_m^{227}, instead of micelles, spherical liquid globules of the solvent are packed in a cubic symmetry. This phase has been observed, e.g., in the phase diagram of a mixture of phosphatidylcholine, diacylglycerol, and water [75]. The Q_m^{227} phase is present between inverse hexagonal and inverse micellar isotropic phases. The unit cell of this phase is composed by two types of inverted micelles [75], eight larger micelles in a tetrahedral diamond-like lattice, and 16 smaller micelles in tetrahedral clusters at the four octants of the cell. Large and small micelles have different values of the ratio between area and volume. The lattice parameter of the Q_m^{227} phase in a mixture of PFL, which is a lipid extracted from the *Pseudomonas fluorescens* [76], and water is 14.6 nm, about three times larger than the lattice parameter of the neighboring hexagonal phase.

7.2.3.2 *Three-dimensional micellar phases of lower symmetries* In the hexagonal micellar phase H_m, micelles are packed on a hexagonal three-dimensional lattice (of hcp structure). A lyotropic mixture of octaethylene glycol dodecyl ether ($C_{12}EO_8$) and water presents the H_m phase, of space group $P6_3/mmc$. It is located in the vicinity of Q_m and H_α phases, at concentrations of $C_{12}EO_8$ between about 35 and 39 wt%, in a temperature range from 5 to 14°C [77]. Lattice parameters of this particular mixture, at 38 wt% of $C_{12}EO_8$, are $a = 6.88$ nm and $c = 11.17$ nm (see Fig. 7.11). The aggregation number is approximately 10^2, and micelles are assumed to be almost spherical.

A rhombohedral Rh phase (such that $a = b = c$ and $\alpha = \beta = \gamma \neq \pi/2$, see Fig. 7.11), was observed in a SDS/water mixture (Fig. 3.9), with $a = 9.99$ nm, which is practically constant in almost the entire phase region, far from the phase transitions, and $\alpha = 108.3°$, at SDS concentrations of about 62 wt% [69,70]. Diffraction peaks observed in the Rh phase of the SDS/water mixture correspond to the following sets of $h\,k\,l$ Miller indices, $1\,\bar{1}\,0, \bar{2}\,1\,1, 2\,\bar{2}\,0, 2\,1\,\bar{1}, 1\,1\,1, 3\,\bar{1}\,\bar{1}, 1\,2\,0$, and $\bar{3}\,2\,1$. If we increase the SDS concentration, there is a phase transition to a cubic phase, with a jump in the lattice parameter, $\Delta a \simeq 0.01$ nm, at the transition. If we decrease the SDS concentration, there is a phase transition to a two-dimensional monoclinic phase M_α. Parameters a and α undergo considerable changes in the vicinity of the Rh–M_α transition.

A tetragonal T_α micellar phase, associated with a bcc lattice, has also been observed in SDS/water mixtures [70]. The T_α phase is stable in a narrow region of the phase diagram, at approximately 65.7 wt% SDS, in a temperature range from 328 to 360 K. Diffraction peaks in the T_α phase of the SDS/water mixture correspond to the following sets of *hkl* Miller indices, 110, 101, 200, 002, 112, 202, 301, 103, 222, 400, 004 or 303, and 204 or 323. The lattice parameters are $a = b = 7.61$ nm and $c = 6.72$ nm.

7.2.3.3 *Direct and inverted bicontinuous cubic phases* Direct and inverted bicontinuous cubic phases are optically isotropic. Cubic bicontinuous structures in lyotropic mixtures are associated with distinct space groups of symmetry [78] (Q^{230} with space group $Ia3d$ [56]; Q^{229} with $Im3m$ [79]; Q^{224} with $Pn3m$ [80]) Another cubic phase, Q^{212}, shows a complex intermediate character between a bicontinuous and micellar phases. It has been modeled as a three-dimensional network of rods, with a structure that encloses the micelles [79]. The Bragg peaks of the X-ray diffraction patterns of these micellar phases are related to a cubic symmetry in direct space. The binary mixture of dodecyltrimethilammonium chloride and water, and the ternary mixture of monoolein, oleic acid and water, present these bicontinuous cubic phases [81]. Deuteron magnetic resonance experiments in a mixture of potassium laurate and D_2O in the cubic phase show that the longitudinal relaxation time is intermediate between the values for lamellar and hexagonal phases. On the other hand, the transverse relaxation time is larger than in lamellar and hexagonal phases. Different regimes of time were also observed in this experiment; there is a fast molecular reorientation of the O–D bonds, with a time scale of 10^{-11} s; also, there is slow motion (time scale of 10^{-6} s) and short- and long-range diffusion (time scales of 10^{-9} and 10^{-7}–10^{-8} s, respectively) [82]. Soap molecules in this cubic structure were shown to have a diffusion coefficient $D = 2 \times 10^{-6}\,\mathrm{cm}^2/\mathrm{s}$, at 90°C.

The microstructure of these phases can be explained in terms of three-dimensional *periodic minimal surfaces*, free from self-intersections, which are generically called *triply periodic minimal surfaces* (TPMS). The basic characteristics of this type of surfaces is the vanishing of the mean curvature at every point [83]. Let us give some details about the three fundamental cubic (periodic) minimal surfaces [84], P-surface, D-surface and G-surface. Schwarz [85] constructed the simplest of these structures, the continuous P-surface, shown in Fig. 7.15.

Repetition of this unit along the directions of the three Cartesian axes gives the bicontinuous P-surface, separating two continuous and congruent channel regions. Another surface that can be constructed is of the diamond-type, called D-surface, and sketched in Fig. 7.16. In this type of surface, channel systems on each side form a diamond lattice. The last fundamental surface is the gyroid or G-surface [86], shown in Fig. 7.17.

The Q^{230} phase was observed [87] in anhydrous salts of fatty acids, of divalent cations, with both direct and inverted topology. In Fig. 7.18(a) we sketch the structure of the Q^{230} phase. The direct structure was found in lyotropic mixtures with the amphiphile dodecyltrimethylammonium chloride (DTAC), with lattice parameter $a = 7.96$ nm. The inverted structure was found in a mixture of monoolein (MO), oleic acid, and water, with a lattice parameter

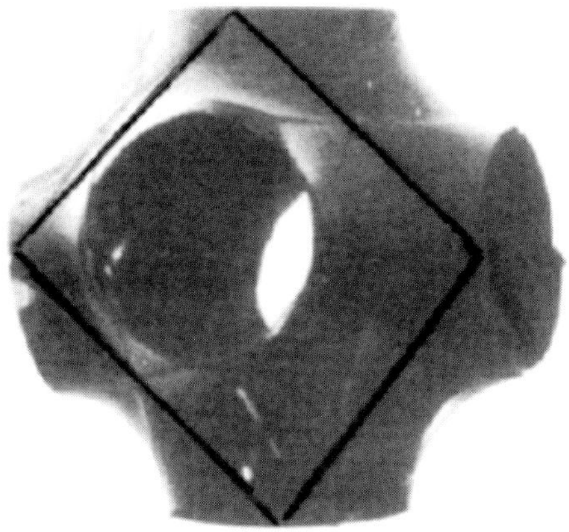

FIG. 7.15. Structure of the Schwarz primitive surface or P-surface [89].

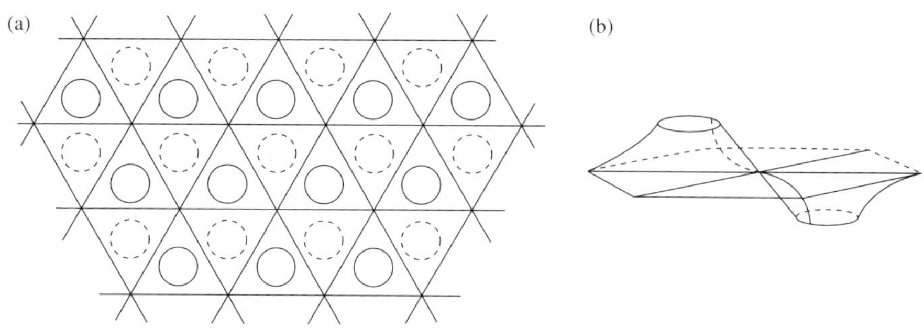

FIG. 7.16. Sketch of the diamond-type surface (D-surface) [89]: (a) lattice formed by two-fold axes; filled and open circles indicate the catenoids going upwards and downwards; (b) illustration of two of these catenoids.

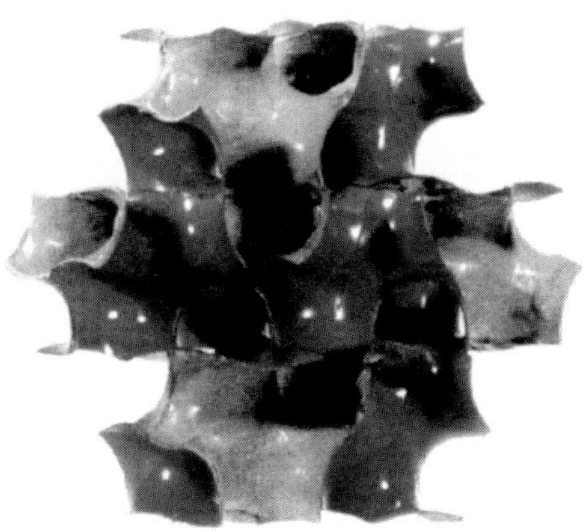

FIG. 7.17. Gyroid or G-surface. The two sides of the bicontinuous structure are represented with gray scales [84].

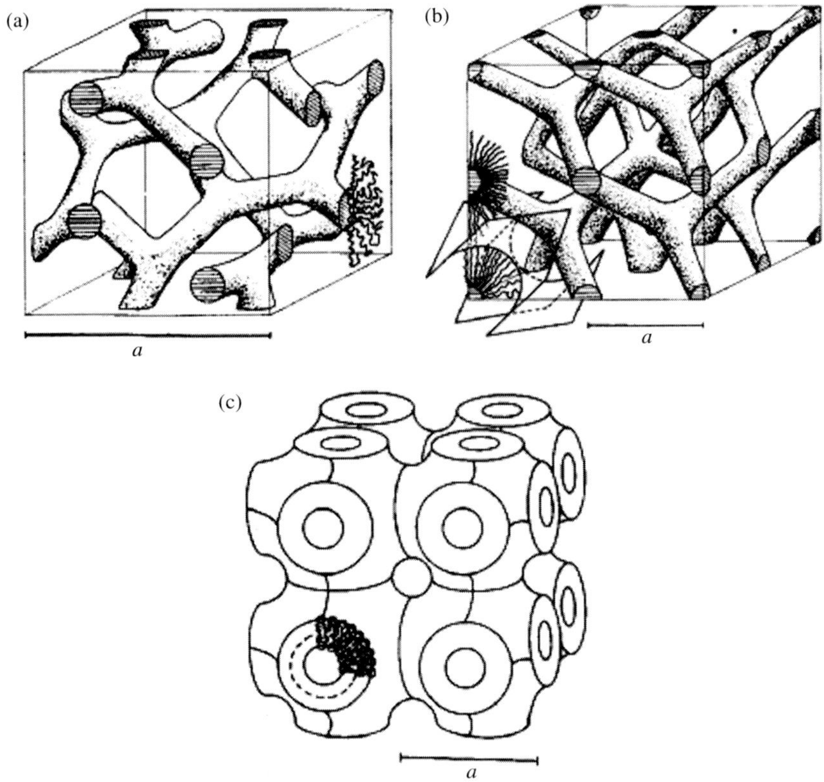

FIG. 7.18. Sketch of bicontinuous cubic structures with a lattice parameter a [91]: (a) Q^{230}, with space group $Ia3d$; (b) Q^{224}, with space group $Pn3m$; (c) Q^{229}, with space group $Im3m$.

$a = 14.30$ nm [73]. This phase consists of interconnected rods, linked three-by-three, forming two three-dimensional networks intertwined and disconnected. In a mixture of $C_{12}EO_8$ and water, the structure of the $Ia\overline{3}d$ phase was described in terms of a complex bicontinuous three-dimensional surface, with convoluted polar–nonpolar interfaces, which has been called *gyroid infinite periodic minimal surface* (G-IPMS or G-surface) [88,89]. Luzzati and coworkers [73] introduced the concept of *chaotic zones*, defined as the loci where the short-range disorder is maximal. The CH_3 end-groups of the hydrocarbon chains are located in the non-polar chaotic zones of the (direct) structure, which is coincident with the G-IPMS.

Q^{224} and Q^{229} phases were observed in inverted topology only. In Fig. 7.18(b) and (c), we sketch the structures of the Q^{224} and Q^{229} phases. The structure of the Q^{224} phase is made of interconnected rods, tetrahedrally linked, in order to form two three-dimensional intertwined and disconnected networks. The Q^{229}

phase seems to be formed by six-by-six linked rods, which are oriented along the sides of a cube. In terms of a structural description on the basis of convoluted surfaces, Q^{224} and Q^{229} phases present a D-surface, with symmetry group $Pn3m$, and a P-surface, with symmetry group $Im3m$, respectively [90].

7.2.3.4 *From micellar to bicontinuous topology* One of the key concepts in order to determine the location of a particular phase in a phase diagram is the preferred interfacial mean curvature (between the amphiphilic aggregate and the solvent). Consider, e.g., a binary mixture of an amphiphile and water. In general, this interfacial curvature increases with hydration, since the area per head group of the amphiphile increases with the amount of water in the mixture. The phase sequence in a hypothetical phase diagram (relative water concentration versus temperature) starts with the inverse micellar phase (at lower water concentrations), going towards the direct micellar phase (at higher water concentrations). If we increase the water concentration in the mixture, there may appear inverse hexagonal, lamellar, and direct hexagonal phases, between the two extrema. Also, there may exist cubic phases in the intermediate regions between each one of the two neighboring phases of this hypothetical sequence [91].

Micellar cubic phases Q_m are usually located between hexagonal (direct and inverse) and isotropic micellar phases in the phase diagrams of lyotropic mixtures. For example, consider the phase diagram sketched in Fig. 7.19, for a mixture of 1-palmitoyllysophosphatidylcholine (PaLPC) and D_2O [74], in which the cubic micellar phase Q_m is located between H_α and L_1 phases.

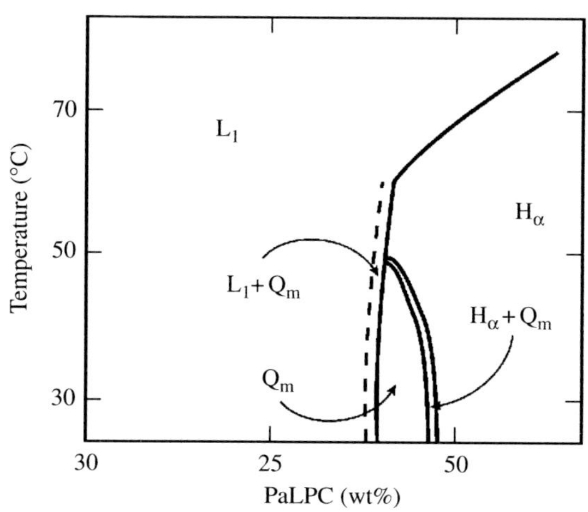

FIG. 7.19. Phase diagram of a mixture of 1-palmitoyllysophosphatidylcholine (PaLPC) and water [74]. The micellar cubic Q_m phase is located between H_α and L_1 phases.

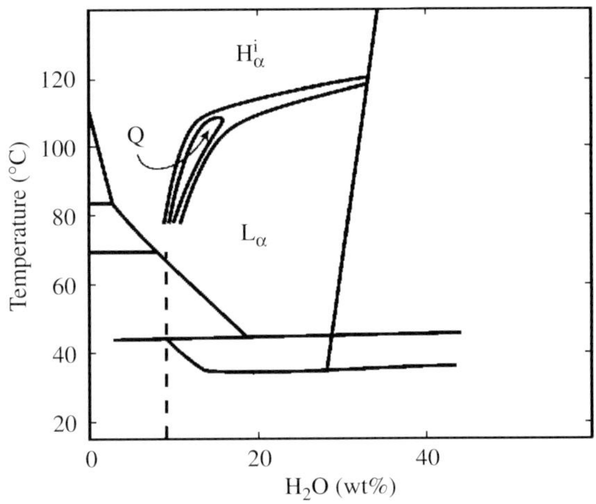

FIG. 7.20. Phase diagram of the lyotropic mixture of (di − C_{12}) dialkyl didodecylphosphatidylethanolamine (DDPE) and water [92]. The bicontinuous cubic Q phase is located between H^i and L_α phases.

The bicontinuous cubic phase Q usually lies between hexagonal and lamellar phases (i.e. between H_α and L_α, and H^i_α and L_α). For example, in the phase diagram of Fig. 7.20, for a mixture of (di-C_{12}) dialkyl didodecylphosphatidylethanolamine (DDPE) and water [92], the bicontinuous cubic Q phase is located between H^i_α and L_α phases. There are no experimental observations of a *direct transition* between micellar and bicontinuous cubic phases.

7.2.4 *The mesh phase*

The mesh phase is found in the phase diagrams of some lyotropic mixtures, between lamellar and hexagonal structures. It seems to be favored if the principal amphiphile has a long chain or if the flexibility of the chain is reduced [93]. Lyotropic mixtures of nonaethylene glycol mono-ether (11-oxa-14,18,22,26-tetramethylheptacosyl), also known as $C_{30}EO_9$, and water [94], and hexadecyl polyethylene oxide, $C_{16}H_{33}O(C_2H_4O)_7H$, also known as $C_{16}E_7$, and water [42,95] present mesh phases in certain regions of the phase diagrams.

In the mesh phase, original flat lamellae have pores (holes) filled with the solvent, organized in a particular two-dimensional symmetry. In Fig. 7.21 we sketch an example of a perforating lamellar structure. The ordering of the holes in the original lamellar plane is not long-ranged. However, it extends to scales which allow the establishment of a clear X-ray diffraction indexation. Bragg peaks in the X-ray diffraction patterns of this phase could be indexed in different ways, since in many cases the number of peaks is not enough to clearly identify the structure. The following structural models have been proposed to account

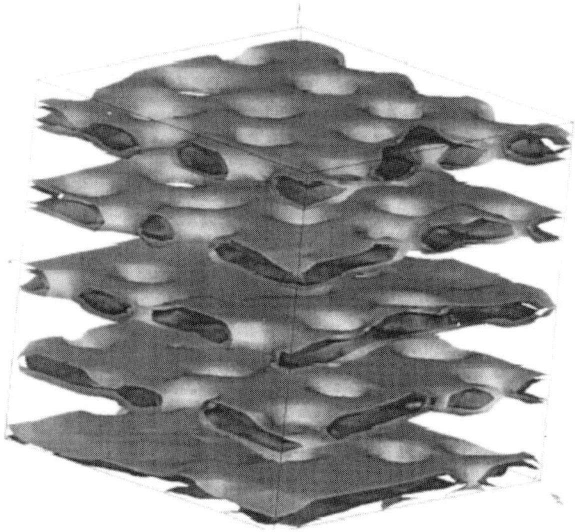

FIG. 7.21. Sketch of a typical perforating lamellar structure of a mesh phase. There are holes filled with the solvent in the original lamellar structure [42].

for these Bragg peaks [94]: (a) a centered tetragonal mesh (see Fig. 1.36(a)); (b) a bicontinuous centered tetragonal structure; (c) a structure with one layer of a *type a* hexagonal mesh from which both a rhombohedral mesh and bicontinuous structures can be constructed; (d) a similar structure, with a layer of *type b* hexagonal. In the particular case of the $C_{30}EO_9$/water mixture, the structure of the mesh phase seems to include layers of *type b* rhombohedral (see Fig. 1.36(b)), with diffraction peaks corresponding to the following set of $h\,k\,l$ Miller indices, 1 1 0, 0 0 2, 1 1 2, 3 0 1, 1 0 3, 2 2 0, and 0 0 4, and lattice parameters $a = 24.4$ nm and $c = 19.6$ nm (the parameter a lies along the corrugated lamellar layer and c is perpendicular to it).

In the $C_{16}E_7$/water mixture, there is a transition from the lamellar to an intermediate phase, of the mesh-type, called *gyroid* by Imai and coworkers [42]. The symmetry group of this phase is $Ia3d$, with a unit-cell dimension of 15.03 nm. The observed diffraction peaks in this gyroid phase correspond to the following set of $h\,k\,l$ Miller indices, 2 1 1, 2 2 0, 3 2 1, 4 0 0, 4 2 0, 3 3 2, 4 2 2, and 4 3 1.

References

[1] V. Luzzati (1968). In *Biological Membranes*, Ed. D. Chapman, Academic Press, New York.
[2] A. Tardieu, V. Luzzati, and F. C. Reman (1973). *J. Mol. Biol.* **75**, 711.

[3] P. Ekwall (1975). In *Advances in Liquid Crystals*, Ed. G. H. Brown, Academic Press, New York, Vol. 1.
[4] J. C. P. Gabriel, F. Camerel, B. J. Lemaire, H. Desvaux, P. Davidson, and P. Batail (2001). *Nature* **413**, 504.
[5] Y. Piffard, A. Verbaere, A. Lachgar, S. Deniard-Courant, and M. Toumoux (1986). *Rev. Chim. Miner.* **23**, 766.
[6] F. Husson, H. Mustacchi, and V. Luzzati (1960). *Acta Crystallogr.* **13**, 668.
[7] P. G. de Gennes and J. Prost (1993). *The Physics of Liquid Crystals*, Clarendon Press, Oxford.
[8] D. Roux and C. M. Knobler (1988). *Phys. Rev. Lett.* **60**, 373.
[9] F. Brochard and P. G. de Gennes (1975). *Pramana Suppl.* **1**, 1.
[10] F. Nallet, D. Roux, and J. Prost (1989). *Phys. Rev. Lett.* **62**, 276.
[11] W. Helfrich (1978). *Z. Naturforsch.* **33a**, 305.
[12] L. Courbin, J. P. Delville, J. Rouch, and P. Panizza (2002). *Phys. Rev. Lett.* **89**, 148305.
[13] O. Diat, D. Roux, and F. Nallet (1993). *J. Phys. II* **3**, 1427.
[14] J. W. McBain and W. C. Sierichs (1948). *J. Am. Oil Chem. Soc.* **25**, 221.
[15] S. Kumar, L. J. Yu, and J. D. Litster (1983). *Phys. Rev. Lett.* **50**, 1672.
[16] M. C. Holmes and J. Charvolin (1984). *J. Phys. Chem.* **88**, 810.
[17] Y. Hendrikx, J. Charvolin, and P. Kekicheff (1987). *Liq. Cryst.* **2**, 677.
[18] J. Rogers and P. A. Winsor (1969). *J. Colloid Interface Sci.* **30**, 247.
[19] R. Strey, R. Schomäcker, D. Roux, F. Nallet, and U. Olsson (1990). *J. Chem. Soc. Faraday Trans.* **86**, 2253.
[20] C. Y. Zhang, S. Sprunt, and J. D. Litster (1993). *Phys. Rev. E* **48**, 2850.
[21] G. S. Smith, C. R. Safinya, D. Roux, and N. A. Clark (1987). *Mol. Cryst. Liq. Cryst.* **144**, 235.
[22] M. P. Hentschel and F. Rustichelli (1991). *Phys. Rev. Lett.* **66**, 903.
[23] J. A. N. Zasadzinski, J. Schneir, J. Gurley, V. Elings, and P. K. Hansma (1988). *Science* **239**, 1013.
[24] D. C. Wack and W. W. Webb (1988). *Phys. Rev. Lett.* **61**, 1210.
[25] P. Ekwall, L. Mandell, and K. Fontell (1968). *Acta. Chem. Scand.* **22**, 365.
[26] P. Ekwall, L. Mandell, and K. Fontell (1969). *J. Colloid Interface Sci.* **29**, 542.
[27] P. Ekwall, L. Mandell, and K. Fontell (1969). *Mol. Cryst. Liq. Cryst.* **8**, 157.
[28] L. Mandell and P. Ekwall (1968). *Acta Polytech. Scand., Chem. Incl. Met. Ser.* **74**, I.
[29] B. Gallot and A. E. Skoulios (1968). *Kolloid-Z.* **208**, 37.
[30] M. Magalhães, A. M. Figueiredo Neto, and P. Tolédano (2000). *Phys. Rev. E* **62**, 5847.
[31] P. Tolédano, C. E. I. Carneiro, and A. M. Figueiredo Neto (2001). *Liq. Cryst.* **28**, 1547.
[32] P. Bassereau, J. Appel, and J. Marignan (1992). *J. Phys. (France) II* **2**, 1257.
[33] R. Schomäcker and R. Strey (1994). *J. Phys. Chem.* **98**, 3908.

[34] M. Mihailescu, M. Monkenbusch, J. Allgaier, H. Frielinghaus, D. Richter, B. Jakobs, and T. Sottmann (2002). *Phys. Rev. E* **66**, 041504.
[35] F. Auguste, P. Barois, L. Fredon, B. Clin, E. J. Dufourc, and A. M. Bellocq (1994). *J. Phys. (France) II* **4**, 2197.
[36] F. Nallet, D. Roux, and J. Pros (1989). *J. Phys. (France)* **50**, 3147.
[37] D. Roux and C. R. Safinya (1988). *J. Phys. (France)* **49**, 307.
[38] J. Yamammoto and H. Tanaka (1995). *Phys. Rev. Lett.* **74**, 932.
[39] W. Helfrich and R. M. Servuss (1984). *Il Nuovo Cimento* **3**, 137.
[40] W. Helfrich (1985). *J. Phys. (France)* **46**, 1263.
[41] G. Bouglet, C. Ligoure, A. M. Bellocq, E. Dufoure, and G. Mosser (1998). *Phys. Rev. E* **57**, 834.
[42] M. Imai, A. Kawaguchi, A. Saeki, K. Nakaya, T. Kato, K. Ito, and Y. Amemiya (2000). *Phys. Rev. E* **62**, 6865.
[43] G. E. S. Toombes, A. C. Finnefrock, M. W. Tate, and S. M. Gruner (2002). *Biophys. J.* **82**, 2504.
[44] M. Rappolt, A. Hickel, F. Bringezu, and K. Lohner (2003). *Biophys. J.* **84**, 3111.
[45] P. T. C. So, S. M. Gruner, and S. Erramilli (1993). *Phys. Rev. Lett.* **70**, 3455.
[46] P. J. Flory (1961). *J. Polym. Sci.* **49**, 105.
[47] V. Luzzati, T. G. Krzywicki, and A. Tardieu (1968). *Nature* **218**, 1031.
[48] V. Luzzati, H. Mustacchi, and A. Skoulios (1958). *Discuss. Faraday Soc.* **25**, 43.
[49] V. Luzzati, H. Mustacchi, A. Skoulios, and F. Husson (1960). *Acta Crystallogr.* **13**, 660.
[50] Y. Hendrikx, J. Charvolin, M. Rawiso, L. Liébert, and M. C. Holmes (1983). *J. Phys. Chem.* **87**, 3991.
[51] S. Alperine, Y. Hendrikx, and J. Charvolin (1985). *J. Phys. (France) Lett.* **46**, L27.
[52] B. Jacrot (1976). *Rep. Prog. Phys.* **39**, 911.
[53] M. M. Woolfson (1970). *X-ray Crystallography*, Cambridge University Press, London.
[54] K. Fontell, L. Mandell, H. Lehtinen, and P. Ekwall (1968). *Acta Polytech. Scand., Chem. Incl. Met. Ser.* **74**, III.
[55] J. Charvolin and B. Mely (1978). *Mol. Cryst. Liq. Cryst.* **41**, 209.
[56] V. Luzzati and A. Spegt (1967). *Nature* **215**, 710.
[57] N. A. Mazer, G. B. Benedek, and M. C. Carey (1976). *J. Phys. Chem.* **80**, 1075.
[58] P. J. Missel, N. A. Mazer, G. B. Benedek, and C. Y. Young (1980). *J. Phys. Chem.* **84**, 1044.
[59] J. Charvolin and P. Rigny (1973). *J. Chem. Phys.* **58**, 3999.
[60] B. Mely, J. Charvolin, and P. Keller (1975). *Chem. Phys. Lipids* **15**, 161.
[61] Y. Hendrikx, J. Charvolin, and M. Rawiso (1984). *J. Colloid Interface Sci.* **100**, 597.

[62] Y. Hendrikx and J. Charvolin (1988). *Liq. Cryst.* **3**, 265.
[63] K. Kratzat and H. Finkelmann (1993). *Liq. Cryst.* **13**, 691.
[64] W. Richtering, J. Läuger, and R. Linemann (1994). *Langmuir* **10**, 4374.
[65] G. Schmidt, S. Müller, P. Lindner, C. Schmidt, and W. Richtering (1998). *J. Phys. Chem. B* **102**, 507.
[66] M. Lukaschek, D. A. Grabowski, and C. Schmidt (1995). *Langmuir* **11**, 3590.
[67] L. Ramos, F. Molino, and G. Porte (2000). *Langmuir* **16**, 5846.
[68] L. Ramos (2001). *Phys. Rev. E* **64**, 061502.
[69] P. Kékicheff and B. Cabane (1987). *J. Phys. (France)* **48**, 1571.
[70] P. Kékicheff and B. Cabane (1988). *Acta Crystallogr. B* **44**, 395.
[71] Y. Hendrikx and J. Charvolin (1981). *J. Phys. (France)* **42**, 1427.
[72] P. Ekwall, L. Mandell, and K. Fontell (1969). *J. Colloid Interface Sci.* **31**, 508.
[73] V. Luzzati, R. Vargas, P. Mariani, A. Gulik, and H. Delacroix (1993). *J. Mol. Biol.* **229**, 540.
[74] P. O. Eriksson, G. Lindblom, and G. Arvidson (1985). *J. Phys. Chem.* **89**, 1050.
[75] J. M. Seddon, E. A. Bartle, and J. Mingins (1990). *J. Phys. Condens. Matter* **2**, SA285.
[76] P. Mariani, E. Rivas, V. Luzzati, and H. Delacroix (1990). *Biochemistry* **29**, 6799.
[77] M. Clerc (1996). *J. Phys. (France) II* **6**, 961.
[78] J. M. Seddon and R. H. Templer (1993). *Philos. Trans. R. Lond. A* **344**, 377.
[79] P. Mariani, V. Luzzati, and H. Delacroix (1988). *J. Mol. Biol.* **204**, 165.
[80] A. Tardieu (1972). PhD. Thesis, Université de Paris-Sud, Orsay, France.
[81] V. Luzzati, R. Vargas, A. Gulik, P. Mariani, J. M. Seddon, and E. Rivas (1992). *Biochemistry* **31**, 279.
[82] J. Charvolin and P. Rigny (1973). *Chem. Phys. Lett.* **18**, 515.
[83] A. Goetz (1970). *Introduction to Differential Geommetry*, Addison Wesley, Reading, MA.
[84] S. Andersson, S. T. Hyde, K. Larsson, and S. Lidin (1988). *Chem. Rev.* **88**, 221.
[85] H. A. Schwarz (1890). *Gesammelte Mathematische Abhandlungen*, Springer, Berlin.
[86] A. H. Schoen (1970). NASA Tech. Rep. no. 05541.
[87] V. Luzzati, A. Tardieu, and T. Gulik-Krzywicki (1968). *Nature* **217**, 1028.
[88] M. Impéror-Clerc and A. M. Levelut (2001). *Eur.Phys. J. E* **4**, 209.
[89] K. Larsson (1989). *J. Phys. Chem.* **93**, 7304.
[90] S. T. Hyde (1990). *J. Phys. (Paris) Colloque* **51**, 209.
[91] J. M. Seddon, J. L. Hogan, N. A. Warrender, and E. P. Peyroula (1990). *Prog. Colloid Polym. Sci.* **81**, 189.
[92] J. M. Seddon, G. Cevec, R. D. Kaye, and D. Marsh (1984). **23**, 2634.

[93] P. Kékicheff and J. G. T. Tiddy (1989). *J. Phys. Chem.* **93**, 2520.
[94] J. Burgoyne, M. C. Holmes, and G. J. T. Tiddy (1995). *J. Phys. Chem.* **99**, 6054.
[95] P. D. T. Huibers, V. S. Lobanov, A. R. Katritzky, D. O. Shah, and M. Karelson (1996). *Langmuir* **12**, 1462.

8

RECENT DEVELOPMENTS AND RELATED AREAS

8.1 Introduction

Interfaces with other fields of chemical physics and interdisciplinary connections are important features of research work on liquid crystals, in particular on lyotropic mixtures. We can give several examples of topics in this interface: magnetic colloids (or *ferrofluids*), microemulsions, Langmuir–Blodgett films, and diblock copolymers.

Ferrofluids are used to dope liquid crystals for different purposes. For example, they may be used for reducing the strength of the magnetic field required to align a certain sample, and for swelling smectic structures in order to promote nanophase segregation. Amphiphilic molecules are also used for coating ferrofluid nanograins in order to avoid their aggregation. In this chapter, we discuss magnetic colloids, in particular the applications in the study of physico-chemical properties of lyotropic mixtures.

As discussed in Chapter 1, there are no sharp conceptual boundaries between lyotropics, in particular the isotropic phases, and microemulsions. In some cases, isotropic phases of the same mixture, in which oil is one of the components, are classified in different sides of this boundary. In this chapter, we discuss microemulsions formed by isolated aggregates and bicontinuous structures. In particular, we look at the topology of the phase diagrams and refer to some of the models used to describe microemulsions.

The study of ultra-thin and highly ordered films (Langmuir and Langmuir–Blodgett films) is also strongly related to the area of lyotropic liquid crystals. Hydrophobic and hydrophilic effects play an important role in the stabilization of these films. In this chapter, we discuss the processes of deposition of thin films, their characterization, and some specific applications.

Diblock copolymers are formed by two independent polymers, with a small affinity, linked by a covalent-type bond. Under some conditions, this system self-organizes in order to increase (decrease) the contacts between similar (antagonic) species. In the discussion of these systems, we emphasize the concepts of affinity, nanoscale segregation and ordering. We also discuss some structural aspects and the phase diagrams of these materials.

Finally, we refer to new lyotropic-type mixtures: chromonic and lyo-banana systems. Chromonics are mixtures of disk or lath-like molecules with solvents. Lyo-bananas are mixtures of a thermotropic banana-type liquid crystal and a solvent.

8.2 Magnetic colloids

There are strong overlaps between the areas of magnetic colloids and lyotropic liquid crystals, from the point of view of the basic mechanisms for the stabilization of colloids and also from purely practical reasons. The introduction of magnetic colloids in a liquid crystalline medium leads to a new class of complex fluids, the so-called *ferroliquid-crystals*, with remarkable properties that are not observed in the undoped system. In this section, we describe magnetic colloids and their use for doping liquid crystals.

8.2.1 *Definition of a ferrofluid*

Ferrofluids, or magnetic fluids, are colloidal suspensions of small magnetic grains, of about 10 nm, dispersed in a liquid carrier. In the reduced volume of a grain, there is just a single magnetic domain. These synthetic materials were produced in the 1960s, in the context of the NASA space program [1–3]. At that time, space scientists faced the problem of transporting fuel from the reservoir to the engines in orbital artificial satellites without the use of heavy and mechanically complex pumps. The creative answer was to dope fuel with ferrofluids and then use small magnetic field gradients, produced by very small coils, in order to move the magnetic grains *and* the fuel through the pipes. The movement of the dissolved ferrofluid grains is then responsible for the collective motion of the fluid. Since this breakthrough, the technological applications of these materials, ranging from industry to medicine, did not stop to increase.

In the absence of external fields, ferrofluids are optically isotropic, with a large magnetic susceptibility. On the other hand, they become birefringent in the presence of magnetic [4] and surface [5,6] fields. Two different types of ferrofluids are available, surfacted and ionic ferrofluids, depending on the physical processes to avoid agglomeration of grains.

8.2.2 *Surfacted ferrofluids*

Surfacted ferrofluids (SFFs) are formed by grains (usually magnetite, Fe_3O_4) coated with surfactant agents (amphiphilic molecules, as oleic acid and aerosol sodium di-2-ethylhexyl-sulfosuccinate) in order to prevent their aggregation. They were the first ferrofluids to be synthesized by the NASA team. Steric repulsion between grains acts as a physical barrier [7] that keeps grains in the solution and stabilizes the colloid. If the grains are dispersed in a polar medium, as water, two layers of surfactant are needed to form an external hydrophilic layer. The polar heads of surfactant molecules can be cationic, anionic or nonionic. If grains are dispersed in a nonpolar medium (as light mineral oil, toluene, octane, hexane, cyclohexane), a single surfactation is enough to form the external layer, with the polar head of the surfactant attached to the surface of the grains and the carbonic chain in contact with the fluid carrier.

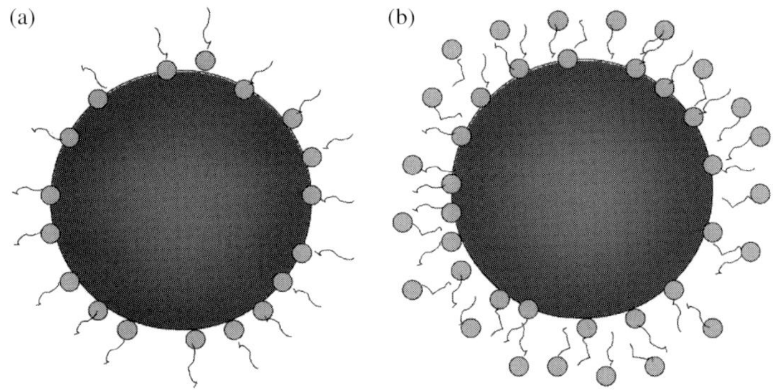

FIG. 8.1. Sketch of surfacted ferrofluid grains: (a) single-layer coated grains; (b) double-layer coated grains.

If a ferrofluid is highly diluted in a solvent, it becomes unstable and the grains flocculate. Different types of liquid carriers are used; for example, liquid metals, as Hg, K, and NaK. In Fig. 8.1, we show the sketch of the grains of a surfacted ferrofluid: (a) single-layer coated grains; (b) double-layer coated grains.

SFFs are commercially available [8,9], but details about manufacturing processes, purity of the fluid carrier and composition of the coating agents, are usually not provided by producers. In some cases, this lack of additional information is a source of difficulties for analyzing experiments with ferrofluids. However, some laboratories in the world are able to produce small quantities of SFFs. Since it is possible to know all the details about the components and manufacturing processes, these homemade materials are the best indicated samples to be used in research laboratories.

8.2.3 *Ionic ferrofluids*

In the ionic ferrofluids (IFF) [10], magnetic grains (usually maghemite, γ-Fe_2O_3, and different ferrites, MFe_2O_4, where M = Mn, Co, Zn, Cu, Ni) are electrically charged to keep the colloidal system stable (by the effect of electric repulsion). Magnetic grains are obtained through a chemical synthesis. An acid–alkaline reaction between grains and the bulk keeps the surface of the grains electrically charged [11,12]. Usually, the liquid carrier is water, and the pH of the solution can vary from about 2 to 12, depending on the sign of the surface charge of the grains. Acid ferrofluids ($pH < 7$) have positively charged grains, and alkaline ferrofluids ($pH > 7$) have negatively charged grains. The charge of the grains changes as a function of the pH of the solution. The *point of zero charge* (PZC) is defined as the value of the pH at which the grains are not charged (in other words, at which the ferrofluid precipitates). The typical surface charge density of a grain, far from the PZC, is about 10 $\mu C/cm^2$ [13]. One of the advantages of this procedure for preparing a ferrofluid is that the sizes of the grains can be better controlled

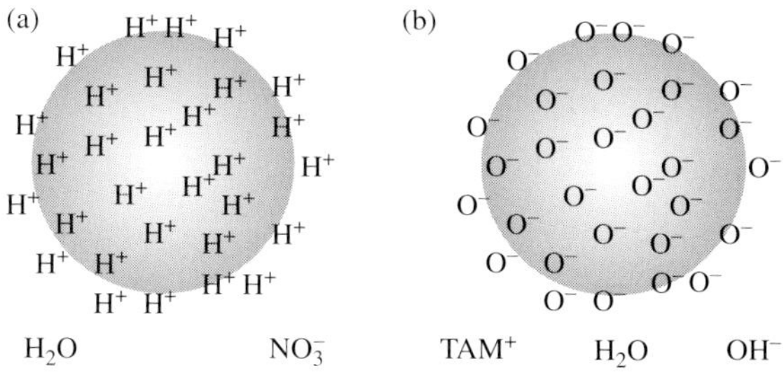

FIG. 8.2. Sketch of ionic ferrofluid grains. The counterions (NO_3^-, tetramethylammonium hydroxide TAM^+, and OH^-) and the water molecules are also represented: (a) acid ferrofluid grain. Positive charges are given by the H^+ located at the surface of the grains; (b) alkaline ferrofluid grain. Negative charges are given by the O^- located at the surface of the grains.

during chemical synthesis. Ionic citrated or tartarated maghemite (γ-Fe_2O_3) ferrofluids have both characteristics (steric and electrostatic repulsion) to prevent aggregation of the magnetic grains. The grains of these ferrofluids are coated with small chain molecules, as $COOH{-}CH_2{-}COH{-}COOH{-}CH_2{-}COOH$ in the case of citrated ferrofluids. In the presence of water, these attached molecules ionize; besides the steric repulsion, there is also an electrostatic interaction. In Fig. 8.2, we show the sketch of IFF grains: (a) acid ferrofluid grain; (b) alkaline ferrofluid grain.

The grains display an almost spherical geometry, which could be sketched as a potato, with a non-uniform shape distribution [14]. The distribution $P(D)$ of sizes of the grains, in terms of the diameter D, is given by a log-normal function [15,16],

$$P(D) = \frac{1}{D\sigma\sqrt{2\pi}} \exp\left(-\frac{\ln^2(D/D_0)}{2\sigma^2}\right), \tag{8.1}$$

where D_0 and σ are the mean diameter and standard deviation, respectively. In Fig. 8.3, we show a plot of $P(D)$ versus D, with $\sigma = 0.2$ and $D_0 = 10$ nm.

The internal crystalline structure of the grains corresponds to the mineral *spinel*, $MgAl_2O_4$, in which divalent ions (Mn^{2+} and Co^{2+}, for example) replace ions of Mg, and ions of Fe^{3+} replace ions of Al^{3+} [17]. In this structure, the bcc primitive unit cell has 32 oxygen atoms, with 64 and 32 interstices of tetrahedral and octahedral symmetries, respectively. In direct spinel structures, ions M and Fe^{2+} occupy 1/8 and 1/2 of the tetrahedral and octahedral sites, respectively. In inverse spinel structures, ions M occupy part of the octahedral sites and ions Fe^{2+} occupy the rest of the octahedral and the tetrahedral sites. Globally, the nanocrystal is ferrimagnetic; the typical lattice parameter of maghemite, Co and

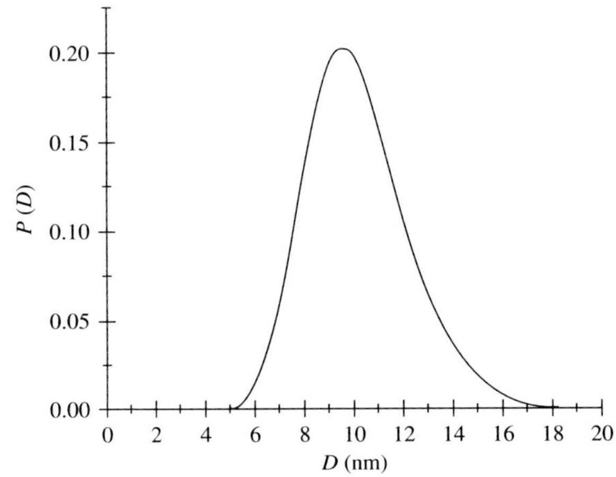

FIG. 8.3. Log-normal plot of size distribution of grains $P(D)$ with $\sigma = 0.2$ and $D_0 = 10$ nm.

Ni ferrites is about 0.83 nm. At saturation, the magnetization of maghemite, Co and Ni ferrites is given by $m_s = 400$, 400 and 270 kA/m, respectively. The magnetic moment of a grain ranges from about 10^3 μ_B to 10^5 μ_B, where μ_B is the Bohr magneton [18].

8.2.4 *Stability of the colloid*

The stability of the magnetic colloid depends on the thermal contribution and on the balance between attractive (van der Waals and dipole–dipole) and repulsive (steric and electrostatic) interactions. In order to avoid sedimentation, the size of the grains cannot be larger than a few nanometers.

8.2.4.1 *Attractive interactions* There are essentially two main attractive interactions between magnetic grains in a ferrofluid, the van der Waals–London and the dipole–dipole interactions.

The van der Waals–London interaction, U_{Aw}, between two spherical grains of diameter D_0 separated by a distance r is written as [19]

$$U_{Aw} = -\frac{A}{6}\left[\frac{2}{\alpha^2 - 4} + \frac{2}{\alpha^2} + \ln\left(\frac{\alpha^2 - 4}{\alpha^2}\right)\right], \tag{8.2}$$

where $\alpha = 2r/D_0$ and A is the Hamaker constant ($A \sim 10^{-19}$ J for ferrite grains [12]). This is a short-range interaction; the attractive force increases with the size of the grains.

The interaction energy between two magnetic dipoles, $\boldsymbol{\mu}_1$ and $\boldsymbol{\mu}_2$, separated by a distance r, is given by

$$U_{\text{Ad}} = \frac{\mu_0}{4\pi r^3}\left[\boldsymbol{\mu}_1 \cdot \boldsymbol{\mu}_2 - 3\left(\boldsymbol{\mu}_1 \cdot \frac{\mathbf{r}}{r}\right)\left(\boldsymbol{\mu}_2 \cdot \frac{\mathbf{r}}{r}\right)\right], \tag{8.3}$$

where $\mathbf{r}$ is the relative position of the grains, and μ_0 is the permeability of vacuum. In the ionic water-based ferrofluid $MnFe_2O_4$, with $D_0 = 10.7$ nm, the magnetic moment of a grain is about 10^{-19} Am^2/grain (or 10^{-16} in CGS units). This interaction is partly responsible for the formation of chains in actual ferrofluids. The formation of chains can also be attributed to the action of an external magnetic field. In this case, there is a reduction of the rotational degrees of freedom, and the grains tend to agglomerate in linear or neck-ring chains.

8.2.4.2 *Repulsive interactions* In SFFs, there are sterical repulsion forces, of a short-range nature. On the other hand, in IFFs, long-range electrostatic interactions between charged grains give rise to repulsive interactions, which guarantee colloidal stability. As the bulk of IFFs is electrically neutral, there are counterions in the bulk of the suspension in order to compensate the surface charged grains (which can be regarded as macro-ions). These counterions are driven to the surface of the grains, but are also subjected to electrostatic repulsion. The calculation of the electrostatic repulsion between grains has to take into account this complex distribution of counterions. In the double-layer model [20], there is a first layer of counterions, called Stern layer, some angstroms thick, which involves the grain, and a second diffuse layer. These layers are separated by a Helmholtz plane. The diffuse layer is characterized by a Debye length, $l_{\text{D}} \equiv \kappa^{-1}$, which is defined as the distance, from the Helmholtz plane, at which the electrostatic potential is 37% smaller than the value at the plane.

The interaction between two electrically charged spherical grains of diameter D_0, separated by a distance r, is written as [20]

$$U_R = \frac{D_0 \pi \sigma^2}{\varepsilon_0 \varepsilon_r \kappa^2} \exp[-\kappa(r - D_0)], \tag{8.4}$$

where $\sigma = \varepsilon_0 \varepsilon_r \kappa \psi_0$ is the surface charge density, $\varepsilon = \varepsilon_0 \varepsilon_r$ is the electric permittivity of the fluid carrier, and ψ_0 is the surface potential of the charged grain at the Helmholtz plane (with typical values $\psi_0 < 50$ mV).

8.2.4.3 *Total interaction potential* At short intergrain distances, the total interaction potential U_{T} between magnetic grains of a ferrofluid as a function of the intergrain distance r has a cutoff due to hard-sphere interactions. There is a potential barrier if we increase the intergrain distances beyond this cutoff. For larger values of r, there appears a secondary minimum, at $r = r_{\text{e}}$. This is the equilibrium distance between grains (mean intergrain distance) in the stable colloid. Larger values of ψ_0 lead to higher potential barriers and smaller values

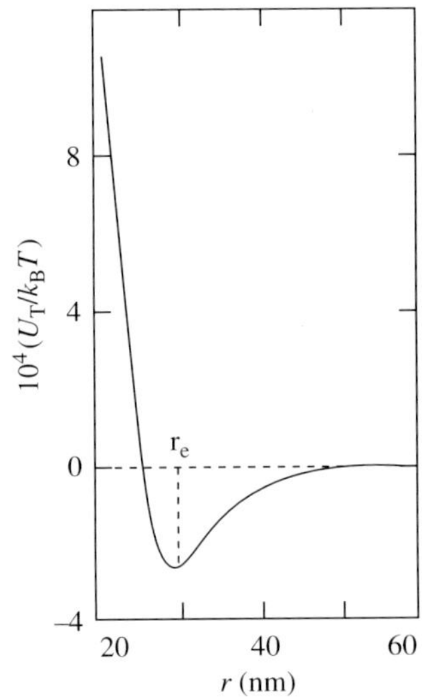

FIG. 8.4. Sketch of the total interaction potential U_T between magnetic grains in a ferrofluid as a function of the intergrain distance r, for $\psi_0 = 15$ mV, pH=2.5, and $D = 8.2$ nm [13].

of the depth of the secondary minimum. In Fig. 8.4 we show a sketch of U_T as a function of r in the vicinity of r_e.

8.2.5 *The mechanisms of rotation of the magnetic moment*

In the absence of external fields, the magnetic moment $\boldsymbol{\mu}$ of a grain is oriented along the direction of easy magnetization, defined by the vector $\mathbf{n}_f$. The magnetic moment is kept along this direction by an anisotropic energy barrier of strength E_a. The Néel time to go through this barrier is given by

$$\tau_N \sim 10^{-9} \exp\left(-\frac{E_a}{k_B T}\right),$$

where τ_N is expressed in seconds. In the absence of a small magnetic field, the ferrofluid behaves as a paramagnetic system formed by nano-ions, which has also been called superparamagnetic.

At a given temperature, and with $E_a > k_B T$, $\boldsymbol{\mu}$ remains parallel to $\mathbf{n}_f$. On the other hand, with $E_a < k_B T$, $\boldsymbol{\mu}$ can flip continuously between $\mathbf{n}_f$ and $-\mathbf{n}_f$, according to a process that is called *Néel rotation.*

Another process of rotation of $\boldsymbol{\mu}$ is a type of *Brownian rotation*, in which grains are supposed to physically rotate with respect to the environment. It is characterized by the typical time

$$\tau_{\mathrm{B}} \simeq \frac{3\eta V_{\mathrm{h}}}{k_{\mathrm{B}}T},$$

where η is the viscosity of the fluid carrier and V_{h} is the hydrodynamic volume of the magnetic grain. In a solution of water, $\tau_{\mathrm{B}} \sim 10^{-6}$ s.

In summary, if a ferrofluid is in the presence of a small magnetic field $\mathbf{H}$, the couplings between the individual magnetic moments $\boldsymbol{\mu}$ of the grains and the field $\mathbf{H}$ depend on different parameters, according to two basic processes (the rotation of the magnetic moment without the physical rotation of the grain, and a physical rotation of the hole grain). If $E_{\mathrm{a}} > k_{\mathrm{B}}T$, $\boldsymbol{\mu}$ orients parallel to $\mathbf{H}$; in this limit of a rigid dipole, there is a physical rotation of the hole grain. In the other limit ($E_{\mathrm{a}} < k_{\mathrm{B}}T$), the torque of $\mathbf{H}$ on $\boldsymbol{\mu}$ gives rise to a rotation of the magnetic moment but not necessarily of $\mathbf{n}_f$ (i.e. of the grain itself). This process depends on the viscosity and on the elastic properties of the liquid carrier in which the grains are placed.

8.2.6 *Thermodiffusion in ferrofluids: The Soret effect*

Thermodiffusion, also called *Soret effect* [21–24], is one of the most interesting phenomena observed in the area of condensed matter. In order to provide a simple explanation of this effect, consider a single-phase solution, formed by two different types of basic units (e.g. a colloidal system in which grains or nanoparticles are dispersed in a fluid carrier), characterized by a volume fraction of one of the components of the solution. If this originally homogeneous material is subjected to a thermal gradient, there is sometimes a concentration current parallel to the direction of the thermal gradient. This is the *Soret effect*, which has also been observed in some gases and solids [23,25,26]. This effect is characterized by the measurement of the thermodiffusion coefficient S_T, also known as *Soret coefficient* [27]. S_T represents the coupling between current of mass and temperature gradient. In gases and ordinary liquids, we have $10^{-5} < S_T < 10^{-3}$ K^{-1} [28]; in metals, $S_T \sim 10^{-3}$ K^{-1} [29]; in polymeric solutions, $S_T \sim 1$ K^{-1} [30]. Recently, values of S_T, from 10^{-3} up to 10^{-1} K^{-1}, have been reported in magnetic colloidal (ferrofluid) systems [31,32]. It is interesting to note that, with the *same* temperature gradient, but depending on the particular ferrofluid under study, there are observations of *positive* or *negative* concentration gradients. It has been proposed [33] the following terminology in order to characterize different magnetic colloids with respect to the sign of S_T: if the grains of a colloid tend to go away from the hottest region, it will be called *thermophobic* ($S_T > 0$); if the grains tend to concentrate in the hottest region, the colloid will be called *thermophilic* ($S_T < 0$). Although the Soret effect has been discovered more than a century ago, we still lack a better explanation of the associated physical mechanisms.

Table 8.1 Sign of the Soret coefficient (last column) as a function of constitutive parameters in maghemite-based noncoated ferrofluids. The label SC refers to the sign of the surface charge (the carrier liquid is always water)

pH	ϕ (%)	Experiment	SC	S_T
12.87	9.93	FRS	−	+
12.56	4.96	FRS	−	+
12.4	3.97	FRS	−	+
$\ll 7$	6.7	FRS	+	−
$\ll 7$	3.3	FRS	+	−
2.46	0.32	Z-scan	+	−
2.91	0.32	Z-scan	+	−
3.49	0.32	Z-scan	+	−
3.49	0.32	Z-scan	+	−

Several experimental methods have been used to measure the Soret coefficient: the column method [34], forced Rayleigh scattering (FRS) experiments [32], and the Z-scan technique [35]. The experimental results are summarized in Tables 8.1 and 8.2:

- the sign of S_T *depends* on the sign of the charge of the surface grains;
- for water-based SFF, the thermodiffusive behavior is *opposite* to IFF; in other words, grains coated with cationic surfactants behave as negatively charged IFF (alkaline) grains, and grains coated with anionic surfactants behave as positively charged IFF (acid) grains;
- SFF with grains coated with nonionic surfactants dispersed in non-polar fluid carriers behave as SFF with grains coated with cationic surfactants;
- the nature of the liquid carrier itself is not the only determinant factor of the sign of S_T, except maybe in the case of the nonpolar fluids, where S_T seems to be always positive.

We now give a qualitative discussion of these results [33]. In IFF, the colloidal stability comes from an acid–alkaline equilibrium reaction between the surface of the grain and the bulk, which is responsible for keeping the surface of the grains electrically charged. If we change some physico-chemical parameters of the solution, so that the system tends to the PZC, the surface charge density decreases and the system tends to flocculate. In other words, the distance between grains decreases until phase separation. The acid–alkaline equilibrium reaction in these colloids depends on temperature. A mechanism that might account for the sign of the Soret coefficient in IFF could be related to

Table 8.2 Sign of the Soret coefficient (last column) as a function of constitutive parameters in coated ferrofluids. The label SC refers to the sign of the surface charge. Labels Ferrotec surfact. and BNE surfact. refer to the commercially available products of Ferrotec Inc. and Beycostatne, respectively. All the grains of the FF in this table have a maghemite core, except the samples of the bottom row, which have a cobalt-ferrite core

Coating	Carrier liquid	pH	ϕ (%)	Experiment	SC	S_T
BNE surfact.	Cyclohexane		6.8	FRS	No	+
BNE surfact.	Cyclohexane		2.5	FRS	No	+
BNE surfact.	Toluene		3.3	FRS	No	+
Oleic acid surfact.	Hexane		6.88	FRS	No	+
Oleic acid surfact.	Octane		19.6	FRS	No	+
Ferrotec surfact.	Light mineral oil		3.6	FRS	No	+
Ferrotec surfact.	Water	9–10	1.8	FRS	+	+
Ferrotec surfact.	Water	9–10	0.45	Z-scan	+	+
Ferrotec surfact.	Water	8–9	1.1	FRS	−	−
Ferrotec surfact.	Water	8–9	0.45	Z-scan	−	−
Citrate ion	Water	7	4.65	FRS	−	−
Citrate ion	Water	7	3.30	FRS	−	−
Citrate ion	Water	7.32	0.09	Z-Scan	−	−
Citrate ion	Water	7.3	0.16	Z-Scan	−	−
Citrate ion	Water	7.44	0.47	Z-Scan	−	−
Citrate ion	Water	7.57	0.79	Z-Scan	−	−
Tartarate ion	Water	8.93	7.23	FRS	−	−
Cobalt-ferrite citrate ion	Water	7	6	FRS	−	−

a temperature unbalance of this reaction; if the charge of the grains increases with the heating of the solution, they move away with respect to each other ($S_T > 0$); if the charge of the grains decreases with heating, they tend to flocculate ($S_T < 0$). In SFF, however, the mechanism described above does not apply since the stability of the colloid is not due to electrostatic repulsions but to the steric interactions. In order to assume a similar mechanism (i.e. in order to reduce or increase the intergrain interactions), we have to consider the modification of the surfactant coating around the grains as a function of temperature.

These results still lack a theoretical explanation. Maybe different mechanisms are related to the thermodiffusive behavior of these complex fluids [33]. Recently, Bringuier and Bourdon [36] prosed a kinetic theory, based on the analysis of a

Brownian motion in a nonuniform temperature profile, in order to predict both signs of the Soret coefficient.

8.2.7 *Doping of liquid crystals with ferrofluids*

In 1970, Brochard and de Gennes [37] proposed an elegant theory of magnetic suspensions in liquid crystals. The system is formed by a liquid crystalline matrix with sticky nanograins of large shape anisotropy ($L/d \sim 10$, where L and d are the length and the diameter of the cylindrical grains, respectively). In order to avoid modifications of the liquid crystalline properties, only a small quantity of grains should be added to the liquid crystal. From this point of view, this procedure is better characterized as a doping [38]. One of the goals of this type of doping is to reduce the magnetic field necessary to orient liquid crystals (from a usual value of order 10 kG to a smaller value by a factor of 10^3 [39]). According to experimental observations for lyotropic nematic liquid crystals [40], above a minimum value of the concentration of grains, c_{m}, the liquid crystalline matrix collectively follows the orientation of the grains. Collective response of the liquid crystalline matrix doped with ferrofluids, in samples of about 10^2 μm of thickness, have been observed with magnetic fields of 5 G.

The first successful observation of macroscopic collective behavior in a nematic liquid crystal doped with magnetic grains was reported in 1970 by Rault and coworkers [41]. In this experiment, however, magnetic grains were much bigger than in actual ferrofluids. One of the main difficulties of this doping is the solubilization of ferrofluids in a thermotropic liquid crystal; despite many attempts, thermotropic liquid crystals doped with ferrofluids are not stable. On the other hand, the doping of lyotropic liquid crystals with water-based surfacted ferrofluids was achieved in 1979 by Liébert and Martinet [42], and since then this method has been used to investigate physico-chemical properties of lyotropics [38,40,43–45]. The occurrence of aggregation of grains depends on the magnetic field gradients in the doped samples. The formation of needles of about 10 μm of length can be easily observed [43]. The presence of these needles can be used to investigate dynamic aspects in the physics of liquid crystals, as relaxation processes and thermal induced hydrodynamic instabilities. There were also observations of depletion layers, that is, regions in which magnetic grains are segregated due to topological configurations of the director [37,43].

Dynamic processes in lyotropic ferronematics, in particular the response of the nematic matrix to pulsed magnetic fields, have been investigated by several authors [46,47]. Later, we will discuss an unusual dynamical behavior of ferronematics as compared to undoped nematics [48].

Due to the new features of ferronematics, in particular the remarkable response to low magnetic fields, and after the work of Brochard and de Gennes, there were several theoretical approaches to the problem of the field–director coupling [49–53].

An interesting question that arises in the physics of ferrofluid-doped mesophases refers to the mechanism responsible for the liquid crystal orientation. Two possibilities have been considered: (a) a magnetic coupling between the local field of the individual grains and the director; and (b) a mechanical coupling between anisometric grains and the director. These questions will be discussed in the following sections.

8.2.7.1 *Mechanical versus magnetic coupling* Consider the effects of a magnetic field in a grain inside a nematic liquid crystal, characterized by a director **n**. The contribution to the free energy due to the magnetic coupling is written as [39]

$$\Delta F = -\frac{1}{2}\chi_{\mathrm{a}} \int [\mathbf{n}(\mathbf{r}) \cdot \mathbf{H}(\mathbf{r})]^2 \, \mathrm{d}^3 r, \tag{8.5}$$

where $\chi_{\mathrm{a}} \sim 10^{-7}$ is the anisotropy of the diamagnetic susceptibility of the liquid crystal [39,54]. In order to obtain the magnetic coupling, we compare the energy necessary to change the configuration from $\mathbf{n} \parallel \mathbf{H}$ to $\mathbf{n} \perp \mathbf{H}$. As $H \alpha \mu/r^3$, this energy is given by $E_{\mathrm{mag}} \sim \chi_{\mathrm{a}}\mu^2/V$, where V is the typical volume of a grain. Using typical values, $V \sim 10^{-21}$ cm^3 and $\mu \sim 10^{-16}$ CGS, we have $E_{\mathrm{mag}} \sim 10^{-21}$ erg, which is much smaller than $k_{\mathrm{B}}T \sim 300$ K $\sim 10^{-14}$ erg. This result clearly shows that thermal fluctuations are enough to overcome an eventual magnetic orientation of the director due to the magnetic field of individual grains.

In order to calculate the order of magnitude of the mechanical coupling between the anisometric (stick-like of length L) grains and the nematic matrix, it is enough to compare the elastic energy of a grain in the configurations with its long axis parallel and perpendicular to **n**. In the minimum energy configuration, the long axis of the grain is parallel to **n**. Using the mean Frank elastic constant [55] of the nematic liquid crystal, given by $k \sim (k_{11} + k_{22} + k_{33})/3 \sim 10^{-6}$ dyn [54], the energy necessary to change the orientation of the long axis of the grain from a parallel to a perpendicular orientation with respect to **n** is of the order of $kL \sim 10^{-12}$ erg, with $L \sim 10$ nm, which is much larger than $k_{\mathrm{B}}T$ (at 300 K).

In conclusion, the mechanical coupling between anisometric grains and the local director of nematics is the mechanism responsible for the orientation of ferrofluid-doped liquid crystals in a small magnetic field.

8.2.7.2 *Doping of thermotropic liquid crystals* A thermotropic ferronematic suspension was obtained in 1970 [41] by mixing large γ-Fe_2O_3 magnetic stick-like grains ($L = 0.35$ μm, with diameter $d = 0.04$ mm), coated with surfactant agents, to MBBA (*p*-methoxybenzilidene-*p*-*n*-butylaniline). Due to their large dimensions, these grains do not fulfill the requirements to be classified as basic units of a ferrofluid. One of the major difficulties of the thermotropic liquid crystal doping with magnetic grains is the stability of the solution. At long times, there appear large aggregates of grains, and the solution is no longer homogeneous. Magnetization experiments performed with this mixture showed that, in the isotropic phase, there is no remanent magnetization; on the other hand, a large remanent magnetization was observed in the nematic phase.

Chen and Amer [56] investigated the mixture of MBBA with similar grains ($L = 0.5$ μm, $L/d = 7$), coated with dimethyl octadecyl aminopropyl trimethoxysilyl chloride (DMOAP). Liquid crystal molecules are expected to anchor on the large grains, which is compatible with the observation of a macroscopic collective behavior of the nematic matrix. Formation of aggregates was also observed in this experiment, specially in "high" fields ($H > 20$ G), giving rise to unusual microscopic textures (called *cellular* by these authors).

Our experience in mixing different types of ferrofluids, with different fluid carriers, to thermotropic liquid crystals show that it is a difficult task to reach a homogeneous mixture at nano-scales. In the most favorable cases, MBBA with nonpolar fluid-based ferrofluid, after some minutes, the mixture presented grain clusters of micrometric dimensions, which indicates demixing.

8.2.7.3 *Doping of lyotropic liquid crystals* Since one of the basic substances of lyotropics is water, their doping with water-based ferrofluids is straightforward. As there is a minimum concentration of grains necessary to produce a collective response of the medium to external magnetic fields, the doping has to be done with grain concentrations larger than $c_{\mathrm{m}} \sim 1/(LD_{\mathrm{s}}^2)$ [37], where D_{s} is the thickness of the sample. On the other hand, experimentalists have to take care in order not to introduce too much ferrofluid in lyotropics in order to avoid undesired modifications of some physico-chemical properties of the liquid crystal. Typical concentrations used in lyotropics, for many practical purposes, are $c \sim 10^{13}$ grains/cm^3, which corresponds to adding about 1 μl of ferrofluid ($\varphi \sim 2\%$, where φ is the volume fraction of magnetic material in the magnetic colloidal solution) to 1 ml of the lyotropic mixture. At this concentration, there were observations of no significative changes in the transition temperatures, birefringence, and elastic constants [38]. The stability of ferronematic mixtures was also reported at large ferrofluid concentration, under some particular conditions, specially related to the cosurfactant relative concentration [57].

Both ionic and surfacted ferrofluids have been used to dope lyotropics. If a large amount of water-based ferrofluid is added to a lyotropic nematic mixture, the first effect can be a shift towards domains of the $\mathrm{N_D}$ and isotropic phases in the phase diagrams, due to the addition of excess water. Usually, the formation of large (micrometer and even millimeter sizes) clusters of grains can be observed in the test tube in which the doped lyotropic mixture is prepared. This condition has to avoided in the experimental investigation of lyotropic mixtures by means of ferrofluid doping.

8.2.7.4 *Ferronematic lyomesophases* Liébert and Martinet reported a pioneering experiment in which a water-based surfacted ferrofluid (Fe_3O_4 grains, with a mean diameter of 15.4 nm) was mixed to a lyotropic nematic mixture of sodium decylsulphate, decanol and water, in the $\mathrm{N_D}$ phase [42]. These authors describe the formation of different types of clusters of grains. The biggest aggregates were removed by centrifugation, but small clusters remain in the mixture. The remaining mixture behaved as a stable system for a long time, without any further aggregation. The collective behavior of the nematic matrix was observed with magnetic fields of about 20 G.

Experimental determinations of c_{m} in lyotropic ferronematics [40,58] confirmed Brochard's prediction, $c_{\mathrm{m}} \sim 1/(LD_{\mathrm{s}}^2)$. Observations of depletion layers, also predicted by Brochard and de Gennes [37], were reported for calamitic ferronematics [43], with a mixture of potassium laurate, decanol and water, doped with surfacted ferrofluids. In this experiment, the formation of clusters of magnetic grains was clearly observed in regions of the sample with a larger concentration of grains, inside domains closed by depletion layers. The use of ferrofluids to orient lyotropics was essential in experiments in which the reciprocal structure of the biaxial nematic phase was determined [44,59–61]. Combination of magnetic orientation and spinning of the sample (see the experimental procedure discussed in Chapter 1) lead to a good orientation of the sample, which is necessary for obtaining the reciprocal space image of the $\mathrm{N_B}$ phase.

An interesting application of the ferrofluid doping of nematics is the possibility to obtain the bend elastic constant, k_{33}, and the anisotropy of the diamagnetic susceptibility, χ_{a}, of lyotropic liquid crystals. There appears a metastable situation if an initial calamitic liquid crystalline film, in a planar alignment geometry, is subjected to a magnetic field perpendicular to the original planar orientation [62]. A periodic bend structure, with inversion walls, is then observed. These walls are oriented parallel to the direction of the magnetic field [54,63,64]. Depending on the field strength, concentration doping and anchoring conditions, the typical time necessary for the walls to relax (disappear), and for the setting of the new planar alignment, can be of the order of some hours. In this case, the process can be analyzed in a quasi-static framework, specially in the beginning (first minutes), and the measurements of the periodicity of the bend distortion allow estimates of k_{33} and χ_{a}. The effective splay-bend elastic constant, k_{13}, was also investigated using ferrofluid doped lyotropic samples [65]. The ratio between k_{13} and the usual Frank elastic constant found in the potassium laurate/decanol/water mixture was positive and of the order of 1 [66,67].

The dynamical behavior of ferronematics as a function of time and strength of the magnetic field was investigated by using optical techniques [46,47]. Assuming a nematic sample characterized by an elastic constant k, doped with magnetic grains of typical dimension D and magnetization at saturation m_{s}, with concentration of grains φ, in a magnetic field $\mathbf{H}$, the torque balance equation per unit area is written as

$$\gamma D\dot{\theta} = -\frac{k}{D}\theta - Dm_{\mathrm{s}}H\theta\varphi, \tag{8.6}$$

where θ is the angle between $\mathbf{H}$ and $\boldsymbol{\mu}$, and γ is the rotational viscosity of the liquid crystal. The solution of Eq. (8.6) is given by

$$\tau^{-1} = \frac{k}{\gamma D^2} + \frac{m_s \varphi}{\gamma} H.$$

In this picture, if samples subjected to a magnetic field are left to relax back to their original conditions, the dynamical behavior of ferronematics is different from undoped nematics. In usual nematics, the relaxation time is proportional to H^{-2}; in ferronematics, the relaxation time is proportional to H^{-1} [46,48]. The critical field of the Fréedericksz transition [55] of a discotic ferronematic sample was found to be two orders of magnitude smaller than its value in undoped samples [68]. In contrast to the work of Brochard and de Gennes [37], this result was theoretically explained by assuming a coupling between anisometric micelles and small clusters of magnetic grains in the mixture [69].

More recently, the Z-scan technique was used to investigate nonlinear properties of ferrofluid-doped lyotropics [70,71]. Values of the nonlinear refraction index n_2 [72] in doped samples were shown to be 10 times larger than those of undoped samples. These results were explained by assuming that the laser beam indirectly heats the sample *via* ferrofluid grains. This mechanism is named *hyperthermia*, being well known in biomedical applications of magnetic fluids [73]. In this scenario, magnetic grains absorb heat from the laser beam and then transfer heat to the lyotropic mixture. Since the absorption of the liquid crystal itself is small, nonlinear effects are also small without the presence of ferrofluid grains. Ferrofluid doping enhances this nonlinear response of the liquid crystalline medium.

It is interesting to note that the typical size of a micelle is about 10 nm, which is also the typical size of a ferrofluid magnetic grain. The packing of micelles and grains in the bulk of ferronematics is not yet well known. Considering the structural data for a mixture of potassium laurate, decanol and water [74], in the nematic phases, the density of micelles is of the order of 10^{18} micelles/cm^3. Comparing this value with the usual ferrofluid doping concentration of about 10^{13} grains/cm^3, we see that, for a fixed volume, there are 10^5 more micelles with respect to the quantity of magnetic grains. In terms of linear dimensions, this corresponds to about 1 grain per 10^2 micelles.

8.2.7.5 *Ferrocholesterics lyomesophases* Ferrocholesteric lyotropic liquid crystals are obtained by doping usual cholesterics with ferrofluids. Both ionic and surfacted ferrofluids can be used for this purpose. All of the requirements presented in the previous section regarding ferronematics also apply to ferrocholesterics.

The first successful attempt to dope a lyocholesteric mixture was also done by Liébert and Martinet [75]. In this experiment, a Ch_D mesophase, made of sodium decylsulfate, decanol, water, and brucine sulfate, was doped with a surfacted water-based ferrofluid, with a concentration of about 1% by weight of ferrofluid.

At this high concentration, large needle-like clusters were formed in the direction parallel to the external magnetic field and perpendicular to the cholesteric planes. It was shown that small magnetic fields, of about 5 G, were already enough to orient samples in the cholesteric phase.

In the case of calamitic ferrocholesterics (Ch_C), small magnetic fields, of about 10 G, were able to unwind the cholesteric helix [76] and produce an oriented planar texture. Another interesting result obtained in this experiment with calamitic ferrocholesterics subjected to a magnetic field is the observation of the relaxation process after the removal of the field. There are observations of hydrodynamic instabilities, at length scales comparable to the sample thickness, that can be followed by inspecting the texture in a polarizing microscope as a function of time. Since the concentration of ferrofluid was large in this experiment, there appeared needle-like clusters of micrometer lengths; after removing the field, these clusters were deformed by the hydrodynamic instabilities, through the appearance of undulations (see Fig. 8.5).

Ferrofluid doping was also done in biaxial cholesterics Ch_B for studying phase transitions and, in general, for establishing different phase diagrams of new cholesteric mixtures [60,77–79]. As discussed before, the doping allows an easy manipulation of small permanent magnets (and coils) in the optical microscope, which helps the observation of textures and the identification of phases.

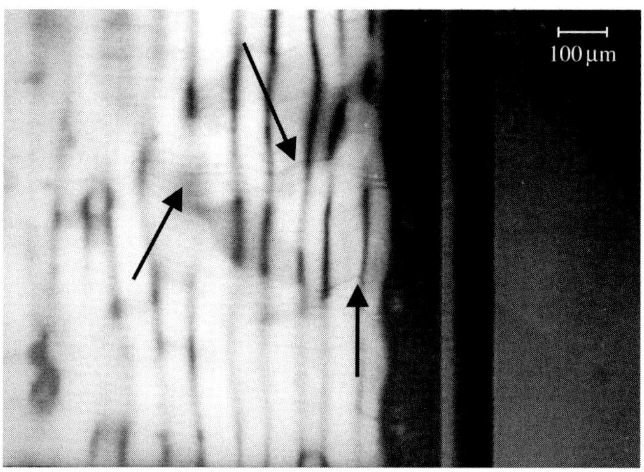

FIG. 8.5. Typical texture of an initially oriented ferrocholesteric Ch_C lyomesophase (unwinded by the magnetic field). The system has been allowed to relax after removing the magnetic field. The texture has been obtained in a polarizing light microscope, between crossed polarizers. Note the presence of large ferrofluid needles (see the indications of the arrows) decorating the hydrodynamic instabilities.

8.2.7.6 *Ferrolamellar and ferrohexagonal lyomesophases* Fabre and coworkers incorporated γ-Fe_2O_3 grains, of a typical diameter of 8 nm, coated with organophosphorated surfactants and dispersed in cyclohexane, in a lyotropic lamellar phase of a mixture of sodium dodecylsulfate (SDS), pentanol (C_5), cyclohexane (C_6), and water [80–82]. The volume fraction of ferrofluid introduced in the lamellar phase was in the range $0.1 < \varphi < 6\%$. The thickness of the lipid bilayer of this lamellar phase ranges from 2 to 40 nm, depending on the cyclohexane content, keeping the water layer almost constant. Stable ferrolamellar phases, without the formation of aggregates and phase separation, were observed under some particular conditions, at $\varphi = 1.5\%$ and with oil layers of thickness 20 and 25 nm [80]. Magnetic grains are expected to accommodate in the hydrophobic region of the lamellar structure. Modifications of the texture (under polarizing microscope) were observed with the direction of the external magnetic field in the plane of the lamellae and perpendicular to it. These results indicate that doping does not change the original lamellar structure.

The flexibility of the lamellae in ferrofluid doped samples was also investigated and compared to the undoped system [83]. It was observed that the value of the bending elastic constant in doped samples is about 10 times larger than in undoped samples [83]. This fact could not be attributed to a hardening of lamellae since the membrane flexibility, described by the mean bending elastic modulus [84], remains the same in both doped and undoped samples. Comparison between lamellar phases doped with magnetic and nonmagnetic grains shows that both systems behave similarly, as far as we consider nonmagnetic effects (spontaneous orientation in the absence of a magnetic field, textures and topology of the phase diagram) [82]. However, these aspects differ significantly from those of undoped samples, indicating that the magnetic interaction between grains does not seem to be relevant for these properties. Apparently, the existence of trapped (magnetic or non-magnetic) grains between the lamellae is the most important feature that determines these properties of the system.

The first realization of a ferrohexagonal phase was obtained by Quilliet and coworkers [85] with the SDS/C_5/water/C_6 lyotropic mixture, which presents a hexagonal phase, doped with the surfacted C_6-based ferrofluid, with maghemite grains of a typical diameter of 7 nm. An interesting feature of this experiment is that a homogeneous phase was not observed. The doped mixture shows a biphasic behavior, with coexisting hexagonal and isotropic phases. In undoped and doped hexagonal phases, on the plane perpendicular to the axes of the cylinders, the lattice parameters were 14.8 and 30.2 nm, respectively. From the analysis of the composition of the sample, it was concluded that there are direct cylinders in the hexagonal phase, which confine the magnetic grains in doped samples. If they are subjected to an external magnetic field, the cylindrical axes of doped samples are oriented along the direction of the field.

8.3 Microemulsions

Many physico-chemical properties of this family of complex fluids have broad overlaps with analogous features of lyotropic liquid mixtures [86,87].

Microemulsions are optically isotropic and thermodynamically stable mixtures of oil (O), water (W) and a surfactant, with the eventual addition of an alcohol [88]. The presence of a surfactant strongly reduces the interfacial tension of the original oil/water mixture, being the main responsible for the formation of microemulsions. More generally, a microemulsion is a thermodynamically stable dispersion of one liquid phase into another. This system is stabilized by an interfacial film of surfactant that separates both liquids.

There are two types of microemulsions, oil-in-water (usually labeled O/W) or water-in-oil (usually labeled W/O). Since the diameters of droplets range from about 10 to 100 nm, these systems are clear solutions, transparent to visible light. The interfacial tension between the two phases is small. A typical composition of a W/O microemulsion, in weight%, (wt%), is given by 28% of water, 16% of pentanol, and 6% of potassium oleate. The nature of a microemulsion does not depend on the way it is prepared. The order of mixing of the different substances is irrelevant. The system reaches an equilibrium state in a longer or shorter time interval, depending on the energy employed in the mixing process.

These systems differ from *emulsions* (sometimes called *macroemulsions*) since they are *kinetically stable*. Emulsions are unstable, with suspended droplets of a typical size of micrometers or even larger; one of the liquids of an emulsion may aggregate and trigger a demixing process. As the large droplets scatter light, the appearance of an emulsion is milky. The nature of an emulsion depends on the way it is prepared, that is, on the order of mixing of different substances and on the (mechanical or thermal) energy of the mixing process.

The conceptual border between lyotropics (in particular the isotropic phases) and microemulsions is not sharp. Isotropic phases of the same mixture, in which oil is one of the components, are sometimes considered at different sides of this frontier. However, an alternative way to differentiate them consists in the characterization of microemulsions as two-phase systems and micellar solutions as one-phase systems. In the literature, there are several reviews of the properties of microemulsions [87,89–92].

From a theoretical point of view, the description of microemulsions has to take into account mechanisms for phase separation and for the self-aggregation of molecules in globules [93]. The pioneer theoretical work of Schulman and Montagne [94] describes the stabilization of a microemulsion on the basis of a model that considers the interfacial tension between water and oil and the surfactant free energy. However, the interaction between globules and fluctuations of shape of a globule (sometimes called "self-interaction") are important ingredients of a theory to describe microemulsions. As discussed in Chapter 1, the hydrophobic–hydrophilic effects are essential for the formation of microemulsions. The presence of a surfactant makes this coexistence compatible and the self-organization mechanism allows the structural organization of the microemulsion.

8.3.1 *Phase diagrams*

Microemulsions are usually formed by a quaternary mixture of a surfactant, alcohol, water and oil. However, some ternary mixtures, without an alcohol, as AOT/oil/water, also form microemulsions [95]. Other mixtures presenting microemulsions are AOT/oil/water/decane and water/dodecane/pentanol/SDS.

We now discuss an introductory example of a mixture with microemulsion phases. In Fig. 8.6, we show a sketch of the isothermal ($T = 25°$C) surface of the phase diagram of the AOT/decane/water mixture [91,92]. Three microemulsion phases have been identified, L_1^M, L_2^M and $L_2'^M$ (note that, in order to avoid any misunderstanding with the isotropic micellar phases, we are using a different nomenclature with respect to Bellocq [91,92]). Phases L_1^M (L_2^M) are direct (inverse) micellar phases in the water-rich (oil-rich) region of this phase diagram. We also have a lyotropic birefringent lamellar phase and several phase coexistence regions. $L_2'^M$ is a microemulsion phase similar to L_2^M, sometimes observed in coexistence regions. A critical point was observed on the demixing curve of the inverted microemulsion region (as well as a line of critical points). Light scattering experiments for different temperatures, at fixed relative concentrations of the components, along a path approaching the critical line, show that the intensity of the scattering follows an Ornstein–Zernike behavior [91]. The correlation length is given by the power law

$$\xi = \xi_0 \left[\frac{T_c - T}{T_c}\right]^{-\nu_t}, \tag{8.7}$$

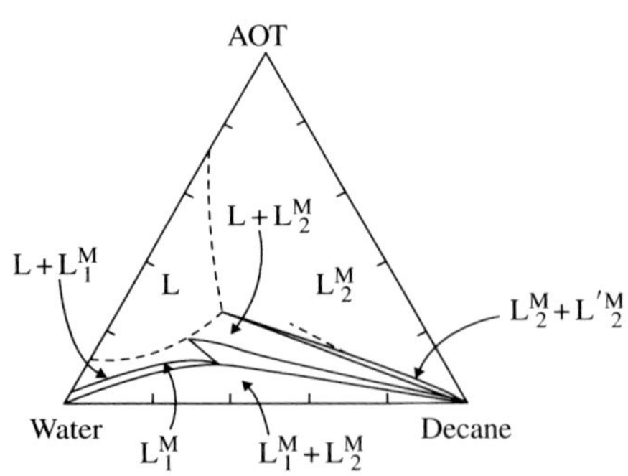

FIG. 8.6. Partial isothermal representation, at $T = 25°$C, of the phase diagram of the mixture AOT/water/decane [92].

where ν_t is a critical exponent at fixed relative concentrations (for fixed temperature, this exponent is called ν_x) and T_c is the critical temperature. It was observed that $\xi_0 \sim 1$ nm, slightly increasing as the temperature increases, and $\nu_t \sim 0.7$. The same experiment, at a fixed temperature and approaching the critical line by changing the relative composition, leads to $\nu_x \sim 0.6$. In contrast to single liquids, besides depending on temperature, the interaction between droplets in microemulsions also depends on parameters related to the relative concentration of the different components of the mixture (and on the physico-chemical properties of these components).

There are two large groups of phase diagrams of microemulsions: with the occurrence of critical points, and without critical points, but with the coexistence of two micellar phases. Besides the coexistence regions discussed above, microemulsions can coexist with an oil-rich phase (which is called Winsor I equilibrium, or WI), with a water-rich phase (WII), or with both phases (WIII).

According to the prevailing experimental picture, in the oil-rich region of the phase diagram of microemulsions, there are almost spherical water droplets in a continuous oil medium, separated from this medium by a surfactant layer, which is assumed to be an incompressible liquid. In water-rich mixtures, the situation is reversed. If water and oil concentrations are comparable, there may appear more complicated topologies, as a bicontinuous structure [96], with continuous water and oil compartments separated by a continuous layer of surfactant [97–100]. The radius of the droplets in mixtures with alcohol slightly increases with the number of carbon atoms in the alcohol chain. However, the spherical micellar shape and the bicontinuous structure were not the unique microscopic pictures proposed to account for experimental observations in microemulsions [101,102]. Disorder lamellar-like and polymer-like phases, with flexible cylindrical molecular aggregates, as well as middle phases [103], in coexistence domains with water and oil, are examples of proposed structures in the literature.

8.3.2 *Models and theoretical approaches*

Due to the different mechanisms responsible for the stability of microemulsions (the role of different components, the relative concentrations, temperature), the theoretical description of these complex fluids is still partial and incomplete. Considerations involving interfacial energy, entropy of mixing of the polar and nonpolar liquids, curvature energy and interactions between different components, must be taken into account. There are different theoretical approaches, sometimes with emphasis on particular systems and particular structural features [94,98,100,104,105].

8.3.2.1 *Droplet-type microemulsions* Microemulsions formed by droplets can be regarded as an intermediate state between a homogeneous mixture of isolated molecules, without the formation of droplets, and a colloidal suspension of hard spheres. A Landau–Ginzburg phenomenological model developed by Safran and coworkers [93,101] assumes that the surfactant (incompressible) layer can be elastically deformed in a splay-type topology, characterized by two orthogonal curvature fields, $\rho_1(\mathbf{r})$ and $\rho_2(\mathbf{r})$. In this context, the free energy per droplet can be written as a function of the mean, H_{m}, and Gaussian, Q, curvatures [93,106],

$$F = K_{\mathrm{sp}} \int (H_{\mathrm{m}} - H_0)^2 \, \mathrm{d}a + K_{\mathrm{ss}} \int Q \, \mathrm{d}a, \tag{8.8}$$

with $H_{\mathrm{m}} = (1/\rho_1 + 1/\rho_2)/2$ and $Q = 1/(\rho_1 \rho_2)$, where K_{sp} and K_{ss} are the energy costs involved in a splay deformation of the interface and the formation of saddles, respectively, and H_0 represents the tendency of the interface to curve towards the polar or non-polar solvent region of the mixture. The next step is the consideration of the molecular packing in the interface (i.e. the interface topology) and the minimization of this free energy with respect to the area per surfactant molecule. The resulting phase diagram [93] shows three distinct regions, in which spheres, rods and lamellae are stabilized, depending on the curvatures and the deformation energies. Rods and lamellae are stabilized at large concentrations of the surfactant, and intermediate forms as ellipsoids were not stabilized. Transitions between phase regions are always first order. Temperature is taken into account by the introduction of the entropy of mixing and the internal entropy of the different types of surfactant aggregates.

Although it is usually assumed that the interactions between surfactant aggregates do not change the overall structure of the phases, these interactions are also an important ingredient of any theoretical approach. Lemaire and coworkers [107] proposed a model in which droplets can interpenetrate forming new aggregates, as dimers, trimers, etc. In this model, droplets can be deformed if they are allowed to physically touch each other. The attractive interaction between droplets includes the London–van der Waals–Hamaker contribution. Additional mechanisms, as the attraction between tips of the tails of the surfactant molecules, have also been taken into account [107].

8.3.2.2 *Bicontinuous-type microemulsions* There are several models for describing microemulsions with bicontinuous structures [100]. For example, we should mention a model of a random distribution of water and oil in microscopic polyhedral cells [98], and a model of superposition of cubes of edges equal to the persistence length ξ_K of the interface [108].

The free energy of the system is usually written in terms of different contributions, including entropic terms, and interfacial and interface curvature terms [92]. The interfacial term takes into account the free energy of the surfactant and the interfacial tension between water and oil without surfactant molecules. The entropic contribution, despite its smallness in absolute values, is essential to allow the stability of microemulsions [109]. The curvature contribution [106] takes into account the energy expense to splay the surfactant interface in orthogonal directions with respect to the spontaneous curvature. This contribution critically depends on the presence of a cosurfactant in the mixture.

Let us discuss a model proposed by Jouffroy and coworkers [108]. The surfactant interface, separating water from oil in the bicontinuous structure of the microemulsion, is assumed to be made of a set of individual pieces, with independent orientations, which turns out to be responsible for the entropy of the film, of area ξ_K^2. Small values of the persistence length ($\xi_K \sim 10$ nm) favor the stabilization of microemulsions, in contrast to the case of liquid crystals, in which $\xi_K \sim 50$ nm. The first contribution to the free energy comes from the surfactant film. The space is divided into cubes of volume ξ_K^3 filled with water or oil. If adjacent cubes have different solvents, they are separated by a surfactant film of area ξ_K^2. This contribution for the total free energy is written as [108]

$$F_1 = \varphi_s \xi_K^3 w^{-1} G(\Sigma), \tag{8.9}$$

where φ_s, w, and $G(\Sigma)$ are the volume fraction of surfactant in the mixture, the volume of the surfactant molecule, and the free energy per surfactant molecule of the film, respectively. The second term in this free energy comes from the entropy of mixing of the set of cubes,

$$F_2 = k_B T[\phi \ln \phi + (1-\phi)\ln(1-\phi)], \tag{8.10}$$

where ϕ is the probability of a cube to be filled with oil. Finally, we have the contribution of the curvature of the surfactant layer. This term requires a more extensive calculation since the details about the topology of the surfactant molecule are essential to describe the interface conformation. This contribution for the free energy is written as

$$F_3 = \lambda f_3(\phi), \tag{8.11}$$

with

$$\lambda \propto \frac{K}{k_B T} H_0 \xi_K, \tag{8.12}$$

where K is the rigidity and $f_3(\phi)$ is a dimensionless function [108]. The total free energy, given by the sum of these terms, depends on two adjustable parameters, ξ_K and λ. The calculated phase diagram displays coexistence regions, with two and three phases, and a microemulsion domain.

As a final remark, we should mention that some authors classify microemulsions as a lyotropic mixture (see, e.g. the definition of a lyotropic mixture in Chapter 1). Indeed, there is a clear overlap between the concepts of a micellar isotropic phase in lyotropic mixtures and the description of microemulsion phases. Bicontinuous isotropic phases (as the sponge phase) are also very close to a microemulsion phase.

8.4 Langmuir–Blodgett films

The investigation of ultra-thin and highly ordered films has many connections with problems in the area of lyotropic liquid crystals. For example, the concepts of mono and multilayers are present and very relevant in both areas of thin films and lyotropic mixtures. Although this may seem a rather abstract idea, we do have a feeling of thickness if we think about pouring oil in a water surface. Traveling waves in water, for example, have much smaller amplitudes in the presence of an oil layer.

After the seminal work of Langmuir [110–113], for which he was awarded the Nobel Prize for Chemistry in 1932, there were many outstanding contributions to the production of stable layer films [114]. Again, the hydrophobic–hydrophilic effects are deeply present in the stabilization of these films. The interaction of liquid crystal molecules or micellar aggregates with these films leads to an interesting field of research about the wetting of substrates by ordered complex fluids. This problem has fundamental implications from the point of view of the understanding of some physico-chemical processes. Also, it has practical implications in the technological manufacturing of devices using liquid crystals. Orientational properties of liquid crystals in layers near a liquid–solid interface can be strongly affected by the deposit of a thin film on the surface.

Langmuir and coworkers developed a technique to form a monolayer film of surfactant molecules at a fluid-gas interface. This type of film is named a *Langmuir film*. Katharine Blodgett, one of the collaborators of Langmuir, developed the technique to transfer this monolayer film to solid substrates and, moreover, to transfer successive layers in order to build a multilayer film over a solid substrate [115,116]. These types of systems are called *Langmuir–Blodgett* (LB) *films*.

8.4.1 *Langmuir films*

The Langmuir monolayer film is usually built on top of a liquid substrate called *subphase*. It should be more appropriate to refer to a liquid–gas interface, with the deposition and organization of molecules of a surfactant. The most commonly used subphase is water, due to its large surface tension ($\gamma = 73$ mN/m, at normal conditions of temperature and pressure). However, other liquids, as mercury [117] and glycerol [118], were also used as subphases.

Langmuir films are formed by surfactant molecules poured on the subphases. The surfactant molecules are dissolved in an organic solvent and spread over the subphase. The solvent evaporates rapidly, and the surfactant molecules remain at the liquid–gas interface.

If the concentration of molecules is small and the distance between them is large, polar heads remain immersed in the liquid while the hydrophobic tail of the molecules remain in the gaseous part of the interface, bent towards the liquid-gas interface (see Fig. 8.7(a)). If the molecules are brought together by physically compressing their polar heads (using a barrier that moves at the liquid-gas interface, and reducing the available total area where the heads are located), the hydrophobic part of the molecules exert a repulsive force and tend to "stand up," with the head still immersed in the liquid (see Fig. 8.7(b)). If we continue to exert a compression of the polar heads, the hydrophobic part of the molecules tends to minimize the available volume, and organizes in a compressed, solid-like,

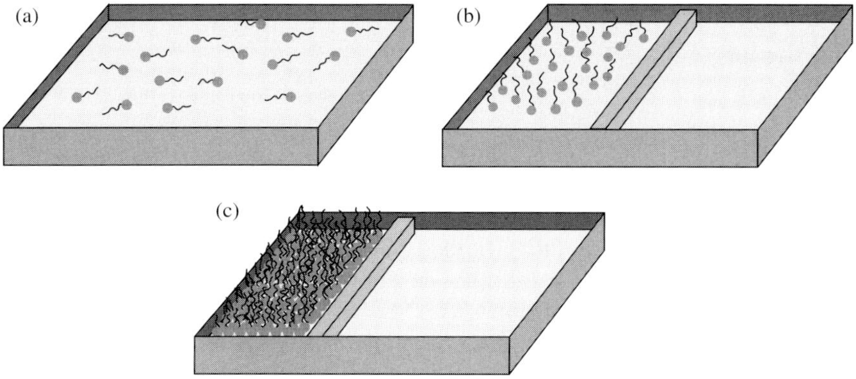

FIG. 8.7. Monolayer of surfactant molecules in a liquid–gas interface: (a) small concentration of surfactant molecules; (b) large concentration of non-compressed surfactant molecules; (c) large concentration of compressed surfactant molecules.

closed-packed layer (see Fig. 8.7(c)) [119] . The surface pressure is defined as

$$\Pi = \gamma - \gamma_0, \tag{8.13}$$

where γ_0 and γ are the surface tension of the subphase with and without the presence of the surfactant molecules. The pressure Π is a measure of the repulsive force between surfactant molecules spread over the subphase.

Usually, we plot Π as a function of the area per surfactant molecule in order to characterize the properties of the surface monolayer spread over a subphase. This plot is called *surface pressure/area isotherm* [119]. As the surface area per surfactant molecule decreases with compression, the typical behavior of Π depends on the surfactant molecule and subphase. Let us describe the case of surfactant molecules as the stearic, $CH_3(CH_2)_{16}COOH$, or oleic, $C_{18}H_{34}O_2$, acids. Initially, Π is a constant; it then increases abruptly, indicating the increasing interaction between surfactant molecules. As the surface per surfactant molecule continuous to decrease, another abrupt increase of Π is observed, indicating a surface phase transition. Surfactant molecules organize in a solid-like structure, with the hydrophobic tails pointing towards the gaseous side of the interface and arranging themselves in a closed-packed structure. The values of the area per surfactant molecule are characteristics of the molecule and the subphase (usually water). If the area per surfactant molecule continuous to decrease, there is a phenomenon of collapse; the monolayer destabilizes, giving rise to more complicated structures as disordered multilayers.

Some polymers, as polyacrylates, polymethacrylates and silicone copolymers [120], can be used for building a Langmuir film. The polar part of the monomer stays at the surface of the subphase, and the hydrophobic part, which usually links to other molecules during the polymerization process, stays at the gaseous part of the interface.

8.4.2 *Deposition of Langmuir–Blodgett films*

The most useful (and practical) application of the Langmuir ideas is perhaps the deposition of successive molecular layers on a solid substrate. The final product of this deposition is a Langmuir–Blodgett film.

The general idea is very simple, as shown schematically in Fig. 8.8. In Fig. 8.8(a), we show a deposition setup with the compression barrier and the displacement transducer. After the Langmuir film is constructed over the substrate, different deposition procedures can be employed by using a dipper that introduces (or removes, depending on the deposition type) the solid substrate in the Langmuir trough [119,121,122]. Typical rates of immersion (or remotion, if the substrate is withdrawn from the liquid side of the interface) are about 1 mm/s.

In the X-type deposition of Fig. 8.8(b), a solid substrate is introduced in the Langmuir trough from the gaseous side of the interface. In this case, the

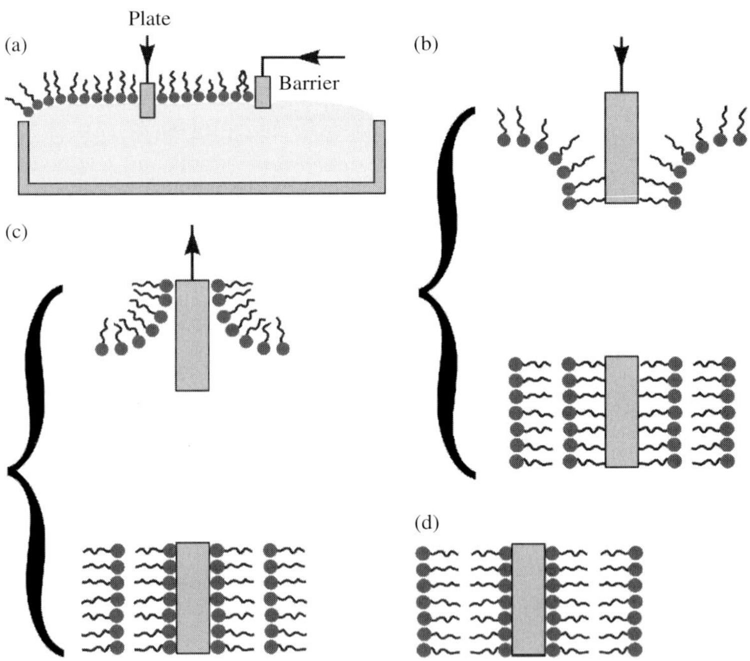

FIG. 8.8. Deposition of a Langmuir–Blodgett film: (a) deposition setup; (b) X-type deposition; (c) Z-type deposition; (d) Y-type deposition.

hydrophobic part of the surfactant molecule is placed in contact with the substrate. Successive layers can be deposited repeating the same procedure. In the Z-type deposit of Fig. 8.8(c), the substrate is withdrawn from the liquid side of the interface towards the gaseous side. In this case, the hydrophilic part of the surfactant molecule is placed in contact with the solid substrate. As in the previous case, successive layers can be deposited onto the substrate, one on top of the other, by repeating the same procedure. Another way to build up an LB film in a symmetrical way consists in the Y-type deposition. In this case, the substrate is initially withdrawn from the liquid (as in the Z-deposition). Then, the substrate with the first deposited layer is immersed into the liquid as in the X-type deposition (see Fig. 8.8(d)). Surfactant molecules can now be successively deposited and are organized with the polar heads of one layer in contact with the heads of the next layer (and with a similar arrangement for hydrophobic parts of the molecules). However, it is important to remark that this simple view of molecular deposit is not the only possibility. In some cases, the type of the deposited film does not depend on the way the substrate is immersed or removed from the subphase. In other words, even if one tries to build up a particular type of film, the final result is always the same, irrespective of the procedure [123]. Substrates can be prepared to present a hydrophobic or hydrophilic character for receiving successive film deposits. One of these possibilities is to deposit onto a substrate one or more monolayers of stearic acid, which gives rise to a good surface, and then depositing the desired molecules.

Film deposition can be characterized by a parameter called *deposition* or *transfer ratio* τ, defined as the ratio between the decrease in the area occupied by the monolayer on the subphase at constant pressure and the coated area of the solid substrate [124]. Another parameter which characterizes the depositions is $\phi_{\mathrm{LB}} = \tau_{\mathrm{e}}/\tau_{\mathrm{i}}$, where τ_{e} and τ_{i} are the remotion and immersion ratios [125]. In this context, X and Y depositions have $\phi_{\mathrm{LB}} = 0, 1$, and Z depositions has $\phi_{\mathrm{LB}} \longrightarrow \infty$.

8.4.3 *Characterization of the film*

Different techniques are used to characterize mono- and multilayer deposited films [126].

An important characteristic of the film is the thickness after the sequence of depositions. The most used techniques to measure film thickness are interference [115], ellipsometry [127], and X-ray and neutron diffraction. The monolayer thickness of fatty acid LB films was found to be a linear function of the number of carbons in the hydrophobic chain [128]. The film structure can be characterized by using X-ray, neutron, and electron diffraction techniques (transmission electron diffraction, TED, and reflection high-energy electron diffraction, RHEED), optical birefringence, infrared and Raman spectroscopy. Fatty acids and other molecules used to build up LB films were shown to present a crystalline structure of subunits of the aliphatic chains [129] . There are experimental evidences that the structure of the first deposited layers may differ from the structure of the subsequent layers [130].

The orientation of the aliphatic chains of the deposited molecules with respect to the surface of the substrate can be determined by infrared and visible spectroscopy techniques. These experiments are particularly sensitive to the relative orientation of the incident electric field of the probe beam and the dipole moments of deposited molecules [131].

The intrinsic geometry of the LB films, in which an inversion symmetry is broken at the interface, leads to some interesting studies of nonlinear properties. If a medium interacts with an electric field E of an electromagnetic wave, the macroscopic polarization P can be written as [132]

$$P = P_0 + \chi^{(1)}E + \chi^{(2)}E^2 + \chi^{(3)}E^3 + \cdots,$$

where P_0 is a constant and $\chi^{(n)}$ is the nth-order susceptibility of the medium. The second-order susceptibility tensor is responsible for the second harmonic generation and the linear electro-optical effect; the third-order susceptibility is related to the third harmonic generation and the Kerr effect [132]. The susceptibility tensors, which reflect the symmetry properties of the films, can be measured by different experimental techniques, as forced Rayleigh scattering (FRS).

8.4.4 *Applications of LB films in the study of lyotropics*

Despite the large number of potential and actual applications of LB films in many fields, the use for investigating properties of lyotropic mixtures, as anchoring and wetting, is still modest. In one of these studies, LB films were used to investigate surface properties of lyotropics [133] . The deposited molecules were lignin, cadmium stearate, and behenic acid, with various numbers of layers, in a film symmetry of Y type. The films were deposited over glass substrates, which were used to build the sample holder for encapsulating the liquid crystal (in this case, a nematic N_D phase). Different optical techniques were used to investigate the effects of the LB film depositions on the anisotropic part of the surface tension of the discotic nematic liquid crystal in contact with the treated substrate. The temperature dependence of the bulk and surface tilt angles of the director with respect to the normal to the surface substrate [67] were measured. Experimental results show that the average director alignment is practically independent of the nature of the molecule deposited onto the substrate. This result is completely different from the observations, under the same conditions, for thermotropic liquid crystals (in this particular study, for 4-heptylcyanobiphenyl, also called K21). These results were explained assuming that the substrate, independently of the deposited LB films, stabilizes a lamellar layer, or bilayer, formed by the surfactant molecules of the lyotropic liquid crystal. This layer screens the effect of the LB films on the alignment of the nematic director. The presence of this layer was previously used to explain the gliding anchoring of the director of a lyotropic N_C phase at amorphous glass surfaces, subjected to an external magnetic field [48].

Even in this situation without the deposition of a particular LB film, the experimental results were explained by assuming the existence of a surfactant layer stabilized by the flat glass surface. In contrast with thermotropics, in which case the individual molecules strongly interact with the substrate and organize a layer that affects the orientation of the bulk molecules, the "carpet" formed by this surfactant deposit in lyotropics screens the influence of surface and allows the gliding of the director under the action of magnetic fields. Although additional experiments are needed to improve this description of the gliding phenomenon, this process may have some connections with the so-called *self-assembling* [134] for building up a thin film over the actual substrate sample-holder.

8.5 Diblock copolymers

The study of polymeric liquid crystals is another interesting area that is strongly related to the investigations of lyotropic mixtures. In particular, the analysis of mesophases formed by basic units of *diblock copolymers* (in short, called " DiCo") is based on concepts as affinity, nanoscale segregation and ordering.

A DiCo is formed by two independent polymers, of species A and B, for example, with a small affinity, linked by a covalent bond. In Fig. 8.9, we show polystyrene (A) and polybutadiene (B), forming a DiCo. Without the link, this system tends to separate into macroscopic segregated states; each type of polymer concentrates in a distinct region of the mixture. On the other hand, if there is a covalent bond between the different species of polymers, the system self-organizes in order to increase the contacts between similar species and decrease the contacts between antagonic species. Even if repulsion between unlike monomers is relatively weak, in a DiCo the repulsion between sequences of unlike species is strong. Examples of this type of system are mixtures of polystyrene and poly-methyl-metacrylate, polystyrene and polybutadiene [135], and polystyrene and polyisoprene [136].

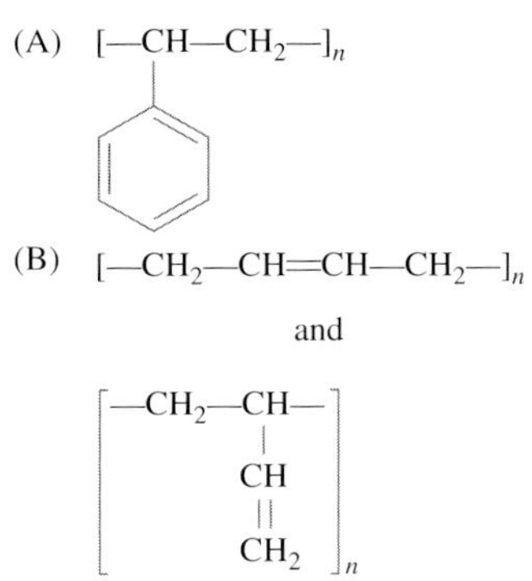

FIG. 8.9. Basic units of the diblock copolymer polystyrene (A) and polybutadiene (B).

The repulsion between unlike species leads to a nanosegregation and, moreover, to a nanoscale structuring into domains of different topologies. Spheres, cylinders and lamellae can be formed and organize themselves as different structures, three-dimensional cubic lattices of spheres, two-dimensional hexagonal lattices of cylinders, and one-dimensional lamellar structures. In a DiCo in which species A and B present a low polymerization index, there may be a stable homogeneous phase; however, there can be transitions to nanosegregated phases by changing some physico-chemical features of the material.

From the experimental point of view, the most used techniques for investigating the structures of a DiCo are small-angle X-ray scattering (SAXS), electron microscopy and electron diffraction.

Different theoretical approaches have been used to describe DiCo systems [137–139]. Microscopic statistical models show that the relevant parameters to describe the nanophase separation in a DiCo are the Flory parameter, which characterizes the A–B interactions, the polymerization index, and the fraction of monomers of one species in the chain [138].

8.5.1 *Structures of diblock copolymers*

Depending on physico-chemical parameters, three types of aggregates were observed in diblock copolymers: spheres, lamellae, and cylinders.

8.5.1.1 *Spherical aggregates* Systems of polystyrene/polyisoprene form spherical aggregates at suitable conditions of temperature and relative concentrations of the different species [136]. Depending on the relative concentrations, spheres with mean radius R_{A} ranging from about 7 to 30 nm, separated by distances D_{di} from about 24 to 106 nm, respectively, were observed by SAXS techniques. The peaks and shoulders in the curves of diffracted intensity indicate a medium-ranged simple cubic lattice, or a cubic close-packed lattice (bcc lattice), which has also been called a paracrystalline macrolattice [140]. The experimental results show that $D_{\mathrm{di}} \propto M_{\mathrm{t}}^{2/3}$ and $R_{\mathrm{A}} \propto M_{\mathrm{A}}^{2/3}$, where M_{t} and M_{A} are the total molecular weight of the DiCo and the molecular weight of the block chain forming the spherical domains (of polyisoprene), respectively. Phases with spherical aggregates were also observed in a system of polystyrene and polybutadiene; the geometrical parameters of the structure are functions of temperature and relative concentrations of the two species, with typical values $D_{\mathrm{di}} \sim 21$ nm and $R_{\mathrm{A}} \sim 7$ nm, and a bcc lattice [135,141]. The interface between the two different species A and B was shown to be well defined at lower temperatures, but it becomes more diffuse as the temperature increases. This behavior indicates a loss of the globular segregation and an almost homogeneous distribution of both species.

A solid–liquid transition has been recently observed in a DiCo, with spherical micellar aggregates, formed by polyoxyethylene and polyoxybutylene in water [142]. If the concentration of micelles is large, they pack in a regular array which corresponds, sometimes, to the formation of a solid or a hard gel phase. Micellar ordering can be of bcc or fcc type, depending on the intermicellar interactions. Rheological experiments, which measure the shear rate and the shear stress of DiCo samples, together with SAXS and light scattering experiments, lead to the characterization of the different phases of the mixtures. The experimental results for the solid–liquid transitions were explained by using a model of hard spheres; the transition is of first order, with a region of phase coexistence between a hard sphere fluid and a solid crystal [142].

8.5.1.2 *Lamellar structure* A diblock copolymer of polystyrene and polyisoprene, with about 50 wt% of polystyrene, forms a lamellar structure, in which species A and B spatially segregate in alternating layers [143,144]. Electron micrograph patterns of this structure show a sequence of black (polyisoprene-rich, usually contrasted by osmium tetroxide) and white (polystyrene-rich) parallel stripes, as a fingerprint pattern, with a periodicity of about 60 nm, which can also be measured by the SAXS technique. The annealing on these DiCo samples with a lamellar structure, at temperatures above the glass transition temperature T_g, increases the periodicity by about 15%. However, annealing at temperatures below T_g does not seem to affect the lamellar periodicity. If copolymers are cast into films of about 10^2 μm thickness, it was observed that the lamellae orient parallel to the surface of the film. The regularity of the lamellar structure seems to be strongly dependent upon the molecular weight of the copolymers.

The morphology of the lamellae near a free surface (e.g. of a DiCo in contact with a gas) displays a microdomain structure [145]. Electron micrograph experiments show a distinct aspect of the lamellar structure in the free surface with respect to the bulk; the lamellae appear somewhat rough and, in some regions, the interfaces of the lamellae orient perpendicular to the free surface. In both cases, the polyisoprene layer separates the DiCo from the free surface. The thickness of this polyisoprene layer is about half of the thickness in the bulk.

There is an interesting behavior if homopolymers are mixed with a DiCo. Consider the case of the same system, polystyrene and polyisoprene, mixed with the corresponding low molecular weight homopolystyrene and homopolyisoprene [146]. Electron micrograph and SAXS results for samples of these homopolymers solubilized in the lamellar structure of the DiCo indicated that the corresponding homopolymers are almost homogeneously distributed inside the lamellae, each one in the region of its similar species (i.e. homopolymer of type A in a region of type A, and the same for species of type B). The corresponding swelling causes an increase of the lamellar periodicity and the chains of the polymers in each domain (A and B) tend to stretch and orient perpendicular to the lamellar interface. Structural phase transitions from lamellae to spheres and cylinders can occur as a function of the content of homopolymers.

If the architecture of a DiCo is of the rod-coil type, there may appear an interesting lamellar topology; the rods can be arranged in a type of smectic structure formed by a monolayer or a bilayer of rigid rods [147]. This liquid crystalline ordering depends on the physico-chemical characteristics of both species of the DiCo and their respective volume fractions. Transitions between smectic-C and smectic-A structures are also predicted as a function of relative volume fractions.

8.5.1.3 *Cylindrical aggregates* A mixture of polybutadiene and polystyrene is an example of a DiCo presenting cylindrical aggregates [148]. Electron micrographs of this system, with a typical content of block styrene of about 68 wt%, show rod-shaped butadiene aggregates (black domains, contrasted by osmium tetroxide) in a homogeneous matrix of styrene. The cylindrical axes are parallel and arranged in a hexagonal close-packed structure. The mean butadiene rod diameter in this particular case is about 19 nm; the lattice parameter of the two-dimensional hexagonal array is about 33 nm. This hexagonal lattice, as in the case of a mosaic crystal, presents small-angle grain boundaries. Regions of lattice imperfections, cracks and defects were also observed in the electron micrographs.

The DiCo poly (ethylenepropylene) and poly (ethylethylene) presents a lamellar structure and a hexagonal phase of cylindrical aggregates [149]. The mechanical properties of diblock copolymers are an interesting subject of research in part due to practical applications. In the particular case of this system, in the presence of applied uniaxial shear, there appear structural instabilities, which can be observed by small-angle scattering techniques. From the theoretical point of view, structural instabilities, as this uniaxial strain-induced instability, were treated by a mean-field approach [150]. In the limit of small strains, two types of instabilities can take place, a modulation of the cylinders along the axes, and a saddle–splay curvature. Uniaxial strains above a critical value should induce a phase transition from the hexagonal closed-packed structure to a modulated cylindrical structure.

8.5.1.4 *Phase transitions* The organization of diblock copolymers in spheres, lamellae, and cylinders depends on temperature T, composition (f is the volume fraction of one species), degree of polymerization N, and a parameter that accounts for the local and nonlocal interactions between polymer species (the Flory–Huggins interaction parameter χ [138]). The DiCo poly(ethylenepropy lene) and poly (ethylethylene) provides an example of a system displaying phase transitions between lamellar and hexagonal structures [149]. As a function of (increasing) temperature, this DiCo presents several structures: lamellae; hexagonally modulated lamellae; hexagonally packed layered channels; and hexagonally packed cylinders. Neutron scattering results show the reversibility of the lamellae modulated–lamellae transitions upon heating or cooling. This phase sequence has also been observed in other polyolefin diblock copolymers.

As a function of temperature, the DiCo polyoxyethylene and polyoxybutylene, mixed with 0.2 M of K_2SO_4, displays two cubic phases formed by spherical aggregates, a bcc and an fcc phase [151]. Leibler developed a microscopic statistical theory of microphase separation in A–B diblock copolymers, using f and the product χN as the relevant parameters [138]. A sketch of the phase diagram in terms of χN and f is shown in Fig. 8.10. This isothermal representation is symmetric with respect to f; at small values of χN, and for all values of f, there is a stable disordered phase (Dis). At a fixed value of f, and increasing χN , there is a cubic (C)–hexagonal (H)–lamellar (L) phase sequence.

The phase diagrams of binary blends of DiCo and homopolymers at constant temperature show a morphology depending on T, f, N, χ, and the concentration of the homopolymer [152]. Blends of homopolystyrene with poly(styrene-*b*-butadiene) and poly(styrene-*b*-isoprene) show isothermal morphology diagrams with several structures (lamellar; ordered bicontinuous double-diamond; cylinders on a hexagonal lattice; spheres on a cubic lattice; disordered aggregates of various shapes) and a macrophase separation.

Matsen and Bates [153] have drawn a mean-field phase diagram for conformationally symmetric DiCo melts using the standard Gaussian polymer model. This phase diagram, sketched in Fig. 8.11, shows regions of stability for disordered (Dis) melts and for ordered structures as lamellae (L), hexagonally packed cylinders (H), body-centered cubic spheres (BCC), close-packed spheres (CPS), and the bicontinuous cubic network (Bic). The CPS phase domains are shown to be very narrow, between the BCC and the disordered phases. This topology of the isothermal phase diagram agrees with many experimental observations.

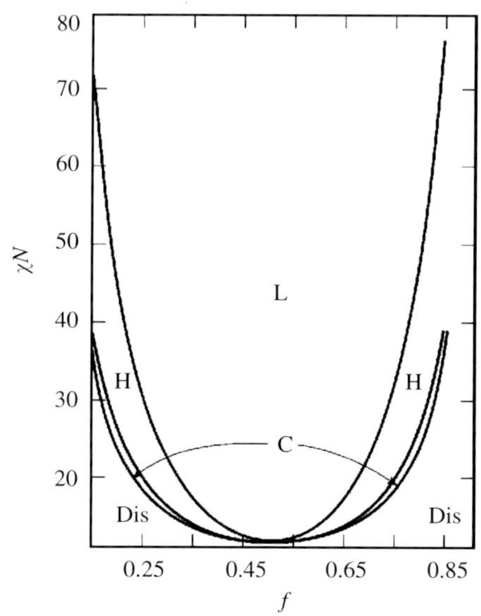

FIG. 8.10. Sketch of the phase diagram of an A–B type of DiCo in the plane χN versus f. Labels Dis, C, H and L refer to the disordered, cubic, hexagonal, and lamellar phases, respectively [138].

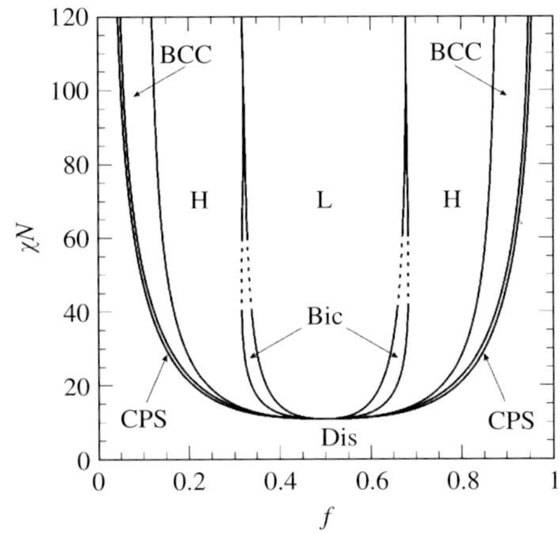

FIG. 8.11. Sketch of a mean-field calculation for the phase diagram of a DiCo in terms of χN and f (see Matsen and Bates [153]). We indicate the following structures: L (lamellar), H (hexagonal cylinders), Bic (bicontinuous cubic), BCC (bcc spheres), CPS (close-packed spheres), and Dis (disordered).

If the DiCo is made of a chain liquid-crystalline block, as a nematic block, and an isotropic block, there appear spherical micelles or cylinders with a nematic core. The phase diagrams, with multiply reentrant sequences, are different from the case of an isotropic DiCo [154]. This type of material is useful for studying the behavior of liquid crystals in confined geometries at nano and micron length scales.

8.6 New lyotropic-type mixtures

As discussed in the previous sections, the presence of sectors with different properties in the same molecule, or the presence of different species in diblock copolymers, with selective affinities, may lead to segregation and self-organization in contact with other molecules. Several examples were discussed in this book and the richness of the polymorphism was sufficiently presented. More recently, there appeared new mixtures of molecules whose interactions involve the concepts of self-aggregation and segregation. Three examples will be discussed: *chromonics*, *lyo-bananas*, and *transparent nematic phases*.

8.6.1 *Chromonics*

There is an increasing interest in the study of the polymorphism of mixtures of thermotropic liquid crystals with nonpolar organic solvents. Smectic thermotropics, of types A and C, have been mixed to organic solvents (n-hexane, n-decane, benzene, and trichloroethylene) [155]. It was shown that solvent molecules

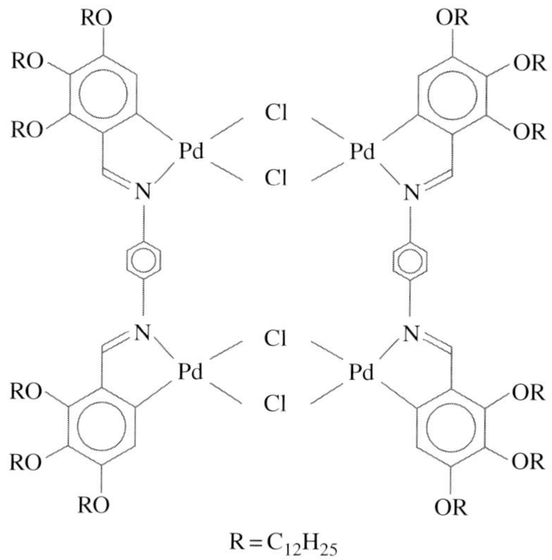

FIG. 8.12. Molecular structure of tetranuclear palladium organyl.

nanosegregate in regions between the smectic layers, forming a lamellar structure. These solvent-rich regions intercalate between the layers of the former smectic structure. Sometimes this process is referred to as a swelling of the smectic structure due to the presence of the solvent, with the increasing of the smectic layer spacing by the addition of the solvent.

Chromonics are made of mixtures of disk or lath-like molecules with solvents [156,157]. In Fig. 8.12, we show the molecular structure of tetranuclear palladium organyl, with a disc-like shape anisotropy. These molecules, which are piled up on top of each other, forming a columnar phase, can pack in a two-dimensional array or even present a uniaxial nematic ordering [158]. The polymorphism of the pure tetranuclear palladium organyl shows a phase transition from a crystalline phase to a discotic columnar phase, at 71°C, and from this structure to an isotropic phase, at 301°C. There is an interesting polymorphism if this type of molecule is mixed with non-polar organic solvents. In Fig. 8.13, we show a sketch of the phase diagram of this mixture (tetranuclear palladium organyl, shown in Figure 8.12, and pentadecane [157]). Besides multiphase domains, there were observations of crystalline (C), nematic (N_1 and N_2), discotic columnar (D), and isotropic (I) phases. Two nematic phases were identified and differentiated by means of their optical microscopic textures. The N_1 phase is preferentially located at the low temperature region of the phase diagram, in contrast to the N_2 phase, which is preferentially located at the high-temperature regions. The presence of nematic phases in this phase diagram is a direct consequence of the introduction of the solvent in the previous packed columnar phase (which is present in the pure material).

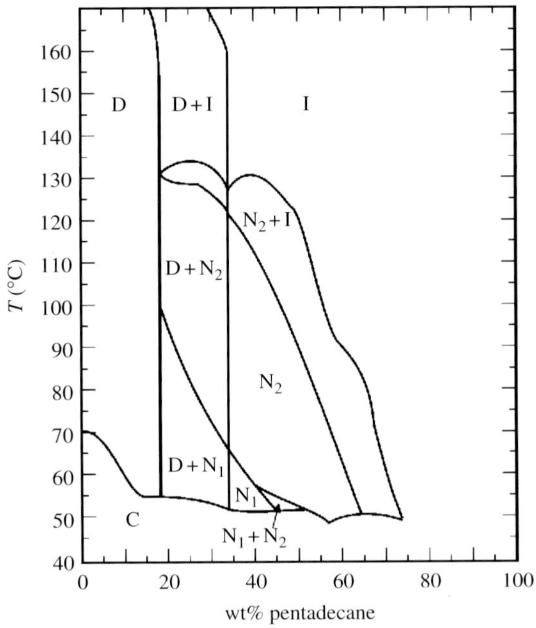

FIG. 8.13. Sketch of the phase diagram of a mixture of tetranuclear palladium organyl (see Fig. 8.12) and pentadecane [157]. We indicate the following structures: crystalline (C), nematic (N_1 and N_2), discotic columnar (D), and isotropic (I).

The ratio between the length of the aliphatic chain of the metal complex and the length of the (unfolded) solvent molecule was shown to affect the polymorphism of the mixtures, at least in some particular cases [159].

Ternary mixtures of tetranuclear palladium organyl (of the type depicted in Fig. 8.12), with enantiomers of the π-acid electron acceptor, (+)-TAPA[(+)-2] and (−)-TAPA[(−)-2], and a nonpolar organic solvent, as pentadecane, have shown a cholesteric phase, besides the nematic and isotropic structures [160]. The charge transfer interactions are assumed to be responsible for the observed cholesteric properties.

The hexakis(alkanoyl)benzene derivatives are another type of molecules with a rich polymorphism in a mixture with cyclic hydrocarbons, as benzene, cyclohexane, and cyclodecane [161]. The pure derivatives present a thermotropic polymorphism, including columnar and cubic mesophases. With the addition of a hydrocarbon, the polymorphism becomes much more interesting, with the stabilization of nematic and isotropic phases in temperatures depending on the amount of solvent.

More recently, carboxyl and alkoxycarbonyl-substituted phthalocyanine copper complexes mixed with different solvents, as nonane, chloroform and benzene, were observed to present a rich lyopolymorphism, with the appearance of nematic and isotropic phases [162].

8.6.2 *The lyo-banana mesophases*

Thermotropic liquid crystals formed by bent core molecules, as the banana-shape systems, have unconventional phases with remarkable properties [163,164]. The possibility of obtaining a chiral mesophase by packing nonchiral molecules is one of these properties. Some of these phases have a ferroelectric polarization, usually at high temperatures, which brings some problems to potential technological applications.

A lyo-banana mesophase can be obtained from the mixture of a thermotropic banana-type liquid crystal with a solvent. Jákli and coworkers mixed a ferroelectric "banana-smectic" material with non-polar xylene as the solvent [165]. They observed that this procedure drastically widens the ferroelectric electro-optical switching range and lowers the clearing point. Due to steric and chemical reasons related to the affinity between species, xylene molecules are assumed to sit between rigid cores of the banana-shaped molecules, weakening the dipole–dipole intermolecular interactions.

Let us look at the steric features of the interactions between "banana" and solvent molecules. If they come into contact, it is known that rigid and flexible molecular segments tend to segregate into packings of similar nature [166]. The extension of this concept to linear and bent segments of molecules leads to a picture in which flexible nonpolar alkyl chains avoid any contacts with the rigid bent core of the molecules, and pack together with aliphatic chains. In the case of organic solvents mixed with thermotropic smectic phases formed by calamitic molecules, it has been observed an increase of the layer spacing with increasing amounts of the solvent [155], which indicates a nanosegregation of the solvent between the smectic layers.

8.6.2.1 *The example of host–C14 guest-hexadecane* As a working example, we now discuss the case of 4-chloro-1,3-phenylinebis [4-(4-*N*-tetradecyloxyphenyliminomethyl) benzoate] [167], also called *C*14 [168] (see Fig. 8.14). As a function of temperature, *C*14 displays a sequence of transitions, from a crystalline phase to a polar smectic C phase, SmCP, at 68°C, and from the SmCP phase to an isotropic ISO phase, at 127°C [169]. Upon cooling from the SmCP phase, there appears a metastable state, SmX, below 60°C. Molecules of *n*-hexadecane [CH_3–$(CH_2)_{14}$–CH_3], also called HEX, are the guest organic solvent mixed to the host *C*14, at different relative concentrations.

If HEX is mixed with *C*14, the clearing point is lowered (e.g. to 91°C at 27 wt%, and to 77°C at 40 wt% HEX concentrations), but the crystallization temperature does not change appreciably. The texture of the crystalline phase changes and there appears an optically isotropic phase at concentrations of HEX larger than 20 wt%. The switchable range does not extend with increasing

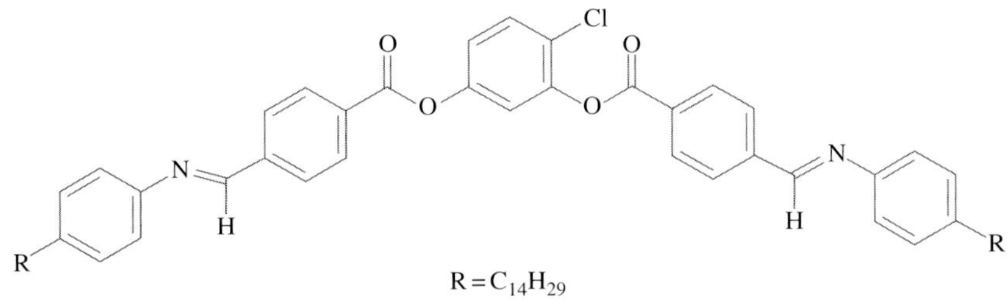

FIG. 8.14. A "banana"-shaped molecule: 4-chloro-1,3-phenylinebis [4-(4-*N*-tetradecyloxyphenyliminomethyl)benzoate].

concentrations of solvent, and non-switchable metastable states appear close to the temperature at which the SmX phase is formed in pure $C14$. Increasing the HEX concentration, the textures of the metastable states gradually lose the features of the SmCP phase and become optically isotropic above 20 wt%, giving rise to a state that has been called X. The small-angle X-ray profiles showed that the layer spacing in the 40 wt% of HEX mixture is about 0.3 nm larger than that in pure $C14$. Differential scanning calorimetetry (DSC) data indicate that the transition enthalpies strongly decrease with increasing concentrations of HEX. Above 45–50 wt% of HEX, the SmCP range disappears, and there is a direct transition between two optically isotropic states (ISO and X phases). At increasing HEX concentrations, there is a decrease of the transition enthalpies, the layer ordering, and the magnitude of the electric polarization. The increase of the layer spacing saturates at 5 wt% of HEX.

The experimental results indicate a nanosegregated structure with the HEX molecules packing in layers which are set by the periodicity of the smectic structure. The driving mechanism for this nanophase segregation are the steric interactions between flexible HEX molecules and rigid bent cores of the liquid crystal molecules. The amount of HEX in the mixture defines two different regimes: (i) at small HEX concentrations, all the HEX molecules are between layers, with uniform layer spacings; (ii) at larger HEX concentrations, the distribution of layer spacings becomes non-uniform. Steric incompatibility between the flexible linear solvent and the rigid bent cores of the $C14$ molecules are supposed to push the HEX molecules to regions with smaller concentrations of $C14$, leading to a sub-micrometer segregation of the solvent.

At increasing HEX concentrations, there appear separated HEX domains of increasing size, and a weakening of the correlations between smectic domains. Due to the isotropic nature of the solvent at sufficiently high concentrations (about 20 wt% of HEX), the alignment of the smectic domains, of sub-micrometer dimensions, becomes uncorrelated and the texture becomes optically isotropic.

The structural model of Fig. 8.15 is based on X-ray scattering observations. In Fig. 8.15(a), we show a typical sketch of the antiferroelectric molecular packing of

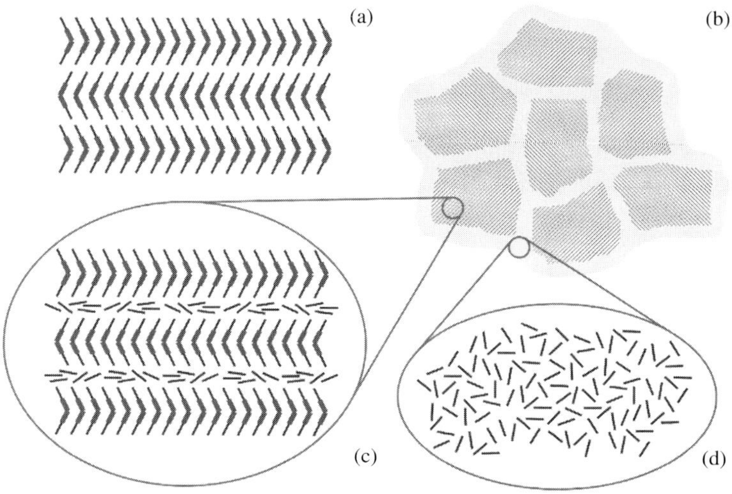

FIG. 8.15. Model for the molecular packing of the pure banana-shaped $C14$ liquid crystal and a mixture containing n-hexadecane [168]: (a) antiferroelectric molecular packing of pure $C14$; (b) submicrometer segregated isotropic HEX domains surrounding the smectic domains of $C14$ with layered nanosegregated HEX molecules; (c) magnified local structure of smectic domains with nanosegregated HEX molecules; (d) magnified local structure of submicrometer segregated HEX molecules.

pure $C14$. In Fig. 8.15(b), we show the sub-micrometer segregated isotropic HEX domains surrounding the smectic regions of $C14$ with the layered nanosegregated HEX molecules. In Fig. 8.15(c) and (d), we show magnified local structures of the smectic domains with nanosegregated and submicron-segregated HEX molecules.

This picture should give hints and indications in order to use the nanosegregation process for designing and producing devices at nanometer length scales.

8.6.3 *Transparent nematic phase*

Yamamoto and Tanaka have recently reported [170] an interesting study of a mixture of water, the thermotropic liquid crystal pentylcyanobiphenyl (5CB) and a double tailed ionic surfactant (didodecyl dimethyl ammonium bromide, DDAB). In the investigated range of molecular concentrations, there were observations of spherical inverted micelles, with a typical radius $a \sim 1.9$ nm, and typical intermicellar distances ranging from about 15 to 4 nm. Water droplets are involved by the polar heads of DDAB molecules, and the hydrophobic parts of those amphiphilic molecules are in contact with 5CB, which plays the role of the non-polar oily medium (5CB has been called oil by some authors). DSC, optical microscopy, and light scattering measurements indicated the existence of an unusual structure between the isotropic phase and a region of phase coexistence. Since it is optically isotropic (it looks like an isotropic phase if it is observed between crossed polarizers), this unusual phase has been called *transparent nematic* (TN). DSC experiments show the presence of two peaks, indicating two thermodynamic transitions. The first transition, at lower temperatures, was associated with the transition between a phase coexistence region and the TN phase; the other transition, at higher temperatures, was associated with the nematic–isotropic transition of 5CB. The nematic features of the TN phase are still present at 10 nm length scales. In the proposed structural model, 5CB molecules are assumed to be strongly anchored perpendicular to the surface of the inverse micelles. At the TN phase, the nematic directors are locally distorted by randomly dispersed micelles. At the macroscopic level, this structure gives rise to the "transparent" aspect of the phase. If we increase the temperature above the nematic–isotropic transition, the distorted nematic structure between micelles is no longer stable and 5CB molecules display an isotropic structure.

More recently, Bellini and coworkers [171] investigated this same mixture by static and dynamic light scattering techniques. Measurements of the intensity autocorrelation function lead to the identification of contributions due to pretransitional paranematic fluctuations (at a nanosecond time scale) and to the scattering by micelles (at a microsecond scale). The fluctuations of correlations of both molecular orientations and micellar concentrations were observed to be enhanced with decreasing temperatures from the isotropic phase towards the demixing transition. This pre-transitional behavior has been explained by a mean-field calculation for a Lebwohl–Lasher model [172]. The effective attractive interactions between micelles are consequences of fluctuations of the 5CB degrees of freedom. Micelles were regarded as holes in the nematic matrix, with free or semifree boundary conditions, in contrast to the strong anchoring proposed by Yamamoto and Tanaka.

References

[1] S. S. Papell (1963). US Patent 3.215.572, October 9.

[2] R. E. Rosensweig and R. Kaiser (1967). NTIS Rep. No. NASW-1219.

[3] R. E. Rosensweig (1985). *Ferrohydrodynamics*, Cambridge University Press, Cambridge.

[4] B. Berkovski and V. Bashtovoy (Eds) (1996). *Magnetic Fluids and Applications Handbook*, Begell House, Wallingford.

[5] C. Matuo, A. Bourdon, A. Bee, and A. M. Figueiredo Neto (1997). *Phys. Rev. E* **56**, R1310.

[6] M. Magalhães, A.M. Figueiredo Neto, A. Bee, and A. Bourdon (2000). *J. Chem. Phys.* **113**, 10246.

[7] S. W. Charles and J. Popplewell (1980). In *Ferromagnetic Material*, Eds E. P. Wohfarth, North-Holland Publishing Company, Amsterdam, Vol. 2.

[8] http://www.ferrotec.com/usa/index.html.
[9] http://www.sigma-hc.co.jp/english/index.html.
[10] R. Massart (1987). *IEEE Trans. Magn.* **17**, 1247 (1981).
[11] R. Massart, E. Dubois, V. Cabuil, and E. Hasmonay (1995). *J. Magn. Magn. Mater.* **149**, 1.
[12] E. Dubois, PhD Thesis (1997). Université Pierre et Marie Curie, Paris 6.
[13] E. Hasmonay, PhD Thesis (1998). Université Pierre et Marie Curie, Paris 6.
[14] A. Bee, R. Massart, and S. Neveu (1995). *J. Magn. Magn. Mater.* **149**, 6.
[15] R. A. Buhrman and C. G. Granqvist (1976). *J. Appl. Phys.* **47**, 2220.
[16] J. C. Bacri, V. Cabuil, R. Massart, R. Perzynski, and D. Salin (1987). *J. Magn. Magn. Mater.* **65**, 285.
[17] ASTM Crystallography International Table, file no. 39–1346.
[18] M. H. Sousa, F. A. Tourinho, J. Depeyrot, G. J. da Silva, and M. C. F. L. Lara (2001). *J. Phys. Chem. B* **105**, 1168.
[19] P. C. Scholten (1978). In *Colloid Chemistry of Magnetic Fluids in Thermomechanics of the Magnetic Fluids*, Ed. B. Berkovski, Hemisphere Publ. Corp., Bristol.
[20] J. N. Israelachvili (1991). *Intermolecular and Surface Forces*, Academic Press, New York.
[21] Ch. Soret (1879). *Arch. Sci. Phys. Nat.*, **2**, 48.
[22] Ch. Soret (1880). *Arch. Sci. Phys. Nat.*, **4**, 209.
[23] S. R. De Groot (1945). *L'Effet Soret, diffusion thermique dans les phases condensées*, Noord-Hollandsche Uitgevers Maatschappij, Amsterdam.
[24] S. R. De Groot and P. Mazur (1962). *Non-Equilibrium Thermodynamics*, North-Holland Publ. Co., Amsterdam.
[25] K. Wada, A. Suzuki, H. Sato, and R. Kikuchi (1985). *J. Phys. Chem. Solids* **46**, 1195.
[26] S. A. Akbar, M. Kaburagi, and H. Sato (1987). *J. Phys. Chem. Solids* **48**, 579.
[27] L. Landau and E. Lifchitz (1971). *Mécanique des Fluides*, Mir, Moscow.
[28] C. Vidal, G. Dewel, and P. Borckmans (1994). *Au-delà de l'équilibre*, Hermann, Paris.
[29] K. E. Grew (1969). In *Transport Phenomena in Fluids*, Ed. H. J. M. Hanley, Marcel Dekker, New York.
[30] W. Köhler (1993). *J. Chem. Phys.* **98**, 660.
[31] J. Lenglet (1996). These de doctorat de l'Université Paris 7 Denis Diderot, Paris, France.
[32] J. Lenglet, A. Bourdon, J. C. Bacri, and G. Demouchy (2002). *Phys. Rev. E* **65**, 031408-1-14.
[33] S. Alves, G. Demouchy, A. Bee, D. Talbot, A. Bourdon, and A. M. Figueiredo Neto (2003). *Philos. Mag.*, **83**, 2059.
[34] O. Ecenarro, J. A. Madariaga, J. Navarro, C. M. Santamaria, J. A. Carrion, and J. M. Saviron (1990). *J. Phys. Condens. Matter* **2**, 2289.

[35] S. Alves, A. Bourdon, and A. M. Figueiredo Neto (2003). *J. Opt. Soc. Am. B* **20**, 713.
[36] E. Bringuier and A. Bourdon (2003). *Phys. Rev. E* **67**, 011404.
[37] F. Brochard and P. G. de Gennes (1970). *J. Phys. (France)* **31**, 691.
[38] A. M. Figueiredo Neto, Y. Galerne, A. M. Levelut, and L. Liébert (1987). In *Physics of Complex and Supermolecular Fluids*, EXXON Monograph Series, Eds. S. A. Safran and N. A. Clark, Wiley, New York, p. 347.
[39] P. G. de Gennes and J. Prost (1993). *The Physics of Liquid Crystals*, Clarendon Press, Oxford.
[40] A. M. Figueiredo Neto and M. M. F. Saba (1986). *Phys. Rev. A* **34**, 3483.
[41] J. Rault, P. E. Cladis, and J. P. Burger (1970). *Phys. Lett.* **32A**, 199.
[42] L. Liébert and A. Martinet (1979). *J. Phys. (Paris) Lett.* **40**, L-363.
[43] L. Liébert and A. M. Figueiredo Neto (1984). *J. Phys. (Paris) Lett.* **45**, L-173.
[44] A. M. Figueiredo Neto, Y. Galerne, A. M. Levelut, and L. Liébert (1985). *J. Phys. (Paris) Lett.* **46**, L-499.
[45] A. M. Figueiredo Neto (1998). In *Phase Transitions in Complex Fluids*, Eds. P. Tolédano and A. M. Figueiredo Neto, World Scientific, Singapore, p. 175.
[46] J. C. Bacri and A. M. Figueiredo Neto (1994). *Phys. Rev. E* **50**, 3860.
[47] C. Y. Matuo and A. M. Figueiredo Neto (1999). *Phys. Rev. E* **60**, 1815.
[48] E. A. Oliveira, A. M. Figueiredo Neto, and G. Durand (1991). *Phys. Rev. A* **44**, R825.
[49] S. V. Burylov and Y. L. Raikher (1995). *Mol. Cryst. Liq. Cryst.* **258**, 107.
[50] Y. L. Raikher and V. I. Stepanov (1996). *J. Intell. Mater. Syst. Struc.* **7**, 550.
[51] A. Yu. Zubarev and L. Yu. Iskakova (1996). *Physica A* **229**, 203.
[52] G. Barbero, A. M. Figueiredo Neto, and N. V. Madhusudana (1997). *Phys. Lett. A* **233**, 105.
[53] G. Barbero, A. M. Figueiredo Neto, and N. V. Madhusudana (1999). *Phys. Lett. A* **251**, 373.
[54] T. Kroin, A. J. Palangana, and A. M. Figueiredo Neto (1989). *Phys. Rev. A* **39**, 5373.
[55] E. B. Priestley (1979). In *Introduction to Liquid Crystals*, Eds E. B. Priestley, P. I. Wojtowicz, and P. Sheng, Plenum Press, New York.
[56] S. H. Chen and N. M. Amer (1983). *Phys. Rev. Lett.* **51**, 2298.
[57] V. Berejnov, Y. Raikher, V. Cabuil, J. C. Bacri, and R. Perzynski (1998). *J. Colloid Interface Sci.* **199**, 215.
[58] C. Y. Matuo and A. M. Figueiredo Neto (1993). *J. Magn. Magn. Mater.* **122**, 53.
[59] A. M. Figueiredo Neto, A. M. Levelut, L. Liébert, and Y. Galerne (1985). *Mol. Cryst. Liq. Cryst.* **129**, 191.

[60] A. M. Figueiredo Neto, L. Liébert, and Y. Galerne (1985). *J. Phys. Chem.* **89**, 3737.
[61] Y. Galerne, A. M. Figueiredo Neto, and L. Liébert (1987). *J. Chem. Phys.* **87**, 1851.
[62] E. A. Oliveira, G. Barbero, and A. M. Figueiredo Neto (1996). *Phys. Rev. E* **54**, 5830.
[63] T. Kroin and A. M. Figueiredo Neto (1987). *Phys. Rev. A* **36**, 2987.
[64] A. J. Palangana and A. M. Figueiredo Neto (1990). *Phys. Rev. A* **41**, 7053.
[65] J. Nehring and A. Saupe (1972). *J. Chem. Phys.* **54**, 337.
[66] S. Fontanini, G. Barbero, and A. M. Figueiredo Neto (1996). *Phys. Rev. E* **53**, 2454.
[67] S. Fontanini, A. L. Alexe-Ionescu, G. Barbero, and A. M. Figueiredo Neto (1997). *J. Chem. Phys.* **106**, 6187.
[68] V. Berejnov, J. C. Bacri, V. Cabuil, R. Perzynski, and Y. Raikher (1998). *Europhys. Lett.* **41**, 507.
[69] K. I. Morozov (2002). *Phys. Rev. E* **66**, 011704.
[70] S. L. Gómez, F. L. S. Cuppo, A. M. Figueiredo Neto, T. Kosa, M. Muramatsu, and R. Horowicz (1999). *Phys. Rev. E* **59**, 3059.
[71] S. L. Gómez and A. M. Figueiredo Neto (2000). *Phys. Rev. E* **62**, 675.
[72] I. C. Khoo (1987). *Liquid Crystals: Physical Properties and Nonlinear Optical Phenomena*, Wiley, New York.
[73] A. Jordan, P. Wust, R. Scholz, B. Tesche, H. Fähling, T. Mitrovics, T. Vogl, J. Cervós-Navarro, and R. Felix (1993). *Int. J. Hyperthermia* **12**, 705.
[74] Y. Galerne, A. M. Figueiredo Neto, and L. Liébert (1987). *J. Chem. Phys.* **87**, 1851.
[75] L. Liébert and A. Martinet (1980). *IEEE Trans. Magn.* **MAG-16**, 266.
[76] A. M. Figueiredo Neto, L. Liébert, and A. M. Levelut (1984). *J. Phys. (France)* **45**, 1505.
[77] A. M. Figueiredo Neto, Y. Galerne, and L. Liébert (1985). *J. Phys. Chem.* **89**, 3939.
[78] A. M. Figueiredo Neto and M. E. Marcondes Helene (1987). *J. Phys. Chem.* **91**, 1466.
[79] M. E. Marcondes Helene and A. M. Figueiredo Neto (1988). *Mol. Cryst. Liq. Cryst.* **162B**, 127.
[80] P. Fabre, C. Casagrande, M. Veyssie, V. Cabuil, and R. Massart (1990). *Phys. Rev. Lett.* **64**, 539.
[81] P. Fabre (1990). *La Recherche* **21**, 1424.
[82] C. Quilliet, P. Fabre, and V. Cabuil (1993). *J. Phys. Chem.* **97**, 287.
[83] V. Ponsinet and P. Fabre (1996). *J. Phys. Chem.* **100**, 5035.
[84] W. Helfrich (1978). *Z. Naturforsch.* **33A**, 305.
[85] C. Quilliet, V. Ponsinet, and V. Cabuil (1994). *J. Phys. Chem.* **98**, 3566.
[86] T. P. Hoar and J. H. Schulman (1943). *Nature (London)* **152**, 102.
[87] L. M. Prince (Ed.) (1976). *Microemulsions. Theory and Pratice*, Academic Press, New York.

[88] I. Danielsson and B. Lindman (1981). *Colloids Surf.* **3**, 391.
[89] K. Shinoda and S. Friberg (1975). *Adv. Colloids Interface Sci.* **4**, 281.
[90] K. L. Mittal (Ed.) (1977). *Micellization, Solubilization and Microemulsions*, Plenum Press, New York.
[91] A. M. Bellocq, J. Biais, P. Bothorel, B. Clin, G. Fourche, P. Lalanne, B. Lamaire, B. Lamanceau, and D. Roux (1984). *Adv. Colloids Interface Sci.* **20**, 167.
[92] A. M. Bellock (1987). In *Physics of Complex and Supermolecular Fluids*, EXXON Monograph, Eds S. A. Safran and N. A. Clark, John Wiley & Sons, New York.
[93] L. A. Turkevich (1987). In *Physics of Complex and Supermolecular Fluids*, EXXON Monograph, Eds S. A. Safran and N. A. Clark, John Wiley & Sons, New York.
[94] J. H. Schulman and J. B. Montagne (1961). *Ann. NY Acad. Sci.* **92**, 366.
[95] P. Honorat, D. Roux, and A. M. Bellocq (1984). *J. Phys. (France) Lett.* **45**, L-961.
[96] S. Friberg, I. Lapeznski, and G. Gillberg (1976). *J. Colloid Interface Sci.* **56**, 19.
[97] L. E. Scriven (1976). *Nature* **263**, 123.
[98] Y. Talmon and S. Prager (1978). *J. Chem. Phys.* **69**, 2984.
[99] P. G. de Gennes and C. Taupin (1982). *J. Phys. Chem.* **86**, 2294.
[100] B. Widon (1984). *J. Chem. Phys.* **81**, 1030.
[101] S. Safran, L. Turkevich, and P. Pincus (1984). *J. Phys. (France) Lett.* **45**, L-69.
[102] D. Roux and C. Coulon (1986). *J. Phys.* **47**, 1257.
[103] S. A. Safran, D. Roux, M. E. Cates, and D. Andelman (1986). *Phys. Rev. Lett.* **57**, 491.
[104] C. A. Miller, R. W. Hwan, W. J. Benton, and T. Tort (1977). *J. Colloid Interface Sci.* **61**, 554.
[105] A. A. Safran and L. E. Turkevich (1983). *Phys. Rev. Lett.* **50**, 1930 (1983).
[106] W. Helfrich (1973). *Z. Naturforsch.* **28**, 6693.
[107] B. Lemaire, P. Bothorel, and D. Roux (1983). *J. Phys. Chem.* **87**, 1023.
[108] J. Jouffroy, P. Levinson, and P. G. de Gennes (1982). *J. Phys. (France)* **43**, 1241.
[109] E. Ruckenstein and J. C. Chi (1975). *J. Chem. Soc. Faraday Trans. 2*, **71**, 1690.
[110] I. Langmuir (1915). *J. Am. Chem. Soc.* **37**, 1139.
[111] I. Langmuir (1916). *J. Am. Chem. Soc.* **38**, 2221.
[112] I. Langmuir (1917). *J. Am. Chem. Soc.* **39**, 1848.
[113] I. Langmuir (1918). *J. Am. Chem. Soc.* **40**, 1361.
[114] C. H. Giles, S. D. Forrester, and G. G. Roberts (1990). In *Langmuir–Blodgett Films*, Ed. G. G. Roberts, Plenum Press, New York.
[115] K. B. Blodgett (1935). *J. Am. Chem. Soc.* **57**, 1007.

[116] K. B. Blodgett (1937). *Phys. Rev.* **51**, 964.
[117] A. H. Ellison (1962). *J. Phys. Chem.* **66**, 1867.
[118] A. Barraud, J. Leloup, and P. Lesieur (1985). *Thin Solid Films* **133**, 113.
[119] R. A. Hann (1990). In *Langmuir–Blodgett Films*, Ed. G. G. Roberts, Plenum Press, New York.
[120] M. Koyama, R. Tomioka, M. Ueno, and K. Meguro (1974). *Colloid Polym. Sci.* **252**, 372.
[121] M. C. Petty and W. A. Barlow (1990). In *Langmuir–Blodgett Films*, Ed. G. G. Roberts, Plenum Press, New York.
[122] O. N. Oliveira Jr. (1992). *Braz. J. Phys.* **22**, 60.
[123] I. Fankuchen (1938). *Phys. Rev.* **53**, 909.
[124] I. Langmuir, V. K. Schaefer, and H. Sobotka (1937). *J. Am. Chem. Soc.* **59**, 1751.
[125] E. P. Honig, J. H. Hengst, and D. den Engelsen (1973). *J. Colloid Interface Sci.* **45**, 92.
[126] G. L. Gaines Jr. (1966). *Insoluble Monolayer at Liquid–Gas Interface*, Wiley-Interscience, New York.
[127] P. Drude (1902). *Theory of Optics*, Longmans, New York.
[128] B. Mann and H. Kuhn (1971). *J. Appl. Phys.* **42**, 4398.
[129] I. R. Peterson and G. J. Russel (1984). *Philos. Mag. A* **49**, 463.
[130] C. A. Jones, J. G. Russel, M. C. Petty, and G. G. Roberts (1986). *Philos. Mag. B* **54**, L89.
[131] P. A. Chollet (1980). *Thin Solid Films* **68**, 13.
[132] Y. R. Shen (1984). *The Principles of Nonlinear Optics*, John Wiley & Sons, New York.
[133] S. Fontanini, A. Strigazzi, G. Barbero, M. C. Salvadori, and A. M. Figueiredo Neto (1998). *Liq. Cryst.* **24**, 793.
[134] O. N. Oliveira Jr., M. Raposo, and A. Dhanabalan (2001). In *Handbook of Surfaces and Interfaces of Materials*, Ed. H. S. Nalwa, Academic Press, San Diego, CA, vol. 40.
[135] F. S. Bates, R. E. Cohen, and C. V. Berney (1982). *Macromolecules* **15**, 589.
[136] T. Hashimoto, M. Fujimura, and H. Kaway (1980). *Macromolecules* **13**, 1660.
[137] E. Helfand and Z. R. Wasserman (1978). *Macromolecules* **11**, 960.
[138] L. Leibler (1980). *Macromolecules* **13**, 1602.
[139] D. R. M. Williams and A. Halperin (1993). *Phys. Rev. Lett.* **71**, 1557.
[140] R. Hosemann and S. N. Bagchi (1962). *Direct Analysis of Matter by Diffraction*, North-Holland, Amsterdam.
[141] R. J. Roe, M. Fishkis, and J. C. Chang (1981). *Macromolecules* **14**, 1091.
[142] V. Castelletto, C. Caillet, J. Fundin, and I. W. Hamley (2002). *J. Chem. Phys.* **116**, 10947.
[143] T. Hashimoto, K. Nagatoshi, A. Todo, H. Hasegawa, and H. Kawai (1974). *Macromolecules* **7**, 364.

[144] T. Hashimoto, A. Todo, H. Itoi, and H. Kawai (1977). *Macromolecules* **10**, 377.
[145] H. Hasegawa and T. Hashimoto (1985). *Macromolecules* **18**, 589.
[146] H. Tanaka, H. Hasegawa, and T. Hashimoto (1991). *Macromolecules* **24**, 240.
[147] M. W. Matsen and C. Barret (1998). *J. Chem. Phys.* **109**, 4108.
[148] G. Kämpf, H. Krömer, and M. Hoffmann (1972). *J. Macromol. Sci.–Phys.* **B6**, 167.
[149] I. W. Hamley, K. A. Koppi, J. H. Rosedale, F. S. Bates, K. Almdal, and K. Mortensen (1993). *Macromolecules* **26**, 5959.
[150] I. W. Hanley (1994). *Phys. Rev. E* **50**, 2872.
[151] I. W. Hamley, J. A. Pople, J. P. A. Fairclough, N. J. Terrill, A. J. Ryan, C. Booth, G. E. Yu, O. Diat, K. Almdal, K. Mortensen, and M. Vigild (1998). *J. Chem. Phys.* **108**, 6929.
[152] K. I. Winey, E. L. Thomas, and L. J. Fetters (1992). *Macromolecules* **25**, 2645.
[153] M. W. Matsen and F. S. Bates (1996). *Macromolecules* **29**, 1091.
[154] D. R. M. Willians and A. Halperin (1993). *Phys. Rev. Lett.* **71**, 1557.
[155] T. Rieker (1995). *Liq. Cryst.* **19**, 497.
[156] B. Kohne, K. Praefcke, T. Derz, H. Hoffmann, and B. Schwandner (1986). *Chimia* **40**, 171.
[157] N. Usoltseva, K. Praefcke, D. Singer, and B. Gündogan (1994). *Liq. Cryst.* **16**, 601.
[158] S. Gaspard, P. Maillard, and J. Billard (1985). *Mol. Cryst. Liq. Cryst.* **123**, 369.
[159] N. Usoltseva, P. Espinet, J. Buey, K. Praefcke, and D. Blunk (1997). *Mol. Cryst. Liq. Cryst.* **299**, 457.
[160] N. Usoltseva, K. Praefcke, D. Singer, and B. Gündogan (1994). *Liq. Cryst.* **16**, 617.
[161] N. Usoltseva, K. Praefcke, A. Smirnova, and D. Blunk (1999). *Liq. Cryst.* **26**, 1723.
[162] N. Usoltseva, V. Bykova, G. Anajeva, A. Smirnova, G. Shaposhnikov, V. Maizlish, E. Kudrik, and A. Shrikov (2000). *Mol. Cryst. Liq. Cryst.* **352**, 45.
[163] T. Niori, T. Sekine, J. Watanabe, T. Furukawa, and H. Takezoe (1996). *J. Mater. Chem.* **6**, 1231.
[164] G. Pelzl, S. Diele, and W. Weissflog (1999). *Adv. Mater.* **11**, 9.
[165] A. Jákli, W. Cao, Y. Huang, C. K. Lee, and L. C. Chien (2001). *Liq. Cryst.* **28**, 1279.
[166] F. Dowell (1983). *Phys. Rev. A* **28**, 3520.
[167] W. Weissflog, C. Lischka, S. Diele, G. Pelzl, I. Wirth (1999). *Mol. Cryst., Liq. Cryst.*, **328**, 101.
[168] M. Y. M. Huang, A. M. Pedreira, O. G. Martins, A. M. Figueiredo Neto, and A. Jákli (2002). *Phys. Rev. E* **66**, 031708.

[169] H. Sawade and G. Heppke, private communication.
[170] J. Yamamoto and H. Tanaka (2001). *Nature (London)* **409**, 321.
[171] T. Bellini, M. Caggioni, N. A. Clark, F. Mantegazza, A. Maritan, and A. Pelizzola (2003). *Phys. Rev. Lett.* **91**, 085704.
[172] P. Lebwohl and G. Lasher (1972). *Phys. Rev. A* **6**, 426.

INDEX